Semiconductor Quantum Wells and Superlattices for Long-Wavelength Infrared Detectors

The Artech House Materials Science Library

Semiconductor Quantum Wells and Superlattices for Long-Wavelength Infrared Detectors

M. O. Manasreh
Editor

Artech House
Boston • London

Library of Congress Cataloging-in-Publication Data

Manasreh, M. O.
    Semiconductor quantum wells and superlattices for long-wavelength
infrared detectors / M.O. Manasreh.
    p.   cm.
    Includes bibliographical references and index.
    ISBN 0-89006-603-5
    1. Infrared detectors—Materials. 2. Compound semiconductors.
3. Superlattices as materials. 4. Quantum wells. I. Title.
    TA1570.M33   1993   92-32246
    621.36'2—dc20   CIP

British Library Cataloguing in Publication Data

Manasreh, M. O.
    Semiconductor Quantum Wells and
    Superlattices for Long-wavelength
    Infrared Detectors
    I. Title.
    621.36

    ISBN 0-89006-603-5

© 1993 ARTECH HOUSE, INC.
685 Canton Street
Norwood, MA 02062

International Standard Book Number: 0-89006-603-5
Library of Congress Catalog Card Number: 92-32246

10 9 8 7 6 5 4 3 2 1

To my mother, wife, and two daughters Sarah Margaret
and Hannah Grace

Contents

Foreword

A few decades ago systems designers desiring to use semiconductor photodetectors had limited options available to them. They had to select one of the few types of semiconductors available (e.g., silicon, germanium, indium antimonide, and so forth). Some of these materials offered the designer two further alternatives: intrinsic or extrinsic materials properties. In the intrinsic approach, the band of wavelengths over which the chosen material may be used is determined by the energy separation of the valence and conduction bands, and it cannot be adjusted to fit the requirement. In the extrinsic approach, photon detection is based on the energizing of an electron (hole) from an impurity dopant to the conduction (valence) band. Since the various dopant and material combinations have different ionization energies, this technique broadens the choice of wavelength bands available to the systems designer. If this were all designers had, however, they would still have only the ability to select from a variety of discrete wavelength bands, bands which might or might not be optimal for achieving the requirements of the system being designed.

With the development of mercury cadmium telluride technology, it has become possible, at least in principle, to select a wavelength band, particularly in the infrared region, to optimize system performance and to build the detector or focal plane array to suit the need. This additional degree of freedom derives from the fact that the energy gap of this ternary material system can be varied all the way from the mercury telluride value (near zero) to the cadmium telluride value (1.44 eV) by varying the mercury/cadmium ratio within the crystal. The flexibility offered by this additional degree of freedom is in large part responsible for the enormous investment that has been made by military departments around the world since the 1960s. Unfortunately, the technology required for longer wavelength bands and higher mercury concentrations has proven to be extremely difficult to harness, and costs have remained high. Hence the quest for additional alternatives continues.

With the maturing of the compound semiconductor technologies (in particular, the gallium arsenide technology) came a new capability: the development of machines and techniques that could grow ultrathin layers of dissimilar compounds. These were applied not only to gallium arsenide and other III–V compounds, but to II–VIs and group IV structures as well. As with any new development, this led to the birth of new concepts and the reevaluation of older ideas that had heretofore been technologically unrealizable. It is from this capability base that the subject of this book, multiquantum well and superlattice detector technology, is derived. This new approach offers many of the same flexibilities theoretically offered by mercury cadmium telluride, but with the added advantage of being based upon a more broadly applicable technology, thus offering the hope that the cost and availability of detectors and focal-plane arrays will be significantly improved. In addition, this quantum technology brings with it some additional degrees of freedom. Now, in addition to the ability for only the desired wavelength response, one can think in terms of using the properties brought on by crystallographic strain to design in optimized electron or hole transport properties as well. And since many other electronic devices and integrated circuits are already produced from compound semiconductors, perhaps the monolithic integration of the detector arrays with their associated readout electronics will be facilitated. There are, of course, disadvantages as well. An extended discussion of the theory, fabrication, and potential of these detectors is the subject of this book.

Gary L. McCoy
Wright Laboratory
June, 1992

Preface

The search continues for novel approaches to the design of semiconductor materials with tailored electronics and optical properties for a new generation of long-wavelength infrared quantum detectors. The precise control over the properties of materials by growth techniques such as molecular beam epitaxy have made it possible to grow novel structures like quantum wells and superlattices. The potential of these structures for long-wavelength infrared detectors has not been fully exploited. This book is devoted to developing an understanding of the basic properties of semiconductor quantum wells and superlattices and how they can be used for long-wavelength infrared detectors and imaging arrays. Recently, infrared detectors have enabled a new and a wide range of applications, including global and space surveillance, detection of cold objects, astronomy and deep space exploration, medical imaging, and remote monitoring of plants and the environment. Therefore, a comprehensive investigation of the infrared properties of quantum wells and superlattices is needed. This book is merely a step, in a fast-changing field, toward achieving an understanding of novel structures that could be used for high-performance infrared detectors and imaging arrays.

The main objective of this book is to provide the reader with a basic understanding of how semiconductor quantum wells and superlattices can be used for long-wavelength infrared detectors and related systems. The material in this book is oriented toward researchers involved in the field of semiconductor infrared materials and devices. Advanced graduate students majoring in electrical engineering and solid state physics can also benefit from this book.

Chapter 1 provides the reader with a general background of quantum detectors and the advanced semiconductor structures (quantum wells and superlattices) that can be used for long-wavelength infrared detectors. There are a few success stories using GaAs/AlGaAs multiple quantum wells for long-wavelength infrared detectors. In Chapter 2 we present a theoretical discussion supported by experimental observation on the modeling of intersubband transitions in GaAs/AlGaAs multiple-quantum wells and similar structures. Chapter 3 is focused on device parameters such as the detectivity, dark current, and other figures of merit of long-wavelength infrared detectors fabricated from GaAs/AlGaAs multiple-quantum wells. The discussion and analysis in Chapter 3 can also be extended to other systems.

Strained-layer superlattices, known as *type II superlattices*, can also be used for long-wavelength infrared detectors. An overview of the modern theoretical methods that serve as the basis for the calculations of the electronic structures of type II superlattices is given in Chapter 4. InAs/GaInSb was taken as an example in this chapter. The results indicate that this system exhibits overall materials properties for long-wavelength infrared detection applications superior to the more conventional bulk HgCdTe alloys.

Other semiconductor systems can also be used for long-wavelength infrared detection; for example, intersubband transitions in both *p*-type and *n*-type Si/SiGe multiple-quantum wells. This system is proposed in Chapter 5. One advantage of *p*-type Si/SiGe multiple quantum wells is that a Brewster's angle incidence of incoming photons is not required. HgTe/CdTe superlattices provide another system that can be used for long-wavelength infrared detectors using normal incidence light. It should be pointed out that this system may have superior properties as compared to bulk HgCdTe systems for long-wavelength infrared detectors. This system, known as *type III superlattices*, is discussed in a great detail in Chapter 6.

The contributions of G.J. Brown, J.P. Loehr, R.L. Whitney, K.F. Cuff, F.W. Adams, C. Mailhiot, K.L. Wang, R.P.G. Karunasiri, T.H. Myers, J.R. Meyer, and C.A. Hoffman are greatfully appreciated. This book could not have been written without the great efforts of these outstanding scientists and engineers.

M.O. Manasreh
Wright Laboratory
June, 1992

Chapter 1

Introduction to Long-Wavelength Infrared Quantum Detectors

M.O. Manasreh and G.J. Brown

Wright Laboratory
Wright-Patterson Air Force Base

1.1 BACKGROUND

Infrared (IR) detectors can be thought of as transducers that convert IR radiation into an electrical signal. The electromagnetic energy that covers the wavelength region of ~1 to 1000 μm is often called *infrared radiation*. Basic characteristics of this type of radiation are that it does not penetrate metals and passes through many crystalline, plastic, and gaseous materials including the earth's atmosphere [1]. Infrared radiation is usually divided into a long-wavelength infrared region (1–30 μm) and a far-infrared region (>30 μm). Infrared detectors are usually classified as either thermal or photon devices [2]. The absorption of light in thermal detectors raises the temperature of the device, which in turn changes some temperature-dependent parameters such as electrical conductivity. Thermal detectors may be thermopile, bolometer, Golay cell detectors, or pyroelectric detectors.

The absorption of long-wavelength radiation in photon (quantum) detectors results directly in some specific quantum event, such as the photoelectric emission of electrons from a surface, or electronic interband transitions in semiconductor materials. Therefore, the output of quantum detectors is governed by the rate of absorption of photons and not directly on the photons' energy. In addition, quantum detectors operating in the long-wavelength spectral region must be cooled to temperatures ≤80 K to reduce the thermal energy noise. Quantum detectors are further classified into two categories: photoconductors, such as CdS, PbS, PbSe, Si:As, Ge, and Au; and photovoltaics, such as InSb and HgCdTe. Many of the quantum detector materials, such as HgCdTe, and junction detectors can operate in a photovoltaic mode under a zero bias voltage (open circuit) or in a photoconductive mode when a reverse bias voltage is applied (short circuit).

In the case of quantum detectors, it is not necessary to heat the entire device material to sense the radiation, as is necessary in thermal detectors. Hence, these detectors are much faster than thermal detectors. Most quantum detectors are fabricated from semiconductor materials, which are fundamentally distinguished from other solid materials by the presence of an electronic energy bandgap. Semiconductor quantum detectors that rely on interband transitions for their operation are called *intrinsic detectors*. Extrinsic detectors, on the other hand, are fabricated from doped semiconductors, where the transitions occur from impurity levels within the bandgap to the appropriate band edge. The wavelength response of the latter class of detectors can extend from 20 to 1,000 μm, but a major disadvantage is that cooling to liquid helium temperature is essential to reduce background noise.

Mercury cadmium telluride, InSb, and extrinsic Si have been the materials of choice for long-wavelength infrared detectors. However, the slow progress in the development of large photovoltaic bulk infrared imaging arrays and the rapid achievements of novel systems have made it more difficult to predict what types of arrays will be readily available for future systems applications. Both tactical and strategic use of infrared imaging systems requires the availability of high-performance detector arrays operating at wavelengths beyond the 10–18 μm spectral band. Spaceborne surveillance systems, low-background infrared seeker-tracker systems, and astronomical payloads need reliable and affordable sensors with long life that can function effectively at temperatures higher than the 20–30 K currently required by bulk quantum detectors.

1.2 QUANTUM DETECTORS BASED ON SEMICONDUCTOR QUANTUM WELLS AND SUPERLATTICES

In recent years major steps have been made toward a new approach for designing semiconductor structures with tailored electronic and optical properties for a new generation of long-wavelength infrared quantum detectors. This approach has come to be known as bandgap engineering [3, 4]. Central to bandgap engineering is the concept that, by spatially varying the composition and doping of a semiconductor over distances ranging from a few micrometers down to ~2.5 Å (~1 monolayer), we can tailor the electronic band structure in a nearly arbitrary and continuous way. Thus, semiconductor structures with new electronic and optical properties can be custom designed for specific applications.

To implement bandgap engineering of infrared detectors, precise control of the composition, doping, and thickness of the semiconductor layers is required. Modern thin film growth techniques such as molecular beam epitaxy and metalorganic chemical vapor deposition have made it possible to deposit ultrathin (10–100 Å) epitaxial layers of precisely controlled alloy compositions on semiconductor substrates on an atomic layer-by-layer basis [5]. High-quality single crystal epitaxial layers composed of a variety of semiconductor alloy combinations (such as GaAs, $Si_{1-x}Ge_x$, $Al_xGa_{1-x}As$) have been grown by using these techniques. In addition to single composition epitaxial layers, these epitaxial growth techniques can be used to deposit multiple layers of two different semiconductors with atomically abrupt interfaces (heterojunctions) [6]. These alternating semiconductor layers have precisely controlled compositions and can be doped within the individual layers that are only a few tens of angstroms thick. These layers are so thin that the energy levels in them show the quantum effects associated with electron confinement within the layers. This confinement, due to the heterojunctions at the interfaces, causes the continuous allowed energy levels associated with macroscopic thicknesses (a free electron) to become discrete allowed energy levels [7].

When there is a conduction band (or valence band) offset in energy at the heterostructure interface due to a difference in the bandgap between the alternating compositions, quantum wells are formed in the band structure. Typically, quantum wells are created by sandwiching a very thin layer of a small bandgap semiconductor material between two thick layers of a wide bandgap material. The allowed electron (or hole) energies are quantized by the presence of the two potential energy barriers formed by the wide bandgap material; that is, the classic electron in a finite potential well problem (see Figure 1.1(a)). The actual values of the discrete energy levels in the quantum wells are determined by the thickness and the depth of the well. The depth of the well is the band discontinuity of the two materials, conduction band discontinuity ΔE_c for electrons and valence band discontinuity ΔE_v for holes. The allowed electronic transitions are then further tailored by the selection of the thickness of the quantum well. Wider wells contain more bound states that are closer in energy than the bound states in narrower wells.

If many quantum wells are grown on top of one another and the barrier layers are made thick enough that the electrons wave functions in adjacent quantum wells do not overlap, then the structure is called a *multiple quantum well* (Figure 1.1(b)). The electronic properties of multiple quantum well heterostructures are obtained simply as a superposition of the results obtained for a single well. If the barrier layers are so thin that tunneling between wells is significant (thickness <50 Å), the result is a superlattice, first proposed by Esaki and Tsu in 1970 [8]. Superlattices are new materials, with properties that are controlled by the artificial periodicity introduced by the multilayer structure. Because of the overlap of the wave functions of the electron energy states in the wells, the discrete electronic transitions of an isolated well become minibands in a superlattice material as shown in Figure 1.1(c). Properties of superlattices can be varied, not only by the choice of the materials used to make up the heterojunctions, but also by the thicknesses of both the well and barrier layers. Additionally, the internal strain field of a heterostructure constructed from two semiconducting materials with mismatched lattice constants can be used to shift

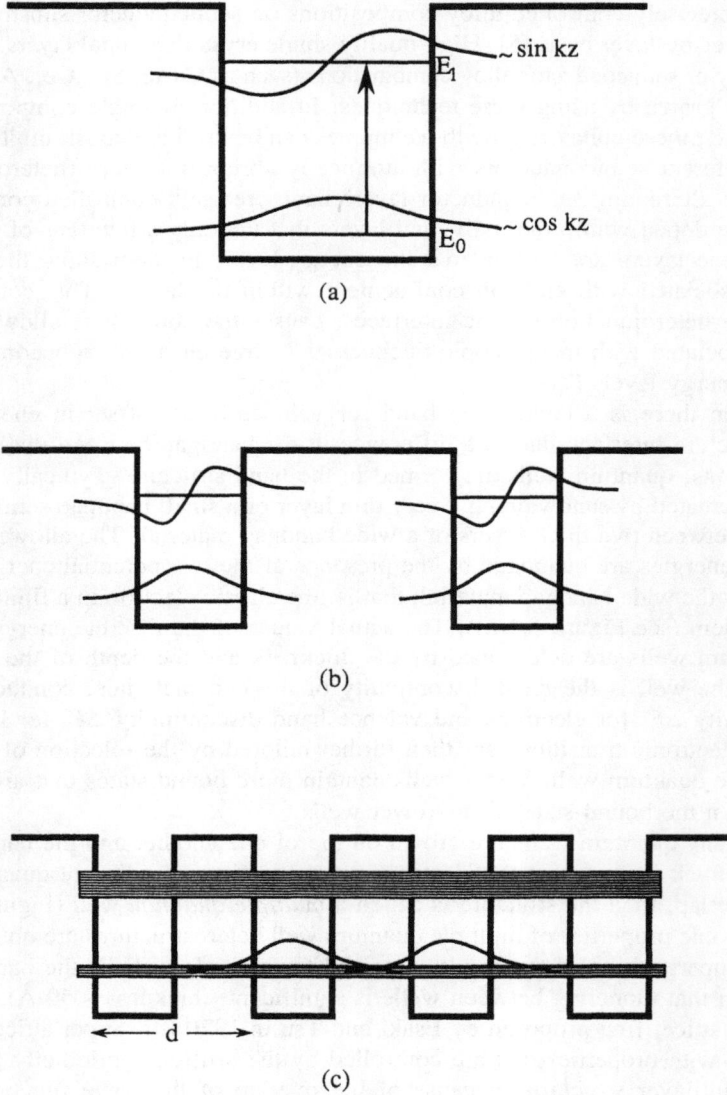

Figure 1.1 A schematic representation of GaAs/AlGaAs conduction bands. (a) Two bound energy levels and wave functions are in a finite isolated quantum well; the vertical arrow indicates a possible transition between these levels, which is known as the *intersubband transition*. (b) A multiple quantum well structure is formed by choosing thick barrier layers. (c) Periodic potential wells of a period *d* superlattice form where the energy levels in each quantum well are overlapped to form minibands.

the band structure of the superlattice [9]. Heterostructures constructed of lattice mismatched materials are referred to as *strained-layer superlattices*. Like the multiple quantum well heterostructures, the appropriate choice of the compositions and structural parameters for the superlattice makes it possible to bandgap engineer a whole range of electronic properties. Superlattices made of III–V semiconductors are classified into two types. In type I superlattices, the band discontinuities are such that both band edges of the smaller gap materials are below those of the wide bandgap material; that is, the valence and conduction band quantum wells are formed in the same layer. The band structure of type II superlattices is such that the top of the valence band of one layer lies above the bottom of the conduction band of the other layer, producing a staggered band offset. This structure causes the valence band quantum wells to be formed in one layer whereas the conduction band quantum wells are formed in the neighboring layer. Thus, the electrons and holes are confined in two separate layers.

One of the uses of multiple quantum well and superlattice heterostructures is the design of novel long-wavelength ($\lambda > 12$ μm) infrared detectors [10]. In bulk semiconductor materials, the selection of compounds or alloys with the appropriate bandgaps and long-wavelength cutoffs for infrared detectors is very limited. For instance, Figure 1.2 summarizes the bandgaps (E_g), cutoffs wavelength (λ_c), and lattice constants (a_0) available with III–V, II–VI, and IV semiconductor materials. It is

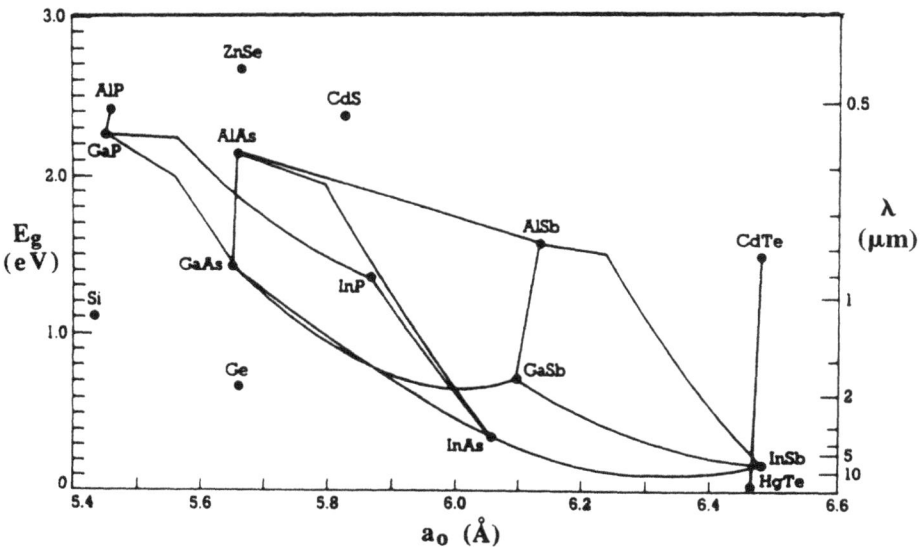

Figure 1.2 Room temperature bandgaps, lattice constants, and wavelength cutoffs for III–V, II–VI, and IV alloy systems.

readily apparent that only the InAsSb system has wavelength cutoffs approaching 12 μm. Therefore, long-wavelength absorption is not possible via valence band to conduction band transitions in any of these semiconductor systems. In the II–VI materials two alloy systems, HgCdTe and HgZnTe, are available for band-to-band infrared absorption at long wavelengths. The main drawback of these alloys for infrared detection is that they are difficult to grow with precisely controlled compositions, which strongly affect the bandgap. Silicon-doped detectors known as *blocked impurity band* (BIB) detectors have also been proposed and used for space surveillance missions. A major disadvantage of BIB detectors is that cooling to temperatures of 20–30 K is essential for satisfactory noise performance. An alternative to using the available bulk semiconductor bandgaps is to use the novel bandgaps created in III–V multilayer heterostructures. Long wavelength quantum detectors made of III–V semiconductor quantum wells and superlattices are expected to have several advantages over bulk HgCdTe detectors as well as other bulk long-wavelength detectors for the following reasons:

1. A higher degree of uniformity, which is of importance for detector arrays;
2. Smaller leakage currents due to the suppression of tunneling available in superlattices;
3. Lower Auger recombination rates due to substantial splitting of the light- and heavy-hole bands;
4. Technologically mature device fabrication and materials processing techniques;
5. Tunability of upper cutoff wavelengths (theoretically, to arbitrarily high values);
6. A large cross section for photon absorption;
7. Radiation hardness;
8. Ease of crystal growth;
9. Specific detectivity (D^*) comparable to that of HgCdTe at temperatures as high as 80 K;
10. Lower cost.

There are two main approaches to fabricating long-wavelength infrared detectors from these III–V semiconductor heterostructures. The first uses the intersubband transitions in III–V semiconductor multiple quantum wells, such as AlGaAs/GaAs and InGaAs/InP as shown in Figure 1.1(a). This type of structure, where both electrons and holes are located in the same layer, is called *type I*. The second uses interband transitions in type II superlattices, such as InAs/InGaSb, as shown in Figure 1.3. In type II superlattices the electrons and holes are located in adjacent layers. Both approaches have been used to produce IR detectors in the spectral range of 8–12 μm [11, 12]. There are two major disadvantages to the first approach. First, the incoming photons must strike the surface at the Brewster's angle of the quantum well materials to allow coupling between the incoming radiation and the electrons that undergo such transitions [13]. Second, to introduce electrons into the quantum well, the materials must be doped with Si or another donor dopant. These devices

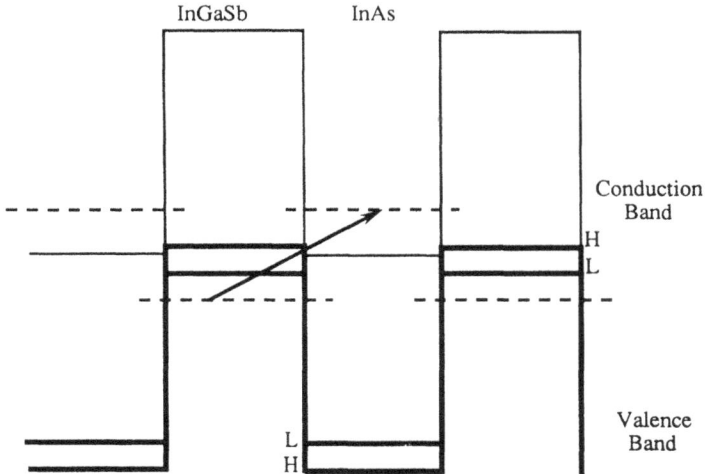

Figure 1.3 Interband transition (arrow) in a type II InGaSb/InAs superlattice. The dashed lines represent the ground states of electrons and holes. The letters H and L indicate heavy and light holes, respectively.

therefore are extrinsic and have limited optical cross sections, because the dopant concentrations are limited by epitaxial growth considerations. In general, theories predict that both intersubband transitions in multiple quantum wells (see Figure 1.1(a)) and interband transitions in type II superlattices (see Figure 1.3) can be continuously adjustable from zero. This possibility has yet to be explored for quantum detectors in the spectral range of 20–30 μm or longer cutoff wavelengths.

In a strained type II superlattice, the degeneracy of the valence band structure is lifted and the light-hole and heavy-hole bands become separated in energy (Figure 1.3). The separation between these bands depends on the amount of strain in the layer and on whether the layer is in biaxial tension or compression. For example, when a layer is in biaxial compression, the heavy-hole band is raised in energy and the light-hole band is lowered. In this class of materials, the optical transitions are between the lowest miniband in the valence band well to the lowest miniband in the neighboring conduction band well. When the quantum confinement effects are small, band offsets can produce band-to-band transition energies less than the bulk bandgap of the layer materials.

As we have seen, a variety of heterostructures can be used to tailor the semi-conductor energy band structure. Research on III–V heterostructures has verified long-wavelength absorption mechanisms in a number of various III–V multiple quantum well and superlattice structures [13–17]. These results have motivated a variety of studies aimed at developing useful heterostructures for long-wavelength infrared detectors [18–23]. Some of the multilayer heterostructures of current interest are GaAs/Al$_x$Ga$_{1-x}$As multiple quantum wells, InAs/Ga$_{1-x}$In$_x$Sb strained-layer super-lattices, Si$_{1-x}$Ge$_x$/Si multiple quantum wells, and HgTe/CdTe type III superlattices. How these novel band structures can be used for infrared detection will be briefly discussed in the next sections and then covered in more depth in the following chapters. The growth issues of GaAs/AlGaAs multiple quantum wells and InAs/Ga$_{1-x}$In$_x$Sb strained-layer superlattices will be discussed in Chapter 4, whereas the growth issues of Si$_{1-x}$Ge$_x$/Si multiple quantum wells and HgTe/CdTe superlattices will be discussed in Chapters 5 and 6, respectively.

1.3 GaAs/Al$_x$Ga$_{1-x}$As MULTIPLE QUANTUM WELLS

In the GaAs/Al$_x$Ga$_{1-x}$As multiple quantum well heterostructure, the mechanism for obtaining long-wavelength infrared radiation absorption is the intersubband transition between the ground state and the first excited state of the quantum wells. The energy separation between these levels is determined by the size of the quantum well, and theoretically goes to zero for sufficiently thick wells. A number of types of GaAs/Al$_x$Ga$_{1-x}$As multiple quantum well detector structures with different Al content in the barriers and various layer thicknesses can be selected to have intersubband transition energies of approximately 12.5 μm [24–27]. Several key experimental and theoretical milestones ranging from the first detection of infrared radiation, fast-response quantum well detectors, 1% quantum efficiency, high-detectivity infrared detectors, to the origin of the blue shift in the intersubband transitions have been achieved during the past several years in these types of infrared detectors [28–40].

The responsivity and detectivity of several of these quantum well detector designs have already been measured [26–27, 41–42]. The most promising design exploits a bound-to-extended state transition as the optical excitation mechanism for the infrared detector, [26, 27, 29, 42, 43]. In the bound-to-extended state multiple quantum well detector, the quantum wells are designed to have two intersubband states. One state is bound in the quantum well and the second state lies slightly above the potential barrier surrounding the well (see Figure 1.4(a)). This is accomplished by narrowing the width of a quantum well that has two bound states. As the well width is decreased, the upper bound state moves toward the top of the well and eventually lies above the potential barrier formed by the Al$_x$Ga$_{1-x}$As conduction band minimum. This state is no longer bound and forms a quantum state in the classical continuum. The significantly stronger dipole oscillator strength of the excited state greatly enhances the bound-to-continuum state absorption. However, if the GaAs wells are undoped, the ground state (lower bound state) is not occupied by electrons, and no photon absorption can occur. The absorption strength thus depends on the number of electrons in the ground state and the number of unoccupied excited states. To fill the ground states, the GaAs layers are heavily doped n-type by the addition

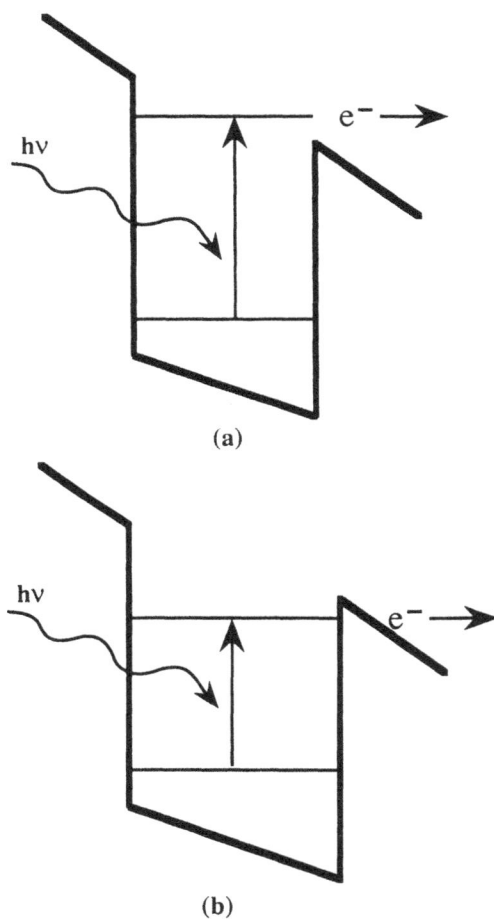

Figure 1.4 Bound-to-continuum transition (a) and bound-to-bound transition (b) in a GaAs/AlGaAs quantum well structure under an applied electric field.

of silicon impurities ($\sim 10^{18}$ cm^{-3}). Infrared response is usually a broad band in the case of bound-to-continuum transition. On the other hand, a narrow band and a stronger photoresponse can be obtained by designing the quantum well so that the excited subband state is just below the top of the barrier conduction band (see Figure 1.4(b)). In the latter case, the excited electrons will tunnel through a small potential barrier.

The absorption spectrum for a bound-to-extended-continuum state GaAs/Al$_{0.2}$Ga$_{0.8}$As multiple quantum well is shown in Figure 1.5. The sharp rise in the

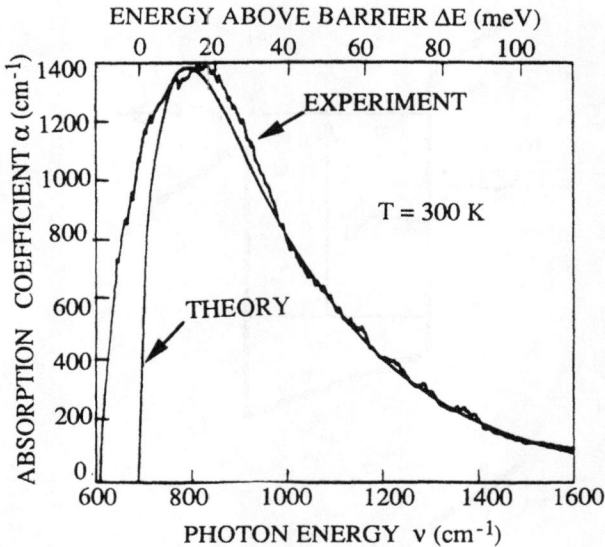

Figure 1.5 Measured and calculated absorption spectrum at 300 K of a bound-to-extended state in a GaAs/AlGaAs multiple quantum well structure. (After Levine et al. [43].)

theoretical spectrum at 680 cm^{-1} corresponds to a photon energy equal to the difference between the bound state energy and top of the Al$_{0.2}$Ga$_{0.8}$As barrier, where the extended continuum states begin. For energies less than this, there are no states in the well and hence no absorption. As the photon energy increases above this threshold value, the measured absorption increases reaching a peak absorption at 830 cm^{-1} (~12 μm) corresponding to an energy of 19 meV above the top of the barrier. At very high energies there is a slow decrease of the absorption because the extended continuum states become more free-electron-like, and hence the dipole matrix element of the transition between the localized bound state in the well and these delocalized continuum states decreases. This absorption spectrum shows the cutoff wavelength (λ_c = 14.3 μm), the peak response wavelength (λ_p = 12 μm), and the photon absorption strength of an infrared detector fabricated from this multiple quantum well structure.

A typical design for a 10 μm infrared detector is given by Levine et al. [27]. First, a heavily doped n^+ GaAs epitaxial contact layer is grown by molecular beam epitaxy on a GaAs substrate. Next, the GaAS/Al$_{0.25}$Ga$_{0.75}$As multiple quantum well structure is grown, which consists of 50 periods of 40 Å thick GaAs wells (doped n-type at 1.5 × 10^{18} cm^{-3}) and 480 Å thick undoped Al$_{0.25}$Ga$_{0.75}$As barriers. Then another heavily doped GaAs epilayer is grown to serve as a top electrical contact. This layered structure is then etched into 200 μm diameter mesas, and gold wires are indium bonded to the top and bottom contact layers. This structure was demonstrated to be a high-sensitivity 10 μm infrared detector at temperatures from 4 to 77 K.

One drawback to the GaAs/Al$_x$Ga$_{1-x}$As multiple quantum well infrared detector is that photons at normal incidence to the multiple quantum well interfaces are not absorbed by this structure. Due to the dipole selection rules associated with the intersubband transitions, the photons must have a component of polarization normal to the multiple quantum well interfaces to excite the transition and be absorbed [15, 44]. For test structures this requirement has been met by back illumination through a 45° polished facet. However, this arrangement is not suitable for two-dimensional infrared detector imaging arrays. One alternative is to etch a grating into the top contact layer of each pixel (mesa) to scatter the incident photons into propagation paths at an angle to the multiple quantum well interfaces [45]. Another alternative is to use a waveguide beneath the multiple quantum well mesas to allow the coupled radiation to enter at an angle and pass through the multiple quantum well stack several times [46]. The most efficient method for optical coupling to the multiple quantum wells is a combination of these techniques, called a *grating coupled waveguide* [47–49]. A grating coupled waveguide is formed by a reflection grating on one side of the multiple quantum well stack and a cladding layer consisting of AlAs on the other. These grating coupled waveguides can have quantum efficiencies of nearly unity (90%–95%) for photons at normal incidence [49].

1.4 InAs/Ga$_{1-x}$In$_x$Sb STRAINED-LAYER SUPERLATTICES

The InAs/Ga$_{1-x}$In$_x$Sb strained-layer superlattice system was proposed by Mailhiot and Smith as an alternate candidate for long-wavelength infrared detection [50, 51]. This material system is a type II superlattice with (staggered) band alignments (see Chapter 4) similar to the InAs/GaSb system [52–54]. In both cases, the InAs conduction band is lower in energy than the GaSb valence band. Because of this unusual band lineup, the superlattice can have a bandgap smaller than that of either constituent material. However, also because of this band lineup, the electrons and holes tend to be localized in different material layers, the electrons in the InAs and the holes in the GaInSb. As a result the electron-hole wave function overlap, that is, the optical matrix element for infrared absorption, decreases rapidly with the increasing superlattice period. In the InAs/GaSb system, the thick layers ($d > 50$ Å) required to reach long-wavelength ($\lambda_c > 10$ μm) sensitivity degrade the optical absorption significantly, due to the increased spatial separation between the electrons and holes in the superlattice. However, by alloying the GaSb with InSb, so that the superlattice is $InAs/Ga_{1-x}In_xSb$, small bandgaps (i.e., long wavelengths) can be achieved with thinner material layers ($d < 25$ Å), thereby increasing the electron-hole overlap. High optical absorption coefficients were calculated for the alloyed $InAs/Ga_{1-x}In_xSb$ superlattice [51].

Alloying GaSb with InSb also increases the lattice constant of the material and causes a small lattice mismatch (<5%) between the $Ga_{1-x}In_xSb$ and InAs layers. Because the lattice mismatch is small, the alternating epilayers can be grown coherently as the layers are strained to the same lattice constant at the growth interface. The thin layers are alternately in biaxial compression and tension, so that the in-plane lattice constants of the individual strained layers are equal. The tetragonal distortions shift the bulk energy levels and split the valence band degeneracies of the light- and heavy-hole energy levels. These strain effects lead to a further enhancement of the $InAs/Ga_{1-x}In_xSb$ superlattice properties. The presence of coherent strain between the InAs and $Ga_{1-x}In_xSb$ layers shifts the band edges such that the superlattice energy gap is reduced. In the strained superlattice, the bandgap forms between electron states split upward from the InAs conduction band and heavy-hole states split downward from the $Ga_{1-x}In_xSb$ valence band. This reduced bandgap is advantageous because longer cutoff wavelengths can be obtained with reduced layer thickness in the strained superlattice, leading to even higher optical absorption coefficients.

Thus, the superlattice bandgap can be tailored to energies suitable for long-wavelength infrared detectors by adjusting the layer thicknesses and the $Ga_{1-x}In_xSb$ alloy composition. The bandgap calculations by Smith and Mailhiot [50, 51] predict that, for a given layer thickness, a superlattice grown along the (111) axis exhibits a smaller energy gap than one grown along the (100) axis, by approximately 40 meV. This difference is large considering that the bandgap of a 10 μm infrared detector is only 100 meV. The larger bandgap of the (100) $InAs/Ga_{1-x}In_xSb$ superlattice would require thicker layers to produce the same long-wavelength cutoff as a (111) superlattice of the same composition. Therefore, the (111) $InAs/Ga_{1-x}In_xSb$ superlattice should have superior optical absorption of the infrared radiation.

At present, there are several reports on the successful growth of $InAs/Ga_{1-x}In_xSb$ superlattices [12, 55–57]. These superlattices were grown along the (100) axis on GaAs substrates by molecular beam epitaxy. Buffer layers were used between the GaAs substrate and $InAs/Ga_{1-x}In_xSb$ superlattice to reduce the mismatch strain between the substrate and the superlattice. The $InAs/Ga_{1-x}In_xSb$ superlattices were grown with layer thicknesses of 25–45 Å and alloy compositions with x values of 0.0 and 0.25. Photoluminescence results for these superlattices verified that the energy gap of the $InAs/Ga_{0.75}In_{0.25}Sb$ superlattice was significantly less than that of an $InAs/GaSb$ superlattice with the same layer thicknesses [55]. In addition, photoconductive devices were fabricated by contacting mesa structures etched into the superlattices. The photoconductivity results agreed closely with the photoluminescence results for these superlattices [12].

1.5 Si/Si$_{1-x}$Ge$_x$ MULTIPLE QUANTUM WELLS

Infrared detectors based on Si/Si$_{1-x}$Ge$_x$ superlattices [58] and multiple quantum wells [59–63] have also been proposed. An important advantage of using Si/Si$_{1-x}$Ge$_x$ multiquantum well or superlattice structures is that a silicon-based infrared detector allows monolithic integration of the detectors and the readout devices currently made of silicon. In the case of Si/Si$_{1-x}$Ge$_x$ multiple quantum wells and superlattices, either conduction band offsets [59–60, 64], where the silicon conduction band is lower than that of the Si$_{1-x}$Ge$_x$, or valence band offsets [61–62], where the silicon valence band is lower than that of the SiGe alloy, can be used for long-wavelength detectors based on intersubband transitions. Due to the relative size of the band offsets, the valence band quantum wells are suitable for 8–14 μm detectors and the conduction band quantum wells are suitable for the 20–30 μm range. Although theoretical modeling of each of these designs has been reported, there have been experimental results only for designs using the intersubband transitions in the valence band quantum wells of Si/Si$_{1-x}$Ge$_x$ structures.

In principle, the infrared detection process of the GaAs/Al$_x$Ga$_{1-x}$As multiple quantum wells can be followed in Si/Si$_{1-x}$Ge$_x$ valence band quantum wells as well. The quantum wells are again designed to have strong ground-to-continuum- state transitions, only now the excited state is in the valence band continuum and the charge carriers are holes instead of electrons. The valence band quantum well is the Si$_{1-x}$Ge$_x$ layer and is doped p type to fill the ground state with holes. The silicon layers form the potential energy barriers. As before, the operating range of the multiple quantum well infrared detector is determined by the choice of well composition (Si$_{1-x}$Ge$_x$) and thickness. However, there is one additional complication when using valence band quantum wells. In the quantum well valence band there is an energy separation between the light-hole band and the heavy-hole band, due to mass different quantization and strain in the Si$_{1-x}$Ge$_x$ layer [63]. This energy separation causes two sets of ground and excited states in the well (Figure 1.6). Therefore, more than one intersubband transition may be possible and must be considered in the design.

For the Si/Si$_{1-x}$Ge$_x$ heterostructures employing conduction band quantum wells, there is the important potential advantage of absorbing radiation at normal incidence. The SiGe alloy has an indirect bandgap that allows the electrons to have an anisotropic effective mass when the conduction band minima are not oriented along the growth axis. The effective mass tensor then has components that can couple the electron motion to the perpendicular motion of radiation at normal incidence; that is, the optical matrix element for the intersubband transition is nonzero. In particular, the (110) and (111) growth orientations will provide large, effective mass components in the growth direction [60]. Therefore, the Si/Si$_{1-x}$Ge$_x$ n-type quantum well detector does not require the GaAs/Al$_x$Ga$_{1-x}$As grating schemes to achieve a large infrared absorption coefficient for radiation at normal incidence. This advantage makes the n-type Si/Si$_{1-x}$Ge$_x$ system a desirable choice for infrared detector arrays.

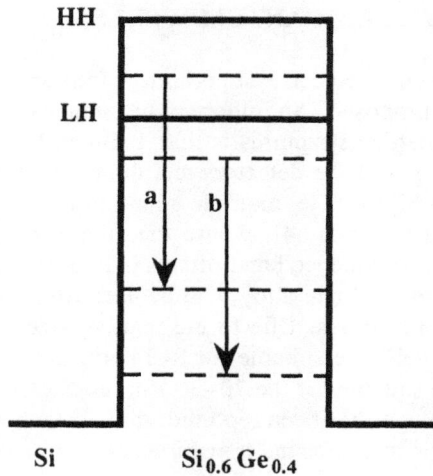

Figure 1.6 Valence band structure of the $Si/Si_{0.6}Ge_{0.4}$ quantum well structure showing the intersubband transitions of both (a) heavy holes and (b) light holes.

The growth of $Si/Si_{1-x}Ge_x$ multiple quantum wells and superlattices has been possible for only a few years and is not as well developed as some of the III–V materials, such as $GaAs/Al_xGa_{1-x}As$. So, much materials research remains to be done on this system. For instance, growth of high-quality (100) material will have to be accomplished before (111) oriented growth is attempted. Also, further modeling of the quantum well compositions and thicknesses required for long-wavelength intersubband transitions needs to be done. This is mainly because the magnitudes of the band offsets needed for the calculations are not well established.

1.6 CONCLUSION

The slow progress in the development of large photovoltaic HgCdTe infrared imaging arrays and the rapid development of other material systems, such as those discussed in the previous sections, have made it difficult to predict what arrays will be available for future applications, such as spaceborne surveillance and infrared seeker-tracker systems. The research on semiconductor heterostructure systems as possible choices for the next generation of long-wavelength infrared detectors is a rapidly advancing field, and the potential of these heterostructures is not yet fully realized and explored (see [65, 66]).

At present at least three semiconductor heterostructures (multiquantum well and superlattice) are competing for selection as the next generation long-wavelength infrared detector materials. These systems are discussed in the next chapters of this book. In addition to having different infrared detection mechanisms, each of these heterostructures has different materials problems to overcome before it can be used in infrared detector imaging arrays. For instance, although the $GaAs/Al_xGa_{1-x}As$ multiquantum well has a mature growth technology and a well-understood optical absorption mechanism, it suffers from an inability to detect radiation at normal incidence. The $InAs/Ga_{1-x}In_xSb$ superlattice growth process is still in the developmental stage, and further research is needed to see if this system can achieve its theoretical potential. The $Si/Si_{1-x}Ge_x$ multiquantum well lacks both a high-quality growth process and a well-developed model of the infrared absorption transitions. In the next few years all of these material problems will have to be addressed and optimized detector structures developed.

REFERENCES

[1]   Vincent, J.D., *Fundamentals of Infrared Detectors Operation and Testing*, John Wiley and Sons, New York, 1990.

[2]   Wilson, J., and J.F.B. Hawkes, *Optoelectronics an Introduction*, Prentice-Hall, Englewood Cliffs, N.J., 1983.

[3]   Capasso, F., *J. Vac Sci. Technol. B*, Vol. 1, 1983, p. 457.

[4]   Capasso, F., *Science*, Vol. 235, 1987, p. 172.

[5]   Dingle, R., "Lightwave Communication Technology," in *Semiconductors and Semimetals*, ed. R.K. Willardson and A.C. Beer, Volume 22, Academic Press, Orlando, Fla., 1985.

[6]   Dingle, R., W. Wiemann, and C.H. Henry, *Phys. Rev. Lett.*, Vol. 33, 1974, p. 827.

[7]   Dingle, R., "Applications of Multiquantum Wells, Selective Doping and Superlattices," in *Semiconductor and Semimetals*, ed. R.K. Willardson and A.C. Beer, Volume 24, Academic Press, San Diego, Calif., 1987.

[8]   Esaki, L., and R. Tsu, *IBM J. Res. Dev.*, Vol. 14, 1970, p. 61.

[9]   O'Reilly, E.P., *Semicond. Sci. Technol.*, Vol. 4, 1989, p. 121.

[10] Osbourn, G.C., *Semicond. Sci. Technol.*, Vol. 5, 1990, p. S5.

[11] Levine, B.F., K.K. Choi, C.G. Bethea, J. Walker, and R.J. Malik, *Appl. Phys. Lett.*, Vol. 50, 1987, p. 1092.

[12] Miles, R.H., D.H. Chow, J.N. Schulman, and T.C. McGill, *Appl. Phys. Lett.*, Vol. 57, 1990, p. 801.

[13] West, L.C., and S.J. Englash, *Appl. Phys. Lett.*, Vol. 46, 1985, p. 1156.

[14] Esaki, L., *IEEE J. Quantum Electron.*, Vol. QE-22, 1986, p. 1611.

[15] Osbourn, G.C., *J. Vac. Sci. Technol. B*, Vol. 2, 1984, p. 176.

[16] Mailhiot, C., and D.L. Smith, *J. Vac. Sci. Technol. A*, Vol. 7, 1989, p. 445.

[17] Dohler, G.H., *IEEE J. Quantum Electron.*, Vol. QE-22, 1986, p. 1682.

[18] Kurtz, S., L.R. Dawson, T.E. Zipperian and S.R. Lee, *Appl. Phys. Lett.*, Vol. 52, 1988, p. 1581.

[19] Kurtz, S., R.M. Bielfeld, L.R. Dawson, I.J. Fritz, and T.E. Zipperian, *Appl. Phys. Lett.*, Vol. 53, 1988, p. 1961.

[20] Levine, B.F., R.J. Malik, J. Walker, K.K. Choi, C.G. Bethea, D.A. Kleinman, and J.M. Vandenburg, *Appl. Phys. Lett.*, Vol. 50, 1987, p. 273.

[21] Choi, K.K., B.F. Levine, C.G. Bethea, J. Walker, and R.J. Malik, *Appl. Phys. Lett.*, Vol. 50, 1987, p. 1814.

[22]   Goosen, K.W., and S.A. Lyon, *Appl. Phys. Lett.*, Vol. 47, 1985, p. 1257.
[23]   Choi, K.K., B.F. Levine, R.J. Malik, J. Walker, and C.G. Bethea, *Phys. Rev. B*, Vol. 15, 1987, p. 4172.
[24]   Levine, B.F., K.K. Choi, C.G. Bethea, J. Walker, and R.J. Malik, *Appl. Phys. Lett.*, Vol. 50, 1987, p. 1092.
[25]   Levine, B.F., C.G. Bethea, K.K. Choi, J. Walker, and R.J. Malik, *Appl. Phys. Lett.*, Vol. 53, 1988, p. 231.
[26]   Levine, B.F., G. Hasnain, C.G. Bethea, and N. Chand, *Appl. Phys. Lett.*, Vol. 54, 1989, p. 2704.
[27]   Levine, B.F., C.G. Bethea, G. Hasnain, V.O. Shen, E. Pelve, R.R. Abbott, and S.J. Hsieh, *Appl. Phys. Lett.*, Vol. 56, 1990, p. 851.
[28]   Chiu, L.C., J.S. Smith, S. Margalit, A. Yariv, and A.Y. Cho, *Infrared Physics*, Vol. 23, 1983, p. 93.
[29]   Coon, D.D., and R.P.G. Karunasiri, *Appl. Phys. Lett.*, Vol. 45, 1984, p. 649.
[30]   Coon, D.D., R.P.G. Karunasiri, and H.C. Liu, *J. Appl. Phys.*, Vol. 60, 1986, p. 2636.
[31]   Herwit, A., and J.S. Harris, Jr., *Appl. Phys. Lett.*, Vol. 50, 1987, p. 685.
[32]   Levine, B.F., C.G. Bethea, G. Hasnain, J. Walker, and R.L. Malik, *Appl. Phys. Lett.*, Vol. 53, 1988, p. 296.
[33]   Goossen, K.W., and S.A. Lyon, *Appl. Phys. Lett.*, Vol. 52, 1988, p. 1701.
[34]   Kastalsky, A., T. Duffield, S.J. Allen, and J. Harbison, *Appl. Phys. Lett.*, Vol. 52, 1988, p. 1320.
[35]   Liu, H.C., D.D. Coon, O. Byungsung, Y.F. Lin, and M.H. Francombe, *Superlattices and Microstructures*, Vol. 4, 1988, p. 343.
[36]   Goosen, K.W., S.A. Lyon, and K. Alavi, *Appl. Phys. Lett.*, Vol. 53, 1988, p. 1027.
[37]   Levine, B.F., C.G. Bethea, K.K. Choi, J. Walker, and R. Malik, *J. Appl. Phys.*, Vol. 64, 1988, p. 1591.
[38]   Bandara, K.M.S.V., D.D. Coon, O. Byungsung, Y.F. Lin, and M.H. Francombe, *Appl. Phys. Lett.*, Vol. 53, 1988, p. 1931.
[39]   Manasreh, M.O., F. Szmulowicz, T. Vaughan, K.R. Evans, C.E. Stutz, and D.W. Fischer, *Phys. Rev. B*, Vol. 43, 1991, p. 9996.
[40]   Manasreh, M.O., F. Szmulowicz, D.W. Fischer, K.R. Evans, and C.E. Stutz, *Appl. Phys. Lett.*, 57, 1990, p. 1790.
[41]   Levine, B.F., C.G. Bethea, G. Hasnain, J. Walker, and R.J. Malik, *Appl. Phys. Lett.*, Vol. 53, 1988, p. 296.
[42]   Janousek, B.K., M.J. Daugherty, W.L. Bloss, M.L., Rosenbluth, M.J. O'Loughlin, H. Kanter, F.J. De Luccia and L.E. Perry, *J. Appl. Phys.*, Vol. 67, 1990, p. 7608.
[43]   Levine, B.F., C.G. Bethea, K.K. Choi, J. Walker, and R.J. Malik, *J. Appl. Phys.*, 64, 1988, p. 1591.
[44]   Allen, S.J., Jr., D.C. Tsui, and B. Vinter, *Solid State Commun.*, Vol. 20, 1976, p. 425.
[45]   G. Hasnain, B.F. Levine, C.G. Bethea, R.A. Logan, J. Walker, and R.J. Malik, *Appl. Phys. Lett.*, Vol. 54, 1989, p. 2515.
[46]   Hasnain, G., B.F. Levine, C.G. Bethea, R.R. Abbott, and S.J. Hsieh, *J. Appl. Phys.*, Vol. 67, 1990, p. 4361.
[47]   Kane, M.J., and N. Apsley, *J. Phys. C*, Vol. 5, 1987, p. 545.
[48]   Andersson, J.Y., L. Lundqvist, and Z.F. Paska, *Appl. Phys. Lett.*, Vol. 58, 1991, p. 2264.
[49]   Andersson, J.Y., and L. Lundqvist, *Appl. Phys. Lett.*, Vol. 59, 1991, p. 857.
[50]   Mailhiot, C., and D.L. Smith, *J. Vac. Sci. Technol. B*, Vol. 5, 1987, p. 1268.
[51]   Smith, D.L., and C. Mailhiot, *J. Appl. Phys.*, Vol. 62, 1987, p. 2545.
[52]   Sai-Halasz, G.A., R. Tsu, and L. Esaki, *Appl. Phys. Lett.*, Vol. 30, 1977, p. 651.

[53]   Sai-Halasz, G.A., L.L. Chang, J.M. Walter, C.A. Chang, and L. Esaki, *Solid State Commun.*, Vol. 27, 1978, p. 935.

[54]   Bleuse, J., P. Voisin, M. Voos, H. Munekata, L.L. Chang, and L. Esaki, *Appl. Phys. Lett.*, Vol. 52, 1988, p. 462.

[55]   Chow, D.H., R.H. Miles, J.R. Soderstrom, and T.C. McGill, *Appl. Phys. Lett.*, Vol. 56, 1990, p. 1418.

[56]   Chow, D.H., R.H. Miles, J.R. Soderstrom, and T.C. McGill, *J. Vac. Sci. Technol. B.*, Vol. 8, 1990, p. 710.

[57]   Campbell, I.H., I. Sela, B.K. Laurich, D.L. Smith, C.R. Bolonesi, L.A. Samoska, A.C. Gossard, and H. Kroemer, *Appl. Phys. Lett.*, Vol. 59, 1991, p. 846.

[58]   Turton, R.J., and M. Jaros, *Appl. Phys. Lett.*, Vol. 56, 1990, p. 767.

[59]   Chang, C.I., D.S. Pan, and R. Somoano, *J. Appl. Phys.*, Vol. 65, 1989, p. 3253.

[60]   Rajakarunanayake, Y. and T.C. McGill, *J. Vac. Sci. Technol. B*, Vol. 8, 1990, p. 929.

[61]   Karunasiri, R.P.G., J.S. Park, K.L. Wang, and L.J. Cheng, *Appl. Phys. Lett.*, Vol. 56, 1990, p. 1342.

[62]   Karunasiri, R.P.G., J.S. Park, Y.J. Mii, and K.L. Wang, *Appl. Phys. Lett.*, Vol. 57, 1990, p. 2585.

[63]   People, R., *Phys. Rev. B*, Vol. 32, 1985, p. 1405.

[64]   Chang, C.I., and D.S. Pan, *J. Appl. Phys.*, Vol. 64, 1988, p. 1573.

[65]   *Semiconductor Science and Technology*, Vol. C6, 1991.

[66]   NATO Advance Research Workshop on Intersubband Transitions in Quantum Wells, Institute d'Etudes Scientifiques de Cargése, France, September, 10–14 1991, NATO Series B., Vol. 288, 1992.

Chapter 2
Theoretical Modeling of the Intersubband Transitions in III–V Semiconductor Multiple Quantum Wells

J.P. Loehr and M.O. Manasreh
Wright Laboratory
Wright-Patterson Air Force Base

2.1 INTRODUCTION

In this chapter we present the basic theoretical modeling necessary to describe the intersubband transitions in type I heterostructure quantum wells. In these structures, the electrons and holes are confined to the *same* layer, as opposed to the type II structures in which the electrons and holes are confined in different layers. We will be interested primarily in transitions between conduction subbands in GaAs/AlGaAs quantum wells and will make extensive comparison with experimental results.

The effective mass approximation underlies almost all device analysis, and this is briefly described in Section 2.2. In Sections 2.3 and 2.4 we apply this approach to solve for the conduction band states in quantum wells. Transitions between these conduction subbands respond only to light that is polarized perpendicular to the layers. The *valence* intersubband transitions, however, can absorb *normally* incident radiation as well, thus offering potentially higher quantum efficiencies. Hence, we discuss the generalized $\mathbf{k} \cdot \mathbf{p}$ method for calculating the quantum well valence band states in Section 2.5. Because this valence bandstructure, as well as the bandgap, can be changed dramatically by applying biaxial strain, we discuss strain effects in Section 2.6.

In Section 2.7 we calculate the optical absorption coefficient for bound-to-bound transitions between conduction subbands in GaAs/AlGaAs structures. In addition, we present detailed comparisons with experimental results for these systems and show that it is important to include nonparabolicity and many-body effects to account for the correct dependence of the absorption coefficient on temperature and doping. When all of these effects are treated, it is possible to use the model to design device structures for a specific interband response. Finally, we conclude in Section 2.8.

2.2 THE EFFECTIVE MASS EQUATION

In any spatially periodic potential the electron states Ψ will take the form

$$\Psi_{n,\mathbf{k}}(\mathbf{r}) = e^{i\mathbf{k}\cdot\mathbf{r}}U_{n,\mathbf{k}}(\mathbf{r}) \qquad (2.1)$$

where $U_{n,\mathbf{k}}(\mathbf{r})$ retains the periodicity of the potential. This discovery, made by Bloch in 1928 [1], commenced the study of modern solid state physics and such functions $\Psi_{n,\mathbf{k}}$ are named after him. The index n is denoted as the *band* index, and the associated collection of energy eigenvalues $\mathcal{E}_n(\mathbf{k})$ is referred to as the *band structure* of the material. The calculation of these eigenvalues is greatly simplified by looking for solutions of the form (2.1). The band structure is extremely important because it is determines the density of states in the material; the wave functions are also significant because they play a part in determining the strength of various scattering processes. Any quantum mechanical study of solids must begin with the band structure, and in principle, any material property can be calculated directly from the bands.

Many methods of varying complexity are used to calculate the energies and wave functions in solids. In this chapter we are interested in primarily the properties of III–V semiconductors; in particular, we wish to investigate the intersubband transitions in multiple quantum wells made from this class of semiconductors. For these materials, the definitive calculation of the bulk band structure is generally taken to be the nonlocal pseudopotential calculation of Chelikowsky and Cohen [2]. This calculation requires extensive computation, however, and is prohibitively time consuming for the quantum well problem. Fortunately, we do not generally require the band structure for all values of $\mathbf{k}$ in the first Brillouin zone. Most of the electronic and long-wavelength optical properties of direct-gap type I semiconductor heterostructures are determined by the conduction and valence band edges. Hence, to calculate these properties we require only the band structure for small values of $\mathbf{k}$ near $\mathbf{k} = 0$. Later, we will discuss a simple perturbation-theory based method, the $\mathbf{k}\cdot\mathbf{p}$ method, used to calculate these band-edge states. Once this bulk band structure has been obtained, the effects of strain, free carriers, electric fields, and heterointerfaces can easily be incorporated within the effective mass approximation.

The effective mass equation (or envelope function approximation) forms the foundation for much of semiconductor and device analysis. It was first suggested by Wannier [3] and elaborated by Slater [4]. Later, a rigorous development of the equation and an extension to multiple bands was given by Luttinger and Kohn [5]. A particularly readable account, directed toward heterostructure applications, may be found in [6]. The approach, which is extremely general, allows one to treat the problem of a crystal in a slowly varying "external" potential $V_{ext}(\mathbf{r})$ rather easily. Instead of solving the full Hamiltonian, including the ion cores, the valence and core electrons, and the external potential, we can solve a reduced Hamiltonian that includes only the band structure $\mathscr{E}_n(\mathbf{k})$ and the external potential. The most common example occurs when only a single nondegenerate band n, such as the conduction band in GaAs, is considered. In this case, the wave function can be expressed as

$$\Psi(\mathbf{r}) = F(\mathbf{r})U_{n,0}(\mathbf{r}) \tag{2.2}$$

where the *envelope function* $F(\mathbf{r})$ and energy eigenvalue E satisfy the *effective mass equation*:

$$[\mathscr{E}_n(-i\nabla) + V_{ext}(\mathbf{r})]F(\mathbf{r}) = EF(\mathbf{r}) \tag{2.3}$$

Here $\mathscr{E}_n(-i\nabla)$ is obtained by evaluating at $\mathbf{k} = -i\nabla$ the second-order Taylor polynomial in $\mathbf{k}$ of $\mathscr{E}_n(\mathbf{k})$. Note that the periodic potential has disappeared, and we are left with only the external potential $V_{ext}(\mathbf{r})$. Of course, the information on the lattice structure has been incorporated via the kinetic energy term $\mathscr{E}_n(-i\nabla)$. It is also important to remember that the full wave function is given by eq. (2.2), the product of the envelope function and the zone center Bloch function.

The simplest case of eq. (2.3) occurs when there is no external potential and the band is spherical; that is,

$$\mathscr{E}_n(\mathbf{k}) = \frac{\hbar^2 k^2}{2m_e} \tag{2.4}$$

In this case the effective mass equation reduces to the Schrödinger equation for a free particle,

$$-\frac{\hbar^2}{2m_e}\nabla^2 F(\mathbf{r}) = EF(\mathbf{r}) \tag{2.5}$$

with the familiar solution

$$F(\mathbf{r}) = e^{i\mathbf{k}\cdot\mathbf{r}}$$

$$E(\mathbf{k}) = \frac{\hbar^2 k^2}{2m_e} \tag{2.6}$$

Hence, the conduction band electron behaves like a free particle with an "effective mass" m_e given by

$$\frac{\hbar^2}{m_e} = \frac{\partial^2 \mathscr{E}_n(0)}{\partial k^2} \qquad (2.7)$$

Although the effective mass m_e appears in an equation that is mathematically identical to the Schrödinger equation for a spinless particle in a scalar potential, it is best to think of m_e as a material property that is analogous to the dielectric constant. The dielectric constant allows us to calculate the response of solids to electric fields by lumping all the effects of the ions and electrons into a material parameter ε. Similarly, the effective mass simplifies the problem of electron motion in a periodic system of ion cores. Naturally, such approximations will break down when the wavelength of the particle (photon or electron) is on the scale of the atomic fluctuations. For example, the problem of x-ray diffraction by crystals is treated quite poorly when the effects of the crystal are represented simply by a static dielectric constant! Similarly, the effective mass equation does not give good results for high-energy electrons, which is equivalent to the observation that the bands become quite nonparabolic for large values of the wavevector (short-wavelength electrons). Fortunately, in direct-gap III–V semiconductors most of the carriers reside in the low-energy band-edge states, and these are represented quite well by the effective mass theory. This is the justification for applying many of the formulas derived for the quantum mechanics of free particles directly to conduction band electrons by making the simple replacement of the free electron mass m_0 by the effective conduction band electron mass m_e. Instead of pursuing this promising line of development, we proceed immediately to the quantum well case.

2.3 CONDUCTION BAND STATES IN QUANTUM WELLS

Modern epitaxial techniques permit us to fabricate structures with extremely sharp interfaces. Because the composition of each monolayer can be controlled, it is possible to grow materials of widely different bandgaps (such as GaAs and $Al_xGa_{1-x}As$) next to each other, thus creating the so-called heterostructure. This produces a very sharp bandgap discontinuity and dramatically affects the carrier behavior. One of the most interesting heterostructures to study is the *quantum well*, in which a thin layer (or well) of a narrow bandgap material is sandwiched between two thicker layers (or barriers) of a wider bandgap material. If the well width is less than the de Broglie wavelength of the carriers in the well material (~ 100 Å in most III–V compounds), the carrier is "quantum" confined, hence the term *quantum well*. Such structures should properly be regarded as completely new materials, with properties wholly distinct from either the well or barrier bulk materials—much in the same way that a crystalline material is completely different from a loose collection of its constituent atoms. Because the well width is smaller than the wavelength of the carriers, quantum wells are *not* truly three-dimensional structures but are instead *quasi-two-dimensional*. This change in the dimensionality of the system results in dramatic variation in many of the material properties of these structures (such as carrier masses, bandgaps, densities of state) and even suggests device concepts that are not possible with simple bulk materials, such as infrared detectors based on intersubband transitions.

The band structure of these quantum wells may easily be addressed within the effective mass equation. If we take the z-axis to be the growth direction, meaning that the material composition will vary in the z direction but will be uniform in the x–y plane, we can again use eq. (2.3), but we must allow the band structure $\mathscr{E}_n(\mathbf{k})$ to depend on z as well. Assuming an arbitrary reference for $\mathscr{E}_n(\mathbf{k}) = 0$, we can take two materials A and B with different dispersion relations and form a *quantum well* of width L by arranging them as

$$\mathscr{E}_n(\mathbf{k}) = \mathscr{E}_A(\mathbf{k}) = \mathscr{E}_A^{\Gamma} + \frac{\hbar^2 k^2}{2m_A} \qquad |z| < L/2$$

$$\mathscr{E}_n(\mathbf{k}) = \mathscr{E}_B(\mathbf{k}) = \mathscr{E}_B^{\Gamma} + \frac{\hbar^2 k^2}{2m_B} \qquad |z| > L/2$$

(2.8)

Thus, when $\mathscr{E}_A^{\Gamma} < \mathscr{E}_B^{\Gamma}$ the electrons see an effective z-dependent "potential" well given by

$$V_0(z) = \mathscr{E}_A^{\Gamma} \qquad |z| < L/2$$

$$V_0(z) = \mathscr{E}_B^{\Gamma} \qquad |z| > L/2$$

(2.9)

At this point we should comment on the significance of the "external" potential $V_{\text{ext}}(\mathbf{r})$ appearing in eq. (2.3), as opposed to the confinement "potential" defined by eq. (2.9). The *confinement* potential of eq. (2.9) arises from the $\mathscr{E}_n(-i\nabla)$ term in the effective mass equation. Hence, it is analogous to the discontinuity in dielectric constant that appears when Maxwell's equations are solved in layered media. The *external* potentials are analogous to the source terms in Maxwell's equations and represent the effects of any fields or charges *that are not already taken into account by the effective masses themselves*. Hence, the external potential must satisfy the Poisson equation, whereas the confinement potential does not. The most common sources of the external potential are applied biases, ionized donors or acceptors, and space charge fields produced by quantum confined carriers. The last may be easily taken into account by solving eq. (2.3) concurrently with the appropriate Poisson equation:

$$\nabla \cdot [\varepsilon(\mathbf{r})\nabla V_{\text{ext}}(\mathbf{r})] = 4\pi \, e\rho(\mathbf{r})$$

(2.10)

where

$$\rho(\mathbf{r}) = \varepsilon[p(\mathbf{r}) - n(\mathbf{r}) + N_D^+ - N_A^-] \tag{2.11}$$

represents the total *excess* charge. Later, we will present a numerical method capable of solving the effective mass equation under these conditions.

One further obstacle in the solution of eq. (2.3) remains; namely, the interpretation of the kinetic energy operator $\mathscr{E}_n(-i\nabla)$ when the material parameters, and hence the band structure $\mathscr{E}_n(\mathbf{k})$, depend on z. The problem arises because ∂/∂_z and $1/m(z)$ do not commute; hence, the ordering is important. Generally, for a single spherical band, the order is taken as

$$\mathscr{E}_n(-i\nabla,z) \rightarrow -\hbar^2 \nabla \cdot \left[\frac{1}{2m_e(z)} \nabla \right] \tag{2.12}$$

It must be admitted that this operator cannot be derived, but this is the preferred quadratic form for a number of reasons. First, the operator is Hermitian, which ensures real energy eigenvalues and conservation of probability for the envelope functions. Second, the operator conserves probability current across the interface, as can be seen by examining the form of the current operator [6]. Last, more recent results [7, 8] show that this is the only physically acceptable quadratic form for a kinetic energy operator with spatially varying mass under very general conditions. Hence, we assume eq. (2.12) and the effective mass equation for conduction electrons in a quantum well becomes

$$\left[-\frac{\hbar^2}{2m_e(z)} \left(\frac{\partial^2}{\partial x^2} + \frac{\partial^2}{\partial y^2} \right) - \frac{\hbar^2}{2} \frac{\partial}{\partial z} \frac{1}{m_e(z)} \frac{\partial}{\partial z} + V_0(z) \right] F(\mathbf{r}) = EF(\mathbf{r}) \tag{2.13}$$

This can be solved in the usual manner; that is,

$$F(\mathbf{r}) = \frac{1}{\sqrt{A}} e^{i(k_x x + k_y y)} f_m(\mathbf{k}_\parallel, z) \tag{2.14}$$

where $f_m(\mathbf{k}_\parallel, z)$ satisfies

$$\left[-\frac{\hbar^2}{2} \frac{d}{dz} \frac{1}{m_e(z)} \frac{d}{dz} + V_0(z) + \frac{\hbar^2 k_\parallel^2}{2m_e(z)} \right] f_m(\mathbf{k}_\parallel, z) = E_m(\mathbf{k}_\parallel) f_m(\mathbf{k}_\parallel, z) \tag{2.15}$$

where $\mathbf{k}_\parallel \equiv k_x \hat{x} + k_y \hat{y}$, m is a subband index, and A is an arbitrary normalization area. Note that the spatially dependent electron mass introduces a $\mathbf{k}_\parallel$-dependent potential term, so that the in-plane and perpendicular motions are not entirely decoupled. This introduces a nonparabolicity to the conduction subbands that broadens the intersubband absorption spectrum. If this effect is ignored, eq. (2.15) becomes the Schrödinger equation for a particle in a finite square well with varying mass. Such an equation can be solved by the usual method of matching solutions in each region with the appropriate current conserving boundary conditions. Numerous examples of this technique may be found in the literature [9]. Of course, this technique works only when the solutions are known in each region, restricting its application to square quantum wells, perhaps with applied electric fields.

When a substantial number of free carriers are present in the well, they will induce band bending. In this case, the effective mass equation will not take the simple form of eq. (2.15), but will contain a space-charge potential as discussed earlier. Hence, the analytic solutions in each region are not known, and we cannot employ the simple method of joining evanescent and oscillatory waves at the boundaries. Instead, some sort of numerical technique must be employed. One method would be to expand the solution $f_m(\mathbf{k}_\parallel, z)$ in terms of a "suitable" set of basis functions. If the basis functions were orthonormal, we would then obtain a secular equation for the expansion coefficients and the energy eigenvalue $E_m(\mathbf{k}_\parallel)$, provided, of course, that we could successfully calculate the matrix elements of the operator $H(z, \partial/\partial z)$ represented in eq. (2.15). This method has been employed successfully [10, 11], but it suffers from a number of drawbacks. In practice, the chosen basis functions are usually *not* orthogonal, resulting in a *generalized* eigenvalue problem instead of the standard secular determinant. Also, a finite number of basis functions must be employed, and this compromises the accuracy of the results. Perhaps the greatest difficulty, though, is that there is no way to determine the accuracy of the solution; moreover, there is no systematic and efficient algorithm that forces the solution to converge. This approach lies within the broad category of *variational* solutions, which are guaranteed to give only the minimal (best) energy eigenvalue within the class of wave functions chosen. Hence, we must possess significant insight to choose solutions that closely resemble the actual wave function. In general, of course, this has been done, and we do not wish to cast doubt on the validity of the variational calculations already reported in the literature. We *do* wish to point out, though, that great care must always be taken when applying the variational technique; hence, it is difficult to arrive at a variational solution that has the necessary flexibility and generality required to address the wide variety of well compositions and doping profiles that we may wish to investigate when designing infrared photodetectors. For this reason, we will present an alternative approach, the finite difference method, that *does* provide sufficient generality, accuracy, and convergence.

2.4 THE FINITE DIFFERENCE METHOD

Once we have arrived at eq. (2.15) for *any* potential $V(z)$, we need only to solve this equation for the energies $E_m(\mathbf{k}_\parallel)$ and wave functions $f_m(\mathbf{k}_\parallel, z)$. A reliable method for solving second-order ordinary differential equations is to replace the differential operators by finite difference approximations [12]. In this approach, the z-axis is subdivided into a finite number, $N - 1$, of intervals of width h; and we solve for the values of the function f at each of the N equally spaced mesh points z_j. This yields, for each value of the in-plane wave vector $\mathbf{k}_\parallel$, an $N \times N$ matrix equation that is diagonalized to obtain the wave functions at each point z_j and the energy eigenvalues. Convergence is then obtained by reducing the step size h. In general, this is a very straightforward procedure. We must take care in the discretization of eq. (2.15), however, because of the spatially inhomogenous effective mass. Because the effective Hamiltonian is Hermitian, a correct discretization will yield a Hermitian matrix, which we now derive.

We start by subdividing the z axis into a finite number of intervals, each of width h. We leave aside for the moment the issues of boundary conditions and the number of intervals chosen. We therefore consider only a set of discrete points z_j and the values of functions $f(z_j) \equiv f_j$ at these points (we leave off writing the explicit subband and $\mathbf{k}_\parallel-$ dependence of f for brevity). Note that $h = z_{j+1} - z_j$. The method is based on the centered-difference approximation to the derivative of a function

$$\left[\frac{\mathrm{d}}{\mathrm{d}z}f(z)\right]_{z_j} \approx \frac{f(z_j + h) - f(z_j - h)}{2h} = \frac{f_{j+1} - f_{j-1}}{2h} \tag{2.16}$$

By successive applications of eq. (2.16), we can discretize the eigenvalue equation. For convenience, let $g(z) \equiv 1/m_e(z)$. Then consider

$$\begin{aligned}\left[\frac{\mathrm{d}}{\mathrm{d}z}g(z)\frac{\mathrm{d}}{\mathrm{d}z}f(z)\right]_{z_j} &= \\ &= \frac{\left[g(z)\frac{\mathrm{d}}{\mathrm{d}z}f(z)\right]_{z_{j+1}} - \left[g(z)\frac{\mathrm{d}}{\mathrm{d}z}f(z)\right]_{z_{j-1}}}{2h} \\ &= \frac{g_{j+1}\left[\dfrac{f_{j+2} - f_j}{2h}\right] - g_{j-1}\left[\dfrac{f_j - f_{j-2}}{2h}\right]}{2h} \\ &= \frac{g_{j+1}f_{j+2} - (g_{j+1} + g_{j-1})f_j + g_{j-1}f_{j-2}}{(2h)^2}\end{aligned} \tag{2.17}$$

Recall that we know the value of $g(z)$ at *any* z; hence, we need solve for only the values of f at the evenly spaced mesh points $z_0 + 2jh$. Because we never evaluate the functions $1/m_e(z)$ and $f(z)$ at the same points, there is no ambiguity when z_j lies directly on the interface.

To write the finite difference matrix for eq. (2.15), we need only determine the boundary conditions on the z axis. In general, there are two conditions that are easy to implement—the superlattice and infinite barrier conditions. If we consider the multiquantum well or superlattice problem, then the wave function must satisfy the Bloch condition in the period d of the structure; that is,

$$f(z \pm d) = e^{\pm ik_z d} f(z) \tag{2.18}$$

This is easily implemented in the finite difference scheme by suitable replacement of f_0 and f_{N+1} according to eq. (2.18).

It is instructive to look at an example of the preceding equations in matrix form. Consider a superlattice consisting of, alternately, three monolayers of A material ($m_e \equiv m_A$) followed by three monolayers of B material ($m_e \equiv m_B$), with each monolayer having width Δ. Then we can write eq. (2.15) as a matrix equation by setting $2h = \Delta = d/(N-1)$ and solving for the values of $f_k = f(z_0 + k\Delta)$. If we take care of the end points with eq. (2.18) and let

$$K_A \equiv -\frac{\hbar^2}{2m_A\Delta^2} \qquad V_A \equiv \mathscr{E}_A^\Gamma + \frac{\hbar^2 k_\parallel^2}{2m_A}$$

$$K_B \equiv -\frac{\hbar^2}{2m_B\Delta^2} \qquad V_B \equiv \mathscr{E}_B^\Gamma + \frac{\hbar^2 k_\parallel^2}{2m_B} \tag{2.19}$$

then we obtain the matrix equation

$$
\begin{bmatrix}
-2K_A + V_A & K_A & 0 & 0 & 0 & K_A\,e^{-ik_z d} \\
K_A & -2K_A + V_A & K_A & 0 & 0 & 0 \\
0 & K_A & -(K_A + K_B) + V_A & K_B & 0 & 0 \\
0 & 0 & K_B & -2K_B + V_B & K_B & 0 \\
0 & 0 & 0 & K_B & -2K_B + V_B & K_B \\
K_A\,e^{ik_z d} & 0 & 0 & 0 & K_B & -(K_B + K_A) + V_B
\end{bmatrix} \times
$$

$$
\begin{bmatrix} f_1 \\ f_2 \\ f_3 \\ f_4 \\ f_5 \\ f_6 \end{bmatrix} = E_m(k_\parallel) \begin{bmatrix} f_1 \\ f_2 \\ f_3 \\ f_4 \\ f_5 \\ f_6 \end{bmatrix} \tag{2.20}
$$

for each value of $k_\parallel$. Note that the matrix is Hermitian, as remarked earlier. The matrix (2.20) would be a real symmetric tridiagonal matrix if not for the "corner terms," $e^{\pm ik_z d}$. Because real symmetric tridiagonal matrices can be diagonalized very quickly, it is attractive to eliminate these terms if possible. Physically, elimination of the corner terms corresponds to placing infinite potential barriers at the end points, forcing $f_0 = f_{N+1} = 0$. For wide barriers ($\sim$100 Å), this is a good approximation, and we can use a simplified tridiagonal form of eq. (2.20). Of course, the interval must be symmetrized to place the well in the middle (which was not done in the version of eq. (2.20) presented previously); otherwise we would force the wave function to zero at one edge of the well, which is not the desirable boundary condition. The *single* quantum well can also be solved with a tridiagonal matrix. In this case, the appropriate boundary condition (for bound states) is that the wave function vanish at $z = \pm\infty$. Of course, we can solve for f only on a finite interval, so we must truncate the wave function prematurely. We can obtain arbitrary accuracy, though, by extending the interval until the eigenvalues converge to the required tolerance.

Finally, we shall comment on the accuracy of the finite difference approach. In principle, eq. (2.15) can be solved to arbitrary accuracy by continually reducing the step size $2h$. It should be realized, however, that eq. (2.15) is itself an approximation and is valid only when the envelope function solution $f(z)$ does not vary appreciably over a unit cell. If convergent results cannot be obtained when $2h$ is set equal to the monolayer thickness, the effective mass approximation is invalid anyway, and in this case a more accurate method such as tight binding method [13] must be employed to calculate the band structure of the quantum well. Therefore, the finite difference solution has the remarkable property that, roughly speaking, it is physically accurate whenever it is numerically accurate, provided that the step size is equal to the monolayer thickness. For GaAs/Al$_x$Ga$_{1-x}$As quantum wells, as long as the well-barrier width is greater than about 30 Å the effective mass equation, and hence the numerical resolution of it presented earlier, gives accurate results.

2.5 VALENCE BAND STATES

As mentioned earlier, many methods can be used to calculate the band structure of solids. Because we are interested in primarily the *optical* properties of lattice matched and strained quantum wells, we require the band structure over only a small region of the Brillouin zone (small $\mathbf{k}$). Therefore, we may address the problem with the $\mathbf{k} \cdot \mathbf{p}$ perturbation theory [14–16]. This theory has the advantage that it can take into account the presence of more than one band, and it can also model the effects of external potentials (heterostructure barriers, applied electric fields, strain potentials, Coulombic potentials) easily through the effective mass approximation. Because we are interested in primarily GaAs, InGaAs, and AlGaAs quantum wells, we have used the four-band $\mathbf{k} \cdot \mathbf{p}$ or "Kohn-Luttinger" Hamiltonian to describe the valence band [5]. This includes the mixing between the heavy holes (HH), which transform under rotations of the cubic group like the $|\, 3/2, \pm 3/2\rangle$ states, and the light holes (LH), which transform like the $|\,3/2,\,\pm1/2\rangle$ states. In bulk semiconductors each of the heavy and light hole states has a spin-degenerate partner, resulting in a total of four bands and hence the 4×4 Kohn-Luttinger Hamiltonian.

It should be noted that in most III–V semiconductors there are two other pairs of bands, the conduction and split-off bands, that could couple to the heavy- and light-hole bands. To include this coupling fully would require a 6×6 or even an 8×8 $\mathbf{k}\cdot\mathbf{p}$ Hamiltonian. We have not done this in this chapter because in InGaAs the conduction bandgap is much larger than the matrix elements of $\mathbf{k}\cdot\mathbf{p}$ between the conduction and valence band states; likewise for the split-off bandgap. In fact, this is the essential approximation made in deriving the 4×4 $\mathbf{k}\cdot\mathbf{p}$ Hamiltonian. In narrow bandgap systems, such as InSb, or in materials with a low split-off bandgap, such as Si, the 4×4 approximation is not a good one and a larger Hamiltonian must be used.

2.5.1 Bulk $\mathbf{k}\cdot\mathbf{p}$ Theory

Before examining the valence bands in quantum wells, it is useful to review the form of the degenerate $\mathbf{k}\cdot\mathbf{p}$ Hamiltonian in bulk material [5]. In the 4×4 basis, the hole wave functions Ψ_k take the form

$$\Psi_{\mathbf{k}}(\mathbf{r}) = e^{i\mathbf{k}\cdot\mathbf{r}}\sum_{\nu} g^{\nu}(\mathbf{k})U_{\nu}^{h}(\mathbf{r}) \tag{2.21}$$

where $\mathbf{k}$ is the three-dimensional wave vector and the U_{ν}^{h} are the zone center Bloch functions having spin symmetry ν. The envelope functions $g^{\nu}(\mathbf{k})$ and energies $\mathscr{E}(\mathbf{k})$ satisfy

$$-\begin{bmatrix} P+Q & L & M & 0 \\ L^{\dagger} & P-Q & 0 & M \\ M^{\dagger} & 0 & P-Q & -L \\ 0 & M^{\dagger} & -L^{\dagger} & P+Q \end{bmatrix}\begin{bmatrix} g^{3/2,3/2}(\mathbf{k}) \\ g^{3/2,1/2}(\mathbf{k}) \\ g^{3/2,-1/2}(\mathbf{k}) \\ g^{3/2,-3/2}(\mathbf{k}) \end{bmatrix} = \mathscr{E}(\mathbf{k})\begin{bmatrix} g^{3/2,3/2}\mathbf{k} \\ g^{3/2,1/2}\mathbf{k} \\ g^{3/2,-1/2}\mathbf{k} \\ g^{3/2,-3/2}\mathbf{k} \end{bmatrix} \tag{2.22}$$

where the matrix elements are given by

$$P = \frac{\hbar^2\gamma_1}{2m_0}k^2 \tag{2.23}$$

$$Q = \frac{\hbar^2\gamma_2}{2m_0}(k_x^2 + k_y^2 - 2k_z^2) \tag{2.24}$$

$$L = \frac{-i\sqrt{3}\hbar^2\,\gamma_3}{m_0}\,(k_x - ik_y)k_z \tag{2.25}$$

$$M = \frac{\sqrt{3}\hbar^2}{4m_0}(\gamma_2 - \gamma_3)(k_x + ik_y)^2 + \frac{\sqrt{3}\hbar^2}{4m_0}(\gamma_2 + \gamma_3)(k_x - ik_y)^2 \tag{2.26}$$

Here m_0 is the free electron mass and the γ_i are the "Luttinger parameters." Lawaetz [17] has shown that for GaAs $\gamma_1 = 7.65$, $\gamma_2 = 2.41$, and $\gamma_3 = 3.28$; for InAs $\gamma_1 = 19.67$, $\gamma_2 = 8.37$, and $\gamma_3 = 9.29$; and for AlAs $\gamma_1 = 4.04$, $\gamma_2 = 0.78$, and $\gamma_3 = 1.57$. Because we are interested in the band structure of the ternary compounds $In_xGa_{1-x}As$ and $Al_xGa_{1-x}As$, we cannot directly use the results for the binary semiconductors. This difficulty is resolved in the virtual crystal approximation, in which the total crystal potential $V_{A_xB_{1-x}C}(\mathbf{r})$ is assumed to be the average potential $xV_{AC}(\mathbf{r}) + (1 - x)V_{BC}(\mathbf{r})$. Because the Luttinger parameters are directly related to matrix elements of the crystal potential, we simply take the average $\gamma_i^{A_xB_{1-x}C} = x\gamma_i^{AC} + (1 - x)\gamma_i^{BC}$ to get the Luttinger parameters for the ternary well material. The effective mass, on the other hand, is *inversely* related to the crystal potential, so we can obtain the mass of the alloy via

$$\left(\frac{1}{m_e}\right)^{A_xB_{1-x}C} = x\left(\frac{1}{m_e}\right)^{AC} + (1 - x)\left(\frac{1}{m_e}\right)^{BC} \tag{2.27}$$

The 4×4 $\mathbf{k}\cdot\mathbf{p}$ Hamiltonian can be solved analytically [18], giving the energies

$$\mathscr{E} = -P \pm (Q^2 + LL^\dagger + MM^\dagger)^{1/2} \tag{2.28}$$

Note that the bands remain parabolic, as $\mathscr{E}$ is quadratic in $|\mathbf{k}|$. It is often convenient to ignore the first term in the expression for M. This *axial approximation* [10] gives a band structure that depends only on the magnitude of the in-plane wave vector $\mathbf{k}$, as can be verified by examining the explicit form of eq. (2.28). This is a good approximation for GaAs/AlGaAs quantum wells and greatly simplifies the calculation of optical and electronic material properties.

2.5.2 Application of the 4×4 $\mathbf{k}\cdot\mathbf{p}$ Theory to the Valence Bands in Quantum Wells

To take into account the effects of confining potentials or applied fields in the z direction, we can take the effective mass approach and replace the quantum number k_z by the differential operator $-i\partial/\partial z$. In this case, the hole state $|m, \mathbf{k}_\parallel\rangle$ is given by

$$\langle r|m, \mathbf{k}_\parallel\rangle = \frac{1}{\sqrt{A}}e^{i\mathbf{k}_\parallel\cdot\mathbf{\rho}} \sum_\nu g_m^\nu (\mathbf{k}_\parallel, z)\, U_\nu^h(\mathbf{r}) \tag{2.29}$$

where $\mathbf{k}_\parallel$ is the in-plane two-dimensional wave vector, $\boldsymbol{\rho} \equiv x\hat{x} + y\hat{y}$ is the in-plane radial coordinate, the U_ν^h are the zone center Bloch functions having spin symmetry ν, m is a subband index, and A is an arbitrary normalization area. The envelope functions $g_m^\nu(\mathbf{k}_\parallel, z)$ and subband energies $E_m^h(\mathbf{k}_\parallel)$ satisfy an equation of the form (2.22), but with matrix entries obtained by replacing k_z in eqs. (2.23)–(2.26) with the operator $-i\partial/\partial z_h$ and by adding in the band-edge potential $V(z)$ on the diagonal; again, we must again be careful to symmetrize any products of k_z and $\gamma_i(z)$. In the axial approximation, this gives

$$P = \frac{\hbar^2\gamma_1(z)}{2m_0}(k_x^2 + k_y^2) - \frac{\hbar^2}{2m_0}\frac{d}{dz}\gamma_1(z)\frac{d}{dz} + V(z) \qquad (2.30)$$

$$Q = \frac{\hbar^2\gamma_2(z)}{2m_0}(k_x^2 + k_y^2) + \frac{\hbar^2}{m_0}\frac{d}{dz}\gamma_2(z)\frac{d}{dz} + V(z) \qquad (2.31)$$

$$L = -\frac{\sqrt{3}\hbar^2}{2m_0}(k_x - ik_y)\left[\gamma_3(z)\frac{d}{dz} + \frac{d}{dz}\gamma_3(z)\right] \qquad (2.32)$$

$$M = \frac{\sqrt{3}\hbar^2}{4m_0}[\gamma_2(z) + \gamma_3(z)](k_x - ik_y)^2 \qquad (2.33)$$

Recall that $\mathbb{O}^\dagger$ represents the *adjoint* of the operator $\mathbb{O}$. If $\mathbb{O}$ is simply a complex number, then $\mathbb{O}^\dagger = \mathbb{O}*$. If $\mathbb{O}$ represents a differential operator, however, we must be careful to take the adjoint properly. In particular, if $\alpha(z)$ is a complex function then it may be shown via integration by parts that

$$\left[\alpha(z)\frac{d}{dz}\right]^\dagger = \frac{d}{dz}[\alpha(z)]* \qquad (2.34)$$

This relation should be used to evaluate $L^\dagger$ in the effective mass approximation. Finally, note that the functions $g_m^\nu(\mathbf{k}_\parallel, z)$ depend on $\mathbf{k}_\parallel$ as well as z and that the energy bands are no longer, in general, parabolic.

2.6 EFFECT OF STRAIN ON THE CONDUCTION AND VALENCE BANDS OF QUANTUM WELLS

Recently, it has become possible to grow thin layers of lattice *mismatched* materials next to each other. The resultant *strain* placed on the material also affects the band structure and allows an additional degree of freedom in designing optoelectronic devices. In particular, because the lattice constant of InAs is 7% larger than that of GaAs, an $In_xGa_{1-x}As$ layer grown between $Al_xGa_{1-x}As$ wells will be placed under *compressive* strain. This will substantially change the band structure of the film and will modify the electronic and optical properties as well. In this section we will discuss a simple model, the *deformation potential* theory, that is often used to describe the effects of strain on the band structure.

To study the effect of strain on the material properties of semiconductors, it is first essential to establish the strain tensor produced by epitaxy. For systems of interest in this chapter, pseudomorphic growth (that is, no dislocations) is assumed. In this case, the epitaxial semiconductor layer is *biaxially* strained in the plane of the substrate by an amount $\varepsilon_\parallel$ and uniaxially strained in the perpendicular (growth) direction by an amount $\varepsilon_\perp$. For a thick substrate, the in-plane strain of the layer is determined from the bulk lattice constants of the substrate material α_S and that of the layer material α_L by

$$\varepsilon_\parallel = \frac{\alpha_S}{\alpha_L} - 1 \equiv \varepsilon \tag{2.35}$$

Because the layer is subjected to no stress in the perpendicular (growth) direction, the perpendicular strain $\varepsilon_\perp$ is simply proportional to $\varepsilon_\parallel$ via

$$\varepsilon_\perp = \frac{-\varepsilon_\parallel}{\sigma} \tag{2.36}$$

where the constant $\sigma \equiv c_{11}/2c_{12}$ is the Poisson ratio: c_{11} and c_{12} are the elastic stiffness constants of the layer. Therefore, for strains achieved by lattice-mismatched epitaxial growth on a (001) substrate, we have shown that the strain tensor is given by $\varepsilon_{xx} = \varepsilon_{yy} = \varepsilon$ and $\varepsilon_{zz} = -(2c_{12}/c_{11})\varepsilon$; all of the off-diagonal strain terms are 0 in this case.

Once the strain tensor ε_{ij} is known, we can apply the deformation potential theory to calculate the effects of strain on various eigenstates in the Brillouin zone. The strain perturbation Hamiltonian $H_\varepsilon^{\alpha\beta}$ is defined and its effects are calculated from first-order perturbation theory. In general we have [19–21]

$$H_\varepsilon^{\alpha\beta} = \sum_{ij} D_{ij}^{\alpha\beta}\, \varepsilon_{ij} \tag{2.37}$$

where $D_{ij}^{\alpha\beta}$ are the elements of the deformation potential operator and transform under symmetry operations as a second-rank tensor. A strain perturbation Hamiltonian for diamond lattices in the $|x\rangle$, $|y\rangle$, $|z\rangle$ basis has been developed by Bir and Pikus [20].

2.6.1 Conduction Band States

In the conduction band of direct bandgap materials, the strain tensor only shifts the position of the band edge by an energy

$$\delta E^{(CB)} = D_{xx} \left(\varepsilon_{xx} + \varepsilon_{yy} + \varepsilon_{zz} \right) \qquad (2.38)$$

and has a rather small effect on the carrier mass. The changes in the effective mass are shown in Figure 2.1. These results are based on calculations done for the lattice matched and strained systems by using the tight binding method adapted to include the deformation potential theory, and they show that the mass decrease occurs due to the addition of indium [22].

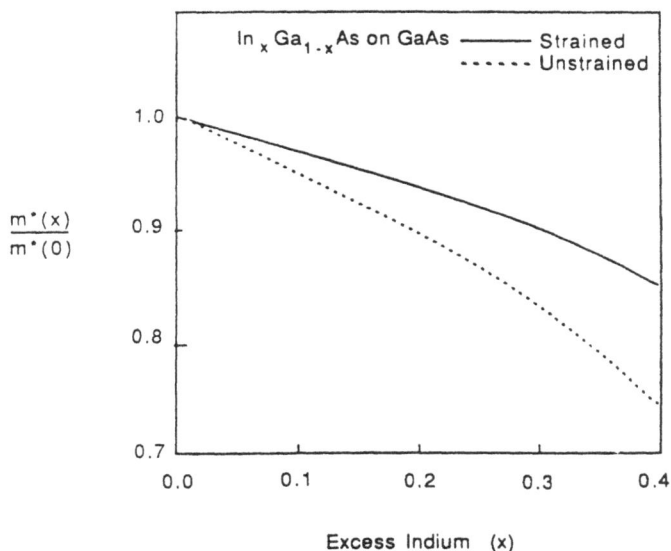

Figure 2.1 Electron effective mass as a function of excess indium mole fraction (x) for In$_x$Ga$_{1-x}$As. Results are given for lattice matched and strained systems. (After Jaffe [22].)

2.6.2 Valence Band States

The effects of strain on the valence band structure are somewhat more complicated, as the HH and LH states are degenerate at the valence band edge. To properly account for the strain it is essential to include full band mixing effects; the changes in the hole masses discussed later will not be given correctly if the HH-LH coupling is ignored. For films grown on an (001) substrate, eq. (2.37) can be used to obtain the strain perturbation Hamiltonian in the $|j, m\rangle$ basis. This gives only the diagonal contributions [23]

$$
\begin{aligned}
\delta E^{(\mathrm{HH})} &= \left[-2a\left(\frac{c_{11} - c_{12}}{c_{11}}\right) + b\left(\frac{c_{11} + 2c_{12}}{c_{11}}\right) \right] \varepsilon \\
\delta E^{(\mathrm{LH})} &= \left[-2a\left(\frac{c_{11} - c_{12}}{c_{11}}\right) - b\left(\frac{c_{11} + 2c_{12}}{c_{11}}\right) \right] \varepsilon
\end{aligned}
\tag{2.39}
$$

to the HH ($|\,3/2, \pm3/2\rangle$) and LH ($|\,3/2, \pm1/2\rangle$) states; here a and b are valence band deformation potentials. Note that there is a common (hydrostatic) term that serves only to shift the bandgap. The difference $\delta \equiv \delta E^{(\mathrm{HH})} - \delta E^{(\mathrm{LH})}$ provides a splitting between the heavy and light holes. For $In_xGa_{1-x}As$ films grown on GaAs, eq. (2.39) gives [23]

$$
\delta = 2b\left(\frac{c_{11} + 2c_{12}}{c_{11}}\right) \varepsilon = -6.4\varepsilon \ (\mathrm{eV})
\tag{2.40}
$$

Here the lattice mismatch ε is related to the excess In mole fraction x by $\varepsilon = -(0.07)x$. This strain perturbation is then inserted into the $4 \times 4\ \mathbf{k} \cdot \mathbf{p}$ theory [5] to obtain the hole dispersion relations in a quantum well. Because the strain affects only the diagonal components, the valence band effective mass Hamiltonian becomes

$$
-\begin{bmatrix} P + Q + \delta/2 & L & M & 0 \\ L^\dagger & P - Q - \delta/2 & 0 & M \\ M^\dagger & 0 & P - Q - \delta/2 & -L \\ 0 & M^\dagger & -L^\dagger & P + Q + \delta/2 \end{bmatrix} \begin{bmatrix} g_m^{3/2,3/2}(\mathbf{k}_\parallel, z) \\ g_m^{3/2,1/2}(\mathbf{k}_\parallel, z) \\ g_m^{3/2,-1/2}(\mathbf{k}_\parallel, z) \\ g_m^{3/2,-3/2}(\mathbf{k}_\parallel, z) \end{bmatrix}
$$

$$
= E_m^h(\mathbf{k}_\parallel) \begin{bmatrix} g_m^{3/2,3/2}(\mathbf{k}_\parallel, z) \\ g_m^{3/2,1/2}(\mathbf{k}_\parallel, z) \\ g_m^{3/2,-1/2}(\mathbf{k}_\parallel, z) \\ g_m^{3/2,-3/2}(\mathbf{k}_\parallel, z) \end{bmatrix}
\tag{2.41}
$$

where all terms have the same significance as in eqs. (2.29)–(2.33), and δ is given by eq. (2.40).

2.6.3 Bulk Results

We now examine the effects of strain on the valence bands in bulk material. In this case, because there is no confining potential in the z direction, it is not necessary to treat k_z as a differential operator, and eq. (2.41) reduces to a simple algebraic equation for the diagonalization of a 4×4 matrix. In fact, because the resulting matrix is so small, it is easy to include the full 6×6 $\mathbf{k} \cdot \mathbf{p}$ Hamiltonian [5] and examine the effect of strain on the split-off bands as well. The resulting bands can then be fit to parabolas to determine the effect of biaxial strain on the valence band effective masses. This has been done in Figure 2.2, which shows the effect of compressive ($\varepsilon < 0$) and tensile ($\varepsilon > 0$) strain on the *in-plane* (x–y) hole masses of bulk GaAs. Note that the split-off band is relatively unaffected by the strain and that compressive strain can reduce the heavy-hole (i.e., the $\mid 3/2, \pm 3/2 \rangle$ states) mass by a factor of three. When tensile strain is applied the character of the band-edge state changes from a heavy hole to a light hole, and therefore the curves are discontinuous at $\mathbf{k}$ = 0. This results because of the *splitting* that the strain introduces via the diagonal term δ, as defined in eq. (2.40). When compressive strain is applied, $\varepsilon < 0$ which makes $\delta > 0$. Thus the heavy hole is raised, the light hole is lowered, and the band-edge state is a heavy hole. When tensile strain is applied, the sign of ε, and hence δ, is changed and the band-edge state becomes a light hole. These effects are shown

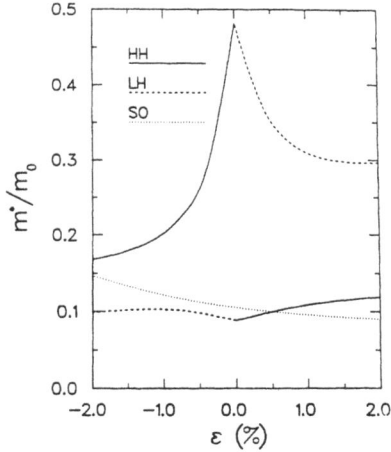

Figure 2.2 Effect of biaxial strain on hole masses in bulk GaAs. Results are obtained from parabolic fits to solutions of the 6×6 $\mathbf{k} \cdot \mathbf{p}$ Hamiltonian.

in Figure 2.3, which shows the effects of strain on the valence bandgaps in bulk GaAs. Note that the bands are degenerate in unstrained material and that strain increases the split-off bandgap as well.

At this point, we discuss briefly the reason why compressive strain reduces the heavy-hole mass. We will confine the discussion of the bulk 4×4 $\mathbf{k} \cdot \mathbf{p}$ Hamiltonian, but the argument will remain true for the 6×6 or quantum well Hamiltonian. Consider a solution to eq. (2.22) via second-order perturbation theory. In this case, we would take the zeroth-order solutions to be simply the diagonal terms, resulting in parabolic bands (as a function of $\mathbf{k}$). Examining the masses of these zeroth-order bands (given by $1/m^*_{\mathrm{HH}_{xx}} = \gamma_1 + \gamma_2$, for example) we would find that they are quite small, much smaller than the observed valence band masses in GaAs. Therefore, it is extremely important to take into account the off-diagonal coupling. We denote the off-diagonal operator by $H'(\mathbf{k})$. If this is treated in the second order, the energies become [24]

$$E_n^{(2)}(\mathbf{k}) = E_n^{(0)}(\mathbf{k}) + \langle n \mid H'(\mathbf{k}) \mid n \rangle + \sum_{n' \neq n} \frac{|\langle n|H'(\mathbf{k})|n'\rangle|^2}{E_n^{(0)}(\mathbf{k}) - E_{n'}^{(0)}(\mathbf{k})} \qquad (2.42)$$

where $E_n^{(0)}(\mathbf{k})$ represents the zeroth-order solution ($n = (3/2, \pm 3/2), (3/2, \pm 1/2)$). Therefore it is evident that the contribution of an off-diagonal matrix element $\langle n|H'(\mathbf{k})|n'\rangle$ is inversely proportional to the energy spacing between the states n and n'. The effect of compressive strain is to *increase* this spacing from $E_n^{(0)} - E_{n'}^{(0)}$ to

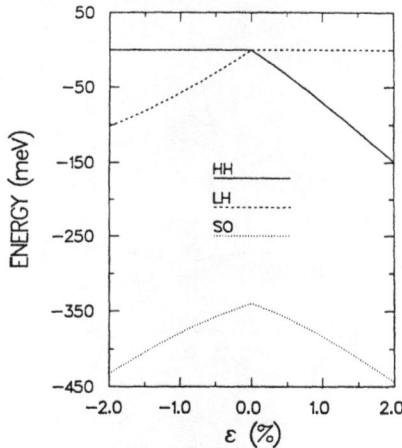

Figure 2.3 Effect of biaxial strain on valence bandgaps in bulk GaAs.

$E_n^{(0)} - E_{n'}^{(0)} + \delta$. Hence, the strain splitting δ reduces the contribution of the off-diagonal terms, allowing the bands to approach the zeroth-order solution, representing a much lower effective mass.

2.7 ABSORPTION COEFFICIENT FOR THE CONDUCTION INTERSUBBAND TRANSITION IN GaAs/AlGaAs MULTIPLE QUANTUM WELLS

We now consider the absorption of light by a multiple quantum well structure. For the wavelengths and intensities of radiation we are interested in, the absorption can be calculated from first-order time-dependent perturbation theory (that is, the "Fermi golden rule") in the dipole approximation [25]. In this case, the transition rate from an initial state $|i\rangle$ to a final state $|f\rangle$ is given by

$$W_{i \to f} = \frac{2\pi}{\hbar} \sum_f |\langle f|H_{\text{int}}|i\rangle|^2 \delta(E_f - E_i - \hbar\omega) \tag{2.43}$$

where the interaction between the incident radiation and the electrons is given in the dipole approximation by (cgs units)

$$H_{\text{int}} = \frac{-e\mathbf{A} \cdot \mathbf{p}}{m_0 c}, \qquad \mathbf{A} = \frac{c}{\eta} \sqrt{\frac{2\pi\hbar}{\omega}} \frac{1}{\sqrt{V_p}} \boldsymbol{\varepsilon} \tag{2.44}$$

Here $E_f(E_i)$ is the energy of the final (initial) state, $\hbar\omega$ is the energy of the incident radiation with polarization $\boldsymbol{\varepsilon}$, η is the index of refraction in the material, $\mathbf{p}$ is the momentum operator, and V_p is the arbitrary normalization volume for the photon.

To determine the selection rules for the transition, we must examine the transition matrix element $\langle f|\boldsymbol{\varepsilon} \cdot \mathbf{p}|i\rangle$. This is accomplished most conveniently by exploiting the following commutation relation, valid for any velocity-independent Hamiltonian H:

$$\frac{\mathbf{p}}{m_0} = \frac{i}{\hbar} [H, \mathbf{r}] \tag{2.45}$$

Substituting for $\mathbf{p}$ via this expression and taking H to be the full electronic Hamiltonian, we obtain through the Hermiticity of H

$$\langle f|\boldsymbol{\varepsilon} \cdot \mathbf{p}|i\rangle = \frac{im_0}{\hbar} (E_f - E_i)\boldsymbol{\varepsilon} \cdot \langle f|\mathbf{r}|i\rangle \tag{2.46}$$

Last, recalling that the energy-conserving delta function forces $E_f - E_i = \hbar\omega$, we see that

$$\langle f|\boldsymbol{\varepsilon} \cdot \mathbf{p}|i\rangle = im_0\omega\boldsymbol{\varepsilon} \cdot \langle f|\mathbf{r}|i\rangle \tag{2.47}$$

We can now evaluate this dipole matrix element for transitions between conduction subbands by using the explicit form of eqs. (2.2) and (2.14). Typically, we consider the initial state to be in the first subband and the final state in the second subband, giving

$$\Psi_i(\mathbf{r}) = \frac{1}{\sqrt{A}} \, e^{i\mathbf{k}_{\|_1}\cdot\boldsymbol{\rho}} f_1(\mathbf{k}_{\|_1}, z) \, U_0(\mathbf{r}) \tag{2.48}$$

$$\Psi_f(\mathbf{r}) = \frac{1}{\sqrt{A}} \, e^{i\mathbf{k}_{\|_2}\cdot\boldsymbol{\rho}} f_2(\mathbf{k}_{\|_2}, z) \, U_0(\mathbf{r}) \tag{2.49}$$

The matrix element is then

$$\langle f|\mathbf{r}|i\rangle = \frac{1}{A} \int d\boldsymbol{\rho} dz \, e^{i\boldsymbol{\rho}\cdot(\mathbf{k}_{\|_1}-\mathbf{k}_{\|_2})} (x\hat{x} + y\hat{y} + z\hat{z}) f_2^*(\mathbf{k}_{\|_2}, z) f_1(\mathbf{k}_{\|_1}, z) \, U_0^*(\mathbf{r})U_0(\mathbf{r}) \tag{2.50}$$

If the barriers on either side of the well are the same, as is usually the case, the Hamiltonian will have inversion symmetry about the center of the well. Hence, the envelope functions $f_m(\mathbf{k}_{\|_m}, z)$ will be either even or odd in z—generally f_1 is even and f_2 is odd. Also, because the cell-periodic portions $U_0(\mathbf{r})$ vary much more rapidly than the rest of the integrand and are orthonormal over a unit cell, they may be effectively dropped from the integrand. This may be shown in two ways: first, by expanding the function $U_0^*(\mathbf{r}) \, U_0(\mathbf{r})$ as a Fourier series in reciprocal lattice vectors $\mathbf{K}$ and keeping only the $\mathbf{K} = 0$ term [5]; second, by breaking the region of integration into unit cells and noting that the rest of the integrand is essentially constant over each cell [26]. Last, if we note that the functions f_m may be chosen real, then the integral becomes

$$\langle f|\mathbf{r}|i\rangle = \frac{1}{A} \int d\boldsymbol{\rho} \, e^{i\boldsymbol{\rho}\cdot(\mathbf{k}_{\|_1}-\mathbf{k}_{\|_2})} (x\hat{x} + y\hat{y}) \int dz \, f_2(\mathbf{k}_{\|_2}, z) f_1(\mathbf{k}_{\|_1}, z)$$
$$+ \frac{1}{A} \int d\boldsymbol{\rho} dz \, e^{i\boldsymbol{\rho}\cdot(\mathbf{k}_{\|_1}-\mathbf{k}_{\|_2})} \int dz f_2(\mathbf{k}_{\|_2}, z) z\hat{z} f_1(\mathbf{k}_{\|_1}, z) \tag{2.51}$$

The first term vanishes because f_1 is even and f_2 is odd. In the second term, the in-plane integral enforces momentum conservation, giving

$$\langle f|\mathbf{r}|i\rangle = \hat{z}\delta_{\mathbf{k}_{\|_1},\mathbf{k}_{\|_2}}M_{21}(\mathbf{k}_{\|_1}, \mathbf{k}_{\|_2}) \tag{2.52}$$

with

$$M_{21}(\mathbf{k}_{\|_1}, \mathbf{k}_{\|_2}) \equiv \int dz f_2(\mathbf{k}_{\|_2}, z) z f_1(\mathbf{k}_{\|_1}, z) \qquad (2.53)$$

Hence, we see that the conduction intersubband transitions respond only to $\hat{z}$-polarized light.

To get the absorption coefficient $\alpha(\hbar\omega)$ for the quantum well, we insert the preceding matrix element into the transition rate (2.43), weight the initial states according to their occupation probability f^0, sum over all initial states, and divide by the velocity c/η of light in the material, giving

$$\alpha(\hbar\omega) = \frac{4\pi^2 e^2(\hbar\omega)}{\eta c\hbar} \frac{1}{V_p} \sum_i f^0(E_i) \sum_f$$

$$(\boldsymbol{\varepsilon} \cdot \hat{z})^2 \delta_{\mathbf{k}_{\|_1}, \mathbf{k}_{\|_2}} |M_{21}(\mathbf{k}_{\|_1}, \mathbf{k}_{\|_2})|^2 \delta(E_f - E_i - \hbar\omega) \qquad (2.54)$$

Note that this expression assumes that the final state remains unoccupied. If we ignore the dependence of the envelope functions f_m on $\mathbf{k}_\|$ and assume parabolic bands, we can simplify further by taking

$$E_f \to E_2(\mathbf{k}_{\|_2}) = E_2(0) + \frac{\hbar^2 k_{\|_2}^2}{2m_e} \qquad (2.55)$$

$$E_i \to E_1(\mathbf{k}_{\|_1}) = E_1(0) + \frac{\hbar^2 k_{\|_1}^2}{2m_e} \qquad (2.56)$$

$$\sum_f \to \sum_{\mathbf{k}_{\|_2}} \qquad (2.57)$$

$$M_{21}(\mathbf{k}_{\|_1}, \mathbf{k}_{\|_2}) \to M_{21} \qquad (2.58)$$

Note also that

$$\sum_i f^0(E_i) \to \sum_{\mathbf{k}_{\|_1}} f^0(E_1(\mathbf{k}_{\|_1})) = N \qquad (2.59)$$

where N is the total number of carriers in a single quantum well; this result is independent of temperature as long as all of the carriers reside in the first subband. With these approximations, the absorption coefficient simplifies to

$$\alpha(\hbar\omega) = \frac{4\pi^2 e^2(\hbar\omega)}{\eta c\hbar} \frac{N}{V_p} (\boldsymbol{\varepsilon} \cdot \hat{z})^2 |M_{21}|^2 \delta(E_2(0) - E_1(0) - \hbar\omega) \qquad (2.60)$$

Now recall that this expression represents the contribution to the optical absorption from a *single* quantum well. If our multiple quantum well structure contains M identical wells, each will contribute to $\alpha(\hbar\omega)$ as previously. Hence, in an incoherent approximation, which will be valid as long as the wells are essentially uncoupled (wide barriers), we simply multiply the preceding result by M to obtain the total absorption for the structure. We can convert this to an *intrinsic* property of the layered material by taking the photon normalization volume V_p equal to the multiple quantum well volume. In this case, we define an average volume carrier density for the structure via

$$\rho_n = \frac{MN}{V_p} \tag{2.61}$$

Finally, we recognize that the transitions will be broadened by such effects as nonparabolicity, phonon scattering, and material inhomogeneities. These effects may be treated phenomenologically by replacing the Dirac delta function in eq. (2.60) with a Lorentzian function of half-width Γ; in practice, the value of Γ is determined by experiment and is on the order of a few meV. With these modifications, the intrinsic absorption of the material becomes

$$\alpha(\hbar\omega) = \frac{4\pi^2 e^2(\hbar\omega)}{\eta c\hbar}\rho_n(\boldsymbol{\varepsilon}\cdot\hat{z})^2|M_{21}|^2\frac{\Gamma}{\pi}\frac{1}{[E_2(0)-E_1(0)-\hbar\omega]^2+\Gamma^2} \tag{2.62}$$

The same kind of development can be applied to determine the absorption due to intersubband transitions between the *valence* bands in quantum wells. In this case, however, the valence band mixing reflected in the 4×4 $\mathbf{k}\cdot\mathbf{p}$ Hamiltonian makes the treatment more complicated. In particular, the first-order $\mathbf{k}\cdot\mathbf{p}$ corrections to the wave-functions must be included in to obtain the proper selection rules; a comprehensive account may be found in [27]. We remark here only that transitions due to *normally* incident light are now allowed because of the strong valence band mixing.

2.7.1 Temperature and Nonparabolicity Effects on the Intersubband Transition

For the sake of simplicity we will consider a $GaAs/Al_{0.7}Ga_{0.3}As$ multiple-quantum well structure grown by the molecular beam epitaxy technique with a 75 Å well width and a 100 Å barrier width. Both the ground and excited states are confined in the well, yielding a bound-to-bound intersubband transition. The infrared absorption spectra of this intersubband transition, recorded experimentally at 296 and 5 K for a sample doped in the barrier, are shown in Figure 2.4 as the solid lines. The dashed lines in this figure are the fitted Lorentzian line shape. The energy position

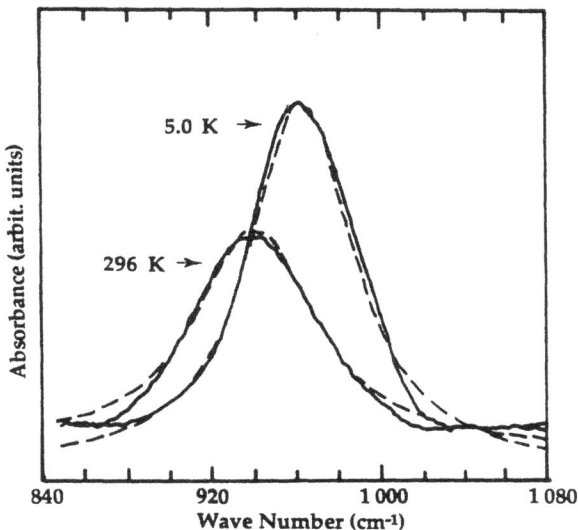

Figure 2.4 Infrared absorption spectra (solid lines) due to the intersubband transition in a GaAs/Al$_{0.3}$Ga$_{0.7}$As quantum well obtained at 296 and 5 K. The dashed lines are the result of a Lorentzian fit.

of the intersubband transition peak is observed to shift to a higher energy (blue shift) as the temperature is reduced. This shift has been reported earlier [28–30] for GaAs/AlGaAs as well as for InGaAs/AlGaAs quantum wells [31]. In addition, the amount of this shift does not seem to be affected either by doping the barrier or the well [28–30].

In Figure 2.5 we plot the peak position (open squares) as a function of temperature. The dashed line is the theoretical peak position energy as a function of temperature obtained from eq. (2.62) and the single particle calculations described in Sections 2.3 and 2.4. In these calculations the effective mass in the well is taken as a function of temperature according to the following expression [32]:

$$m_1^* = \cfrac{1}{1 + 7.51\left(\cfrac{2}{E_{G_2}} + \cfrac{1}{E_{G_2} + 0.343}\right)} \tag{2.63a}$$

and the effective mass in the barrier is taken as

$$m_2^* = \cfrac{1}{1 + C\left(\cfrac{2}{E_{G_1}} + \cfrac{1}{E_{G_1} + 0.343}\right)} \tag{2.63b}$$

42

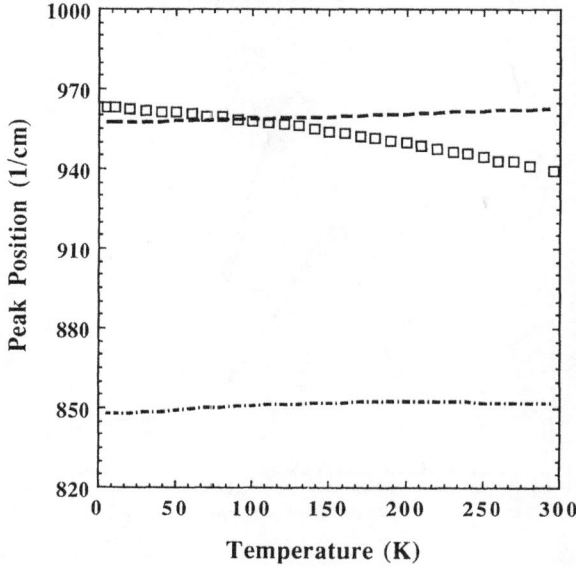

Figure 2.5 The peak position energy of the intersubband transition absorption as a function of temperature. The open squares are the experimental data. The dashed line represents the theoretical calculations using the envelope function approximation method, considering only the electron effective mass as a function of temperature. The dashed-dotted line is the same as the dashed line but nonparabolicity, anisotropy, conduction band offset, Fermi level, and lineshape broadening effects as a function of temperature are included.

where $E_G = 1.519 + 1.247x - (5.405 + 0.694x)10^{-4}T/T + 204$ (eV), E_{G_1} and E_{G_2} are the bandgaps of $Al_xGa_{1-x}As$ and $GaAs$, respectively, and $C = 7.51 - 3.13x$. The conduction band offset was taken as $\Delta E_C = (0.57)(1.247x)$ (eV). The theoretical results exhibit a red shift as a function of temperature. In addition, the nonparabolicity treated by Manasreh et al. [33] using the one-electron energy band structure for the well obtained by Ekenberg's nonparabolic-anisotropic envelope function formalism [34] and given by

$$\mathcal{H} = \alpha_0\frac{\partial^4}{\partial z^4} - \{[\hbar^2/(2m_1^*)] + (2\alpha_0 + \beta_0)(k_x^2 + k_y^2)\}\frac{\partial^2}{\partial z^2} \\ + \hbar^2(k_x^2 + k_y^2)/(2m_1^*) + (2\alpha_0 + \beta_0)\,k_x^2\,k_y^2 + \alpha_0\,(k_x^4 + k_y^4) + V \tag{2.64}$$

does not seem to account for the experimentally observed blue shift in Figure 2.5 (squares). This nonparabolicity result is shown as as dashed-dotted line in Figure 2.5. The latter treatment included temperature-dependent electron effective masses for both the well and the barrier (see eq. (2.63)), which result from their dependence on the bandgap. The bandgap was calculated using the Varshni parameterization [35]. The nonparabolicity parameter was taken as temperature dependent (see eqs. (2.64) and (2.65) of Blakemore, [32]) with the low temperature value of $\alpha' = 0.600$ eV^{-1}; the anisotropy parameter $\beta' = 0.702$ eV^{-1} was kept constant (see Ekenberg [34] for the definitions of α' and β'). The conduction band offset was taken as temperature dependent [36] with the temperature gradient of $(-0.57) \cdot (1.15 \times 10^{-4}) \cdot (x)$ where 0.57 reflects the 57/43 conduction to valence band offset ratio and x is the alloy composition. Our calculations show that the excited state has a lower curvature than the ground state, which contributes to the width of the calculated absorption spectrum. It should be pointed out that the theoretical absorption coefficient given by eq. (2.62) was broadened by using $\Gamma(T) = \gamma_1 + \gamma_2/[\exp(\hbar\omega/kT) - 1]$, where $\hbar\omega$ is the GaAs optical phonon energy ($\sim$36 meV), $\gamma_2 = 2.55$ meV, and $2\gamma_1$ is the full width at half maximum obtained at $T = 4.2$ K and given as a function of the two-dimensional electron gas density (σ) by $2\gamma_1 = 6.2883 + 0.202 \times 10^{-11}\,\sigma + 4.448 \times 10^{-25}\,\sigma^2$ (meV) (see Manasreh et al. [33] and [37] for additional discussion). On the basis of these calculations we conclude that a one-electron envelope function calculations give intersubband transition energies lower than those observed experimentally and predict no significant shift as a function of temperature.

2.7.2 Many-Body Effects on the Intersubband Transition

From the previous section we conclude that the nonparabolicity does not seem to account for the intersubband transition peak in GaAs/AlGaAs multiple-quantum wells. The next step is to include many-body effects in the calculations. Typical bound-to-bound intersubband absorption spectra are shown in Figure 2.6 measured at 5 K for five different samples with different dopant concentrations. Spectrum a is obtained for a barrier-doped sample. It is clear from this figure that the peak position energy is increased (blue shift) as the *two-dimensional electron gas* (2DEG) density (σ) is increased. To analyze the blue shift observed as a function of σ (we will refer to this type of blue shift as a *density blue shift*), the total integrated absorption (I) of the intersubband transition spectrum of each sample is converted to the 2DEG density, σ, according to the relationship [28] $I = N\sigma L\ (e^2h)/(4\varepsilon_0 m^*c)f/(n^2\sqrt{n^2 + 1})$, where N is the number of wells, L is the well width (75 Å), m^* is the electron effective mass in GaAs, n is the refractive index of GaAs ($\sim$3.3), and f is the oscillator strength ($\sim$13.91 for the present multiple-quantum wells, MQW). The results are shown in Figure 2.7 as solid squares. The vertical error bars were introduced to account for a monolayer inhomogeneity in the GaAs quantum well across the sample. All the GaAs/AlGaAs multiple quantum well structures presented in this chapter have the same structure as described in the previous section.

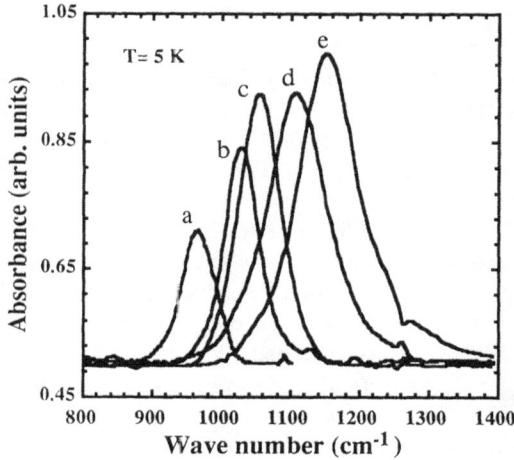

Figure 2.6 Selected intersubband infrared absorption spectra in GaAs/Al$_{0.3}$Ga$_{0.7}$As multiple quantum wells with different dopant concentrations (σ): (a) 3.25×10^{11} cm^{-2}, (b) 1.09×10^{12} cm^{-2}, (c) 1.13×10^{12} cm^{-2}, (d) 1.80×10^{12} cm^{-2}, and (e) 2.29×10^{12} cm^{-2}. All samples were doped in the well except (a), which was doped in the barrier.

To estimate the peak position energies of the intersubband transition spectra, the linear absorption coefficient $\alpha(\omega)$ is first calculated from eq. (2.62). The energies, E_0 and E_1, were calculated as a function of the wave vector [38] by using a single-particle band structure for the well following Ekenberg's nonparabolic-anisotropic envelope function formalism [34] (see eq. (2.64)). In these calculations, the effective masses of both the well and the barrier were obtained from their dependence on the bandgap, which was calculated by using Varshni parameterization, with the conduction to valence-band offset ratio taken as 57/43. The intersubband transition peak position energy (we will refer to the theoretical peak position energy as $h\nu_0$) is then obtained from the broadened optical absorption coefficient (eq. (2.62)). The latter results exhibit a red shift as σ is increased, in disagreement with the experimental measurements (see the short-dashed line in Figure 2.7). However, the behavior of $h\nu_0$ as a function of σ is in good agreement with the depolarized light Raman scattering measurements of single-particle transitions [39].

The next step is to investigate the depolarization (plasmon shift) and excitonlike (interaction of the excited electron with the hole in the ground state, analogous to the exciton associated the the valence-to-conduction band transition) effects on $h\nu_0$. To include these effects we apply the formalism of Ando [40], which gives the shifted intersubband transition energy (ΔE) as

$$\Delta E = h\nu_0\sqrt{1 + \phi - \beta} \qquad (2.65)$$

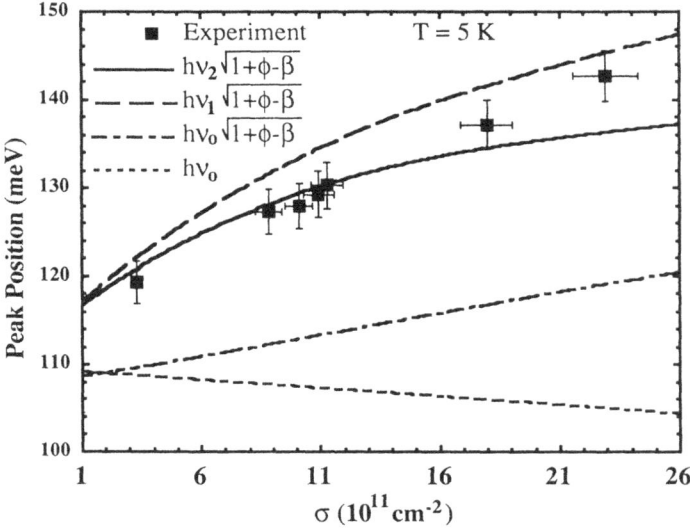

Figure 2.7 The peak position energy of the intersubband transition as a function of the 2DEG density (σ). The closed squares represent the experimental data, the short-dashed line represents the single particle calculations of the intersubband transition energy, that is, $h\nu_0$, the long-short-dashed line represents eq. (2.65), which includes the depolarization and excitonlike effects; the long-dashed line represents eq. (2.65) with $h\nu_0$ replaced by $h\nu_1$, which includes the exchange interaction effect in addition to the depolarization and excitonlike effects; and the solid line represents eq. (2.65) with $h\nu_0$ replaced by eq. (2.72), which includes the direct Coulomb interaction effect in addition to the depolarization, excitonlike, and exchange interaction effects.

where the factor ϕ accounts for the depolarization shift and is given by

$$\phi = 2\frac{4\pi e^2}{\varepsilon}\sigma S_{01}\frac{1}{h\nu_0} \tag{2.66}$$

where ε is the dielectric constant and S_{01} is the Coulomb matrix element given by

$$S_{01} = \int_0^\infty dz\left[\int_0^z \xi_1(z')\xi_0(z')\,dz'\right]^2 \tag{2.67}$$

where ξ_0 and ξ_1 are the wave functions for the ground and excited states, respectively. The factor β in eq. (2.65) accounts for the excitonlike shift given by

$$\beta = -\frac{2\sigma}{h\nu_0}\int_{-\infty}^{+\infty} dz\xi_1(z)^2\xi_0(z)^2\frac{\partial V_{xc}[\sigma(z)]}{\partial\sigma(z)} \tag{2.68}$$

where $V_{xc}[\sigma(z)]$ is the exchange-correlation potential as a function of the local sheet density $\sigma(z)$ and is described elsewhere [41]. The result of eq. (2.65) after calculating eqs. (2.66) through (2.68) is presented in Figure 2.7 as the long-short-dashed line. Three points should be mentioned here. First, the excitonlike shift is very small as compared to the depolarization shift. Second, eq. (2.65) produces a trend similar to that of the experimental data, that is, a blue shift as σ is increased. Third, the experimental peak position energy as a function of σ is much larger than the results obtained from eq. (2.65). Therefore, the depolarization and excitonlike shifts alone do not appear to account for what is observed experimentally, and other effects should be considered.

Because the dopant densities in the the present multiple-quantum well samples are relatively high, we would expect that many-body effects such as electron-electron exchange and direct Coulomb interactions for the ground state are significant. Bandara et al. [42] derived approximate expressions for these interactions and concluded that the electron-electron exchange interaction energy for the ground state cannot be ignored. Manasreh et al. [37] extended the work of Bandara et al. and derived the exchange interaction energy calculated from an expression for the mth subband, which is based on the Hartree-Fock approximation, given by [43]

$$E_{\text{exch}}^{(m)} = -\sum_n \int d\mathbf{r} \int d\mathbf{r}' \frac{e^2}{4\pi\varepsilon} \frac{1}{|\mathbf{r} - \mathbf{r}'|} \Psi_n^*(\mathbf{r}')\Psi_m(\mathbf{r}')\Psi_n(\mathbf{r})\,\Psi_m^*(\mathbf{r}) \quad (2.69)$$

where the summation is over all occupied subbands. To render the model analytical, following Bandara et al. [42], Manasreh et al. [37] used wave functions for a simple parabolic band and an infinite square well. With these simplifications, eq. (2.69) reduces for the ground state (in first order) to

$$E_{\text{exch}}^{(0)}(k) = -\frac{e^2 k_F}{4\pi\varepsilon}\left[\frac{2}{\pi}\mathbf{E}\left(\frac{k}{k_F}\right) - \left(\frac{\pi}{6} - \frac{5}{8\pi}\right)\frac{k_F}{k_L}\right] \quad \text{for} \quad (k < k_F) \quad (2.70a)$$

$$E_{\text{exch}}^{(0)}(k) = -\frac{e^2 k_F}{4\pi\varepsilon}\left[\frac{2}{\pi k_F}\left(k\mathbf{E}\left(\frac{k_F}{k}\right) + \frac{(k_F^2 - k^2)}{k}\mathbf{K}\left(\frac{k_F}{k}\right)\right) - \left(\frac{\pi}{6} - \frac{5}{8\pi}\right)\frac{k_F}{k_L}\right] \quad (2.70b)$$

$$\text{for} \ (k > k_F)$$

For the first excited state eq. (2.69) reduces to

$$E_{\text{exch}}^{(1)}(k) = -\frac{e^2 k_F}{4\pi\varepsilon}\left(\frac{5}{9\pi}\right)\frac{k_F}{k_L} \quad (2.70c)$$

where k is the magnitude of the $k_\parallel$ wave vector, k_F is the Fermi wave vector ($k_F = \sqrt{2\pi\sigma}$), $k_L = \pi/L$, $\mathbf{E}(x)$ is an elliptic integral of the second kind, and $\mathbf{K}(x)$ is an elliptic integral of the first kind. The fact that the ground state exchange energy is k dependent, although while the excited state exchange energy is not, is due to the orthogonality of the ground and excited state wave functions. The Hartree-Fock exchange energy suffers the well-known defect of an infinite slope at the Fermi wave vector. This gives rise to an unphysical "dip" in the absorption spectrum at $k = k_F$ and a zero density of states at the Fermi energy. By adding $E_{exch}^{(0)}(k)$ and $E_{exch}^{(1)}(k)$ to the ground state energy (E_0) and excited state energy (E_1) in eq. (2.62), respectively, we can obtain the new peak position energy of the intersubband transition, which is defined here as $h\nu_1$. The results of eq. (2.65) after replacing $h\nu_0$ in eqs. (2.65), (2.66), and (2.68) by $h\nu_1$, using eq. (2.70), are presented by the long-dashed line in Figure 2.7. Thus far, it is evident that the agreement between theory and experiment is greatly improved by including depolarization, excitonlike, and exchange interaction effects. It should be pointed out that the exchange interaction alone when it is added to the single-particle calculations cannot explain the peak position as a function of σ.

The agreement between theory and experiment can be further improved by adding $E_{dir}^{(0)}$ and $E_{dir}^{(1)}$ to the ground state energy (E_0) and excited state energy (E_1) in eq. (2.62), respectively, where $E_{dir}^{(0)}$ and $E_{dir}^{(1)}$ are the direct Coulomb interaction energies for the ground and excited states, respectively, given by [42]

$$E_{dir}^{(0)} = \frac{3e^2\sigma}{8\varepsilon k_L^2 L} \tag{2.71a}$$

$$E_{dir}^{(1)} = \frac{e^2\sigma}{4\varepsilon k_L^2 L} \tag{2.71b}$$

The new peak position energy of the intersubband transition, which is defined here as $h\nu_2$, can therefore be written as

$$h\nu_2 = h\nu_1 - (E_{dir}^{(0)} - E_{dir}^{(1)}) \tag{2.72}$$

The results of eq. (2.65) after replacing $h\nu_0$ in eqs. (2.65), (2.66), and (2.68) by $h\nu_2$ using eq. (2.72) are presented by the solid line in Figure 2.7. From this figure, we see that the effect of the E_{dir} term on ΔE is small (~ 1 meV at 3×10^{11} cm^{-2}) as compared to that of E_{exch} or the depolarization shift and becomes considerably larger as σ is increased (~ 10 meV at 2.6×10^{12} cm^{-2}). The agreement between experiment and theory when the depolarization shift, excitonlike shift, exchange interaction energy, and direct Coulomb interaction energy are included is remarkable, in particular for $\sigma \leq 16 \times 10^{11}$ cm^{-2}. It is important to note that these effects cannot account for the experimental peak position energy of the intersubband transition as a function of σ when they are considered separately.

The temperature (*T*) and many-body effects on the intersubband transition in a GaAs/Al$_{0.3}$Ga$_{0.7}$As multiple quantum well are also studied. In the following discussion we will show that the intersubband blue shift observed as the temperature decreases is increased as a function of σ. This behavior can be explained by incorporating the electron-electron exchange and direct Coulomb interactions, depolarization (plasmon shift), and excitonlike shift in the nonparabolic-anisotropic envelope function approximation model described previously. In addition, the intersubband transition peak position energy as well as the total integrated area are decreased by illuminating the samples with white light at 5 K. The latter observation provides additional support that the intersubband blue shift, which is observed as σ is increased, is due to many-body effects.

Figure 2.8 shows the infrared absorption spectra for a sample with $\sigma = 2.3 \times 10^{12}$ cm^{-2}. Spectrum *a* was taken after cooling the sample in the dark to 5 K and spectrum *c* was obtained at 298 K. Spectrum *b* was taken at 5 K after illuminating the sample with a secondary white light for 2 min. We will discuss the temperature effect on the intersubband transition, that is, spectra *a* and *c*, first and discuss the illumination results (persistent photo effect) later. It is observed that the total integrated absorption of the intersubband transition of the multiple quantum well samples doped in the well remains constant between 5 and 300 K. The peak position energy, on the other hand, is shifted toward higher energy as *T* is decreased (we will refer to difference between the peak position energies of spectra *a* and *c* as a *temperature blue shift*). The temperature blue shift of this particular sample is ~5.5 meV and is plotted along with temperature blue shifts obtained for samples doped with different

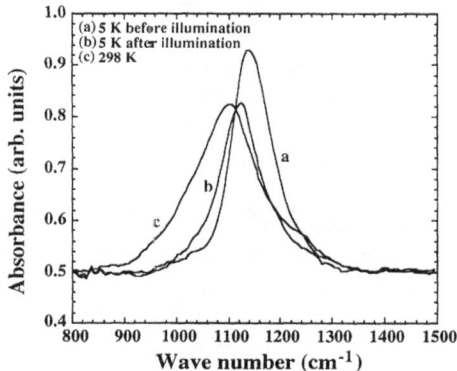

Figure 2.8 Optical absorption of the *IT* in a sample doped in the well with $\sigma = 2.3 \times 10^{12}$ cm Spectrum *a* was taken at 5 K after cooling the sample in the dark. Spectrum *b* is the sa as spectrum *a* but after illuminating the sample with white light for 2 minutes. Spectrur is taken at 298 K.

concentrations in Figure 2.9 as solid squares. It is clear from this figure that the temperature blue shift is decreased as σ decreases.

The theoretical peak position energies of the intersubband transition spectra are obtained from the calculated linear absorption coefficient described by eq. (2.62). The result of single particle calculations using the nonparabolicity approach (see eq. (2.64)) with the effective mass given by eq. (2.63) is presented by the long-dashed line in Figure 2.9. This theoretical result exhibits a trend opposite to that observed experimentally. The short-dashed line represents the theoretical calculations of the temperature blue shift after incorporating the exchange and direct Coulomb interactions described in eqs. (2.69)–(2.71). A further use of screened exchange would improve the calculation, especially at the higher concentrations used in this work, but these calculations are not included here. Furthermore, we neglect the explicit temperature dependence of the exchange interactions which is known to be small [44, 45]. The electron-electron exchange and direct Coulomb interactions do not seem to explain the measured temperature blue shift as shown in Figure 2.9. The next step is to include other effects, such as depolarization and excitonlike shift described by eqs. (2.65)–(2.68). It should be noted that eq. (2.65) is derived under the assumption that the interband density of single particle states is a delta function.

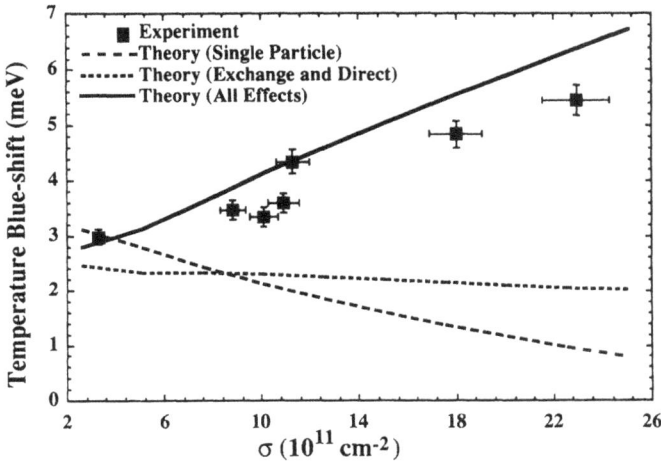

Figure 2.9 The temperature blue shift of the peak position energy as a function of 2DEG concentration. The solid squares were obtained by taking the difference between the peak position energy measured at 5 K and 298 K. The solid line represents the temperature blue shift calculated from eq. (2.65) with all interaction described by eqs. (2.66) through (2.71). The short-dashed line represents the theoretical calculations with only electron-electron exchange (eq. (2.70)) and direct Coulomb (eq. (2.71)) interactions. The long-dashed line represents the single particle calculations.

However, according to Zaluzny [46], Ando's formalism [40] for finite-width inter-band density of states changes the shape but not the peak shift of the calculated absorption spectrum, so that eq. (2.65) is correct for the case treated here as well. For comparison with the experiment, the result of eqs. (2.65) and (2.72) is given by the solid line in Figure 2.9. It is evident from this figure that the agreement between theory and experiment is greatly improved by including depolarization, ex-citonlike, electron-electron exchange, and Coulomb interaction effects. For small dopant concentrations, the many-body effects are small and therefore single particle calculations that include the nonparabolicity and anisotropy can explain the temper-ature blue shift. The theoretical results depicted in Figure 2.9 may not be accurate. Further theoretical derivations are in progress.

It should be pointed out that eq. (2.70) and ϕ and β in eq. (2.65) are derived at 0 K, and they do not depend explicitly on temperature. However, these equations depend implicitly on temperature through the temperature dependence of the band parameters. We should also point out that the strongest temperature dependences reside in the effective masses and in the Fermi occupancy factors, which are correctly reflected in our calculation.

The theory described by the solid line in Figure 2.9 somewhat overestimates the experimental measurements. Using a better treatment of the exchange (screened and temperature-dependent exchange and the higher-order terms in eq. (2.70)) would improve the comparison with experiment. Hence, it is clear that the temperature blue shift observed in the peak position energy of the intersubband transition in GaAs/ AlGaAs is very complicated and requires extensive theoretical modeling.

Additional support for the many-body effects on the intersubband transition optical absorption can be provided by varying σ. One method of varying σ is by illuminating the sample with a secondary white light at $T \leq 100$ K in which the electrons can be excited (removed) out of the well and trapped by defects (persistent photo effect). The results of this test are presented in Figure 2.8 for a sample doped in the well with $\sigma \approx 2.3 \times 10^{12}$ cm^{-2}. Spectrum b in this figure is the intersubband transition optical absorption obtained after cooling the sample in the dark to 5 K and then illuminating the sample with white light for 2 min. It is noted in this figure (spectrum b) that the peak position is shifted to a lower energy and the total integrated absorption area is reduced by about 20% after illumination. This result demonstrates that σ is decreased after illuminating the sample and consequently the many-body effects are reduced, as can be clearly seen by the red shift of the peak position energy. We have verified that the change in carrier density in the well leads to very small (less than 1 meV) Coulombic shifts of the calculated energies and that the density dependence of the exchange is responsible for the shift of the measured spec-trum following white light illumination. It should be noted that the total integrated absorption areas of spectra a and c are the same within an experimental error, sug-gesting that the electrons are confined to the quantum well in the temperature range of 5 to 298 K.

The peak position energy of the intersubband transition observed in the sample used in Figure 2.8 is studied as a function of temperature. The results are shown in Figure 2.10. The solid squares represent the peak position energy of the intersubband transition spectra taken between 5 and 298 K after cooling the sample in the dark to 5 K (this is the peak position energy of spectrum *a* of Figure 2.8). The open circles represent the peak position energy after cooling the sample in the dark followed by white light illumination (this is the peak position energy of spectrum *b* of Figure 2.8). The peak position energy starts to recover at 110 K, and a complete recovery occurs at ~150 K. The temperature at which the recovery occurs is identical to that of the DX center [47]. The latter result is in good agreement with the results reported by Dischler et al. [48] for GaAs/AlGaAs multiple quantum well samples. It should be noted that the amount of the red shift observed after white light illumination is reduced as the dopant concentration is decreased. The exact nature of the electron trapping process is not yet clear. However, we offer the following tentative explanation for this persistent photo effect. Because the trapping effect is observed in the heavily doped multiple quantum well, where the diffusion of silicon from the well to the barrier is known to occur from the capacitance-voltage and secondary ion mass spectroscopy measurements, a DX center is formed in the barrier

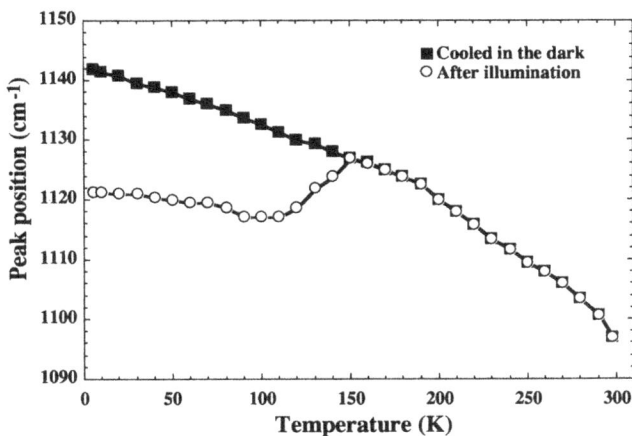

Figure 2.10 The effect of the secondary white light illumination on the peak position energy of the intersubband transition as a function of temperature. The solid squares represent the peak position energy of spectrum *a* of Figure 2.8, which was taken after cooling the sample in the dark to 5 K. The open circles represent the peak position energy of spectrum *b* of Figure 2.8, which was taken after cooling the sample in the dark to 5 K and white light illumination.

materials. The electrons are therefore excited from the well by the white light illumination and then trapped by the DX center. This explanation differs from the Dischler et al. [48] model, which proposes that the persistent photo effect is due to the trapping of electrons by acceptors present in the well material.

2.8 CONCLUSION

The basis of the theoretical modeling of the bound-to-bound intersubband transitions in type I quantum wells made of III–V compound semiconductors, in particular GaAs/ AlGaAs, is presented in this chapter using $\mathbf{k} \cdot \mathbf{p}$ theory. Single particle calculations of the conduction and valence bands states are developed in this chapter. The optical absorption coefficient of the bound-to-bound intersubband transition is derived and used to estimate the peak position energy of the intersubband transition in GaAs/ AlGaAs multiple quantum wells. The nonparabolicity effect, which is included in our calculations, cannot explain the blue shift observed in the peak position energy of the intersubband transition as the temperature is reduced.

In this chapter, we have shown that the blue shift observed in the peak position energy of the intersubband transition in GaAs/AlGaAs multiple quantum wells at 5 K as a function of the 2DEG density, σ, is quantitatively accounted for when the depolarization, excitonlike, the ground state electron-electron exchange interaction, and the ground state direct Coulomb interaction effects are incorporated in the nonparabolic-anisotropic envelope function approximation calculations. All these effects are found to be necessary to account for what is observed experimentally. Single particle calculations, on the other hand, show a red shift as a function of σ in good agreement with Raman scattering measurements [39].

It is also shown that the temperature blue shift obtained by taking the difference between the peak position energies of the intersubband transition in GaAs/AlGaAs MQWs at 5 and 298 K is increased approximately linearly as σ is increased. This blue shift is also explained by incorporating the exchange and direct Coulomb interactions, depolarization, and excitonlike shifts in the calculated peak position energy of the intersubband transition. The persistent photo effect observed after a secondary white light illumination demonstrates that the peak position energy of the intersubband transition can be shifted and that the total integrated absorption can also be changed by varying σ. The latter conclusion provides a strong argument that many-body effects play a significant role in the intersubband transitions in quantum wells.

ACKNOWLEDGMENTS

This work was partially supported by the Air Force Office of Scientific Research. We would like to thank E. Taylor and J. Ehret for the MBE growth. We also would like to thank F. Szmulowicz, T. Vaughan, V. Jogai, and C.E. Stutz for useful discussions.

REFERENCES

[1]   Bloch, F. *Zeits. f. Physik*, Vol. 52, 1928, p. 555.
[2]   Chelikowsky, J., and M. Cohen, *Phys. Rev. B*, Vol. 14, 1976, p. 556.
[3]   Wannier, G.H., *Phys. Rev.*, Vol. 52, 1937, p. 191.
[4]   Slater, J.C., *Phys. Rev.*, Vol. 76, 1949, p. 1592.
[5]   Luttinger, J.M., and W. Kohn, *Phys. Rev.*, Vol. 97, 1955, p. 869.
[6]   Altarelli, M., in *Heterojunctions and Semiconductor Superlattices*, ed. G. Allan, G. Bastard, N. Boccora, and M. Voos, Springer-Verlag, Berlin, 1986.
[7]   Thomsen, J., G.T. Einevoll, and P.C. Hemmer, *Phys. Rev. B*, Vol. 39, 1989, p. 12783.
[8]   Morrow, R.A., *Phys. Rev. B*, Vol. 36, 1987, p. 4836.
[9]   See, for example, Bastard, G., *Wave Mechanics Applied to Semiconductor Heterostructures*, Halsted Press, London, 1988.
[10] Altarelli, M., U. Ekenberg, and A. Fasolino, *Phys. Rev. B*, Vol. 32, 1985, p. 5138.
[11] Sanders, G.D., and Y.C. Chang, *Phys. Rev. B*, Vol. 35, 1987, p. 1300.
[12] See, for example, Burden and Faires, *Numerical Analysis*, Prindle, Weber and Schmidt, Boston, 1985.
[13] Chang, Y.C., and J.N. Schulman, *Phys. Rev. B*, Vol. 31, 1985, p. 2069.
[14] Kane, E.O., *J. Phys. Chem. Solids*, Vol. 1, 1956, p. 82.
[15] Kane, E.O., *J. Phys. Chem. Solids*, Vol. 1, 1957, p. 249.
[16] Kane, E.O., *Physics of III–V Compounds*, Academic Press, New York, 1966, Vol. 1, Ch. 3, pp. 75–100.
[17] Lawaetz, P., *Phys. Rev. B*, Vol. 4, 1971, p. 3460.
[18] Andreani, L.C., A. Pasquarello, and F. Bassasni, *Phys. Rev. B*, Vol. 36, 1987, p. 5887.
[19] Hinckley, J.M., and J. Singh, *Phys. Rev. B*, Vol. 41, 1990, p. 2912, J.M. Hinckley and J. Singh, *Phys. Rev.* Vol. 42, 1990, p. 3546.
[20] Bir, G.L., and G.E. Pikus, *Symmetry and Strain-Induced Effects in Semiconductors*, John Wiley and Sons, New York, 1974.
[21] Ridley, B.K., *Quantum Processes in Semiconductors*, Clarendon Press, Oxford, 1982.
[22] Jaffe, M., Ph.D. thesis, University of Michigan, 1989, and the references therein.
[23] Kato, H., N. Iguchi, S. Chika, M. Nakayama, and N. Sano, *J. Appl. Phys.*, Vol. 59, 1986, p. 588.
[24] See, for example, Schiff, L.I., *Quantum Mechanics*, McGraw-Hill, New York, 1968.
[25] See, for example, Sakurai, J.J., *Advanced Quantum Mechanics*, Addison-Wesley, New York, 1967.
[26] See, for example, Datta, S., *Quantum Phenomena*, Vol. 8 in the Modular Series on Solid State Devices, Addison-Wesley, Reading, MA, 1989.
[27] Chang, Y.C., and R.B. James, *Phys. Rev. B*, Vol. 39, 1989, p. 12672.
[28] West, L.C., and S.J. Eglash, *Appl. Phys. Lett.*, Vol. 46, 1985, p. 1156.
[29] Levine, B.F., C.G. Bethea, K.K. Choi, J. Walker, and R.J. Malik, *Appl. Phys. Lett.*, Vol. 53, 1988, p. 231.
[30] Covington, B.C., C.C. Lee, B.H. Hu, H.F. Taylor, and D.C. Streit, *Appl. Phys. Lett.*, Vol. 54, 1989, p. 2145.
[31] Zhou, X., P.K. Bhattacharya, G. Hugo, S.C. Hong, and E. Gulari, *Appl. Phys. Lett.*, Vol. 54, 1989, p. 855.
[32] Blakemore, J.S., *J. Appl. Phys.*, Vol. 53, 1982, p. R123.

[33]   Manasreh, M.O., F. Szmulowicz, D.W. Fischer, K.R. Evans, and C.E. Stutz, *Appl. Phys. Lett.*, Vol. 57, 1990, p. 1790.

[34]   Ekenberg, U., *Phys. Rev. B*, Vol. 36, 1987, p. 6152.

[35]   Varshni, Y.P., *Physica*, Vol. 34, 1967, p. 149.

[36]   Adachi, S., *J. Appl. Phys.*, Vol. 58, 1985, p. R1.

[37]   Manasreh, M.O., F. Szmulowicz, T. Vaughan, K.R. Evans, and C.E. Stutz, *NATO Series B*, Vol. 288, 1992, p. 287.

[38]   Manasreh, M.O., F. Szmulowicz, T. Vaughan, K.R. Evans, C.E. Stutz, and D.W. Fischer, *Phys. Rev. B*, Vol. 43, 1991, p. 9996.

[39]   Ramsteiner, M., J.D. Ralston, P. Koidl, B. Dischler, H. Beibl, J. Wagner, and H. Ennen, *J. Appl. Phys.*, Vol. 67, 1990, p. 3900.

[40]   Ando, T., *Solid State Commun.*, Vol. 21, 1977, p. 133.

[41]   Bloss, W.L., *J. Appl. Phys.*, Vol. 66, 1989, p. 3639.

[42]   Bandara, K.M.S.V., D.D. Coon, O. Byungsung, Y.F. Lin, and M.H. Francombe, *Appl. Phys. Lett.*, Vol. 53, 1988, p. 1931.

[43]   Ashcroft, N.W., and N.D. Mermin, *Solid State Physics*, Holt, Rinehart and Winston, New York, 1976, Chapter 17.

[44]   Das Sarma, S., R.K. Kalia, M. Nakayama, and J.J. Quinn, *Phys. Rev. B*, Vol. 19, 1979, p. 6397.

[45]   Das Sarma, S., and B. Vinter, *Phys. Rev. B*, Vol. 23, 1981, p. 6823.

[46]   Zaluzny, M., *Phys. Rev. B*, Vol. 43, 1991, p. 4511.

[47]   Theis, T.N., T.N. Morgan, B.D. Parker, and S.L. Wright, *Materials Science Forum*, Vol. 38–41, 1989, p. 1073.

[48]   Dischler, B., J.D. Ralston, P. Koidl, P. Hiesinger, M. Ramsteiner, and M. Maier, in *Materials Research Society*, Vol. 216, 1991, p. 347.

Chapter 3

Long-Wavelength Infrared Photodetectors Based on Intersubband Transitions in III–V Semiconductor Quantum Wells

R.L. Whitney, K.F. Cuff, and F.W. Adams
Lockheed Palo Alto Research Laboratories

3.1 INTRODUCTION

This chapter briefly reviews the development and operating principles of GaAs/AlGaAs multiple quantum well detectors and discusses in detail the important device-optimization issues for application of these devices to staring infrared sensor systems.

3.1.1 Purpose and Scope

Long-wavelength infrared (LWIR) detectors, based on intersubband absorption in III–V *multiple quantum well* (MQW) structures, have been rapidly developed to the point where these photodetectors are becoming attractive for a number of sensor applications. A major impetus for their development has been the move to large-area staring focal-plane arrays in advanced *infrared* (IR) sensor systems. This is because III–V materials and quantum well structures, particularly in the GaAs/AlAs alloy system, can be grown on large-area substrates with excellent uniformity. Recent demonstrations of prototype GaAs/AlGaAs MQW staring focal plane arrays have validated this concept. We think that this technology now deserves serious attention from sensor designers.

On the other hand, the single detector performance of the basic MQW LWIR photoconductive detector is, and for fundamental reasons will remain, inferior to that of ideal HgCdTe infrared detectors under most operating conditions. Whether

MQW infrared detectors will be the appropriate choice for a given application consequently will depend on the application and the resulting sensor system requirements. The potential for low-cost, highly uniform, large-area multiple quantum well detector arrays is an advantage that may allow them to compete successfully with conventional IR detectors in several important applications. Even for such "niche" applications, their performance must be optimized.

Accordingly, our objective in this chapter is to provide the basic framework for evaluating the performance of staring IR sensors using MQW detectors and to give some insight into the optimization of GaAs/AlGaAs MQW detector arrays for specific applications.

Much of our discussion is not specific to quantum-well detectors but can be applied to any staring focal-plane array. However, to our knowledge no such discussion has previously appeared in print. There are at least two reasons for this. First, large staring IR detector arrays have only recently begun to emerge from the laboratory for use in sensor systems, and the factors that limit staring-array performance are different from those that dominate scanning sensor design. The impact of these factors is sharpened by the low contrast of LWIR scenes as compared to shorter wavelengths. (Before the appearance of large LWIR staring arrays, one could argue that it was not necessary to consider how correlated noise, for example, affects *detector* optimization.) Second, the impact of application-specific considerations is more marked for MQW infrared detectors than for conventional photodiode arrays. This is because the single-detector noise performance of the quantum well detectors can be inferior (so that application-specific optimization can make the difference between adequate and unusable devices) and because the control of GaAs/AlGaAs materials and structure is superior (so that one can expect to carry out these optimizations accurately and consistently and to observe their effect on array performance). With such control of detector characteristics possible, sensor designers need no longer simply accept what they are given. And when the effect of optimization is potentially as significant as it is for multiple quantum well detectors, they are motivated to discover how best to accomplish it.

For the purpose of sensor design, it is necessary to have simple, scalable expressions for detector behavior as a function of operating wavelength, temperature, bias, device dimensions, and other physical variables. However, even in the well-studied GaAs/AlGaAs material system, experimental knowledge of some physical properties is incomplete; and even with the simplest device design, some MQW detector properties require evaluating complex calculations or are still incompletely understood. Consequently, to carry out our agenda we must make assumptions concerning physical properties and simplifications in the calculation of detector behavior. Specific numbers generated using our assumptions and simplified expressions will therefore not always be quantitatively precise. We believe, however, that their behavior is qualitatively correct, and that their quantitative accuracy is sufficient to give the reader an understanding of what is most important in optimizing these detectors for use in staring sensors. With these reservations in mind, this chapter can be used to make intelligent judgments of the usefulness of GaAs/AlGaAs quantum well infrared detectors in specific applications.

3.1.2 Quantum Well Detector Basics

Since the first reported detection of infrared radiation using an intersubband transition in a GaAs/AlGaAs *quantum well* (QW) structure, new ideas for detector structures have proliferated. Although the GaAs/AlAs alloy system has been used most frequently, quantum well infrared detectors have been made successfully in other III–V alloy systems. We shall not attempt in this chapter to cover all variations of material and device structures. Instead, we will concentrate on the GaAs/AlGaAs material system and on the simplest device design.

We have found it convenient to use the term *quantum well infrared photodetector* (QWIP) [1] for these devices. Besides conciseness, this term achieves a favorable balance of specificity (QW, IR) and generality (photodetector) that allow its use in distinguishing a whole class of intersubband-absorption detector concepts from more familiar detectors that use conventional infrared-sensitive materials. The focus of this chapter is what we call the basic QWIP: a simple photoconductor in which the IR-sensitive material is a quantum well structure.

Quantum wells are formed by sandwiching thin layers of a narrower-bandgap semiconductor, such as GaAs, between layers of a wider-bandgap semiconductor like AlGaAs. A low-temperature growth technology, such as *molecular beam epitaxy* (MBE) or *metalorganic chemical vapor deposition* (MOCVD), is essential to growing sufficiently thin layers with sharp interfaces. In the GaAs/AlGaAs system, quantum wells are formed in the GaAs layers in both conduction and valence bands. Figure 3.1 shows schematically the valence and conduction band edges of a basic IR-sensitive GaAs/AlGaAs quantum well structure. The AlAs fraction of the barrier alloy, x, and the layer thicknesses indicated there are representative values intended to give an idea of the magnitudes.

If the well material is doped with n-type impurities, electronic transitions between appropriately spaced states within the conduction band can cause significant IR absorption. When the upper electronic state of the transition is in the continuum or near the top of the well, photoexcited carriers have a high probability of leaving the well, and they can then be collected as photocurrent. In Figure 3.1, we have indicated doping of the end contacts by hatching and the location of the confined state in the n-doped conduction-band well with a line; the barriers are left undoped. Since the confined valence-band states in an n-doped structure do not contain a significant number of holes, we have not shown them.

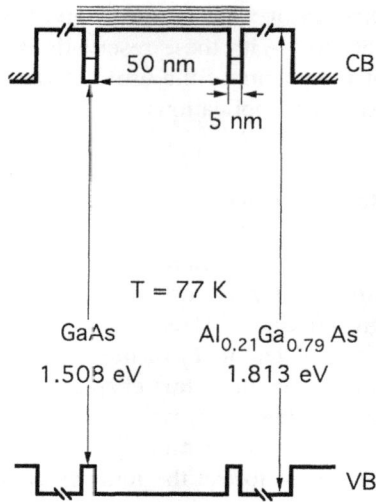

Figure 3.1 Band diagram for GaAs/AlGaAs quantum well infrared detector structure.

In Figure 3.1, we have supposed that the well is so narrow, or, equivalently, that the barrier is so low, that only one confined state is split off from the conduction band. We have indicated by horizontal lines a range of energies in the continuum for which the probability of an optical transition from the confined state is enhanced by a resonance condition. This is the bound-to-continuum transition QWIP, which is the device we shall discuss in this chapter. Alternatively, the upper state may be confined in the quantum well, but with such small confinement energy that photoexcited carriers quickly escape by tunneling; this is the bound-to-bound transition QWIP, which we shall not discuss.

Total absorption can be increased by growing many wells and barriers together in a single structure, forming a multiple quantum well device. Peak response wavelength, bandwidth, and device resistance are adjusted by varying well and barrier widths, barrier height (composition), and doping concentration. Such flexibility in device design, inherent in "bandgap-engineered" [2] devices such as the QWIP, allows tailoring the device structure for specific applications—a very intriguing possibility. However, to optimize properly, the application must first be well understood and the values of important sensor-system parameters must be at least approximately known.

3.1.3 Organization

In this chapter, we first state the basic figures of merit that characterize staring infrared sensors. Then we briefly review the development of GaAs/AlGaAs quantum well infrared photodetectors. We discuss in some detail the simplified parametric models of dark current, absorptance, photoconductive gain, and correlated noise that we use to analyze the performance of GaAs/AlGaAs QWIP focal planes. Finally, we show how to optimize QWIP focal planes by varying detector parameters to optimize a suitable sensor figure of merit and illustrate how the application strongly influences optimization by applying these principles to two typical infrared sensor systems, a thermal imager and a satellite atmospheric sounder.

3.2 FUNDAMENTALS OF INFRARED DETECTION FOR STARING APPLICATIONS

To evaluate staring focal-plane arrays correctly, the figures of merit used for single detectors must be supplemented to include the effects of spatial variations in noise and signal response among the detectors of an array. It is also important to account for the noise of the amplifiers required to read out the signal current. In this section we summarize these considerations. We state the important single-detector figures of merit, referring the reader to one of the excellent reviews available [3, 4] for details. Then we give two basic array figures of merit in the form that we use later.

The most commonly used figure of merit for single infrared detectors is the detectivity, which is a normalized signal-to-noise ratio. The spectral detectivity, D_λ^*, is given by

$$D_\lambda^* = \frac{R_\lambda \sqrt{A_p}}{i_n} \qquad (\text{cm}\sqrt{\text{Hz}}/\text{W}) \qquad (3.1)$$

where A_p is the photosensitive area of the detector, i_n is the noise current (A/$\sqrt{\text{Hz}}$), and R_λ is the spectral responsivity (A/W). For a photoconductor, the noise current is given by

$$i_n = \sqrt{4egI} \qquad \text{A}/\sqrt{\text{Hz}} \qquad (3.2)$$

where I is the total current flowing in the device and g is the photoconductive gain. The current responsivity is defined as the ratio of the *root-mean square* (rms) signal current from a monochromatic source, I_s, to the rms signal power, P_λ, given by

$$R_\lambda = \frac{I_s}{P_\lambda} = \frac{e\lambda\eta g}{hc} \approx 0.8\,\lambda\eta g \ (\text{A}/\text{W}) \qquad (3.3)$$

where e is the electronic charge, λ is the wavelength of the signal source (μm), η is the absorptance of the detector, h is Planck's constant, and c is the speed of light.

An alternate figure of merit is the blackbody detectivity, D_{BB}^*, which can be expressed as

$$D_{\mathrm{BB}}^* = \frac{R_{\mathrm{BB}}\sqrt{A_p}}{i_n} \tag{3.4}$$

where R_{BB} is the blackbody responsivity given by

$$R_{\mathrm{BB}} = \frac{\displaystyle\int_{\Delta\lambda} R(\lambda)S(\lambda, T)\,\mathrm{d}\lambda}{\displaystyle\int_{\Delta\lambda} S(\lambda, T)\,\mathrm{d}\lambda} \tag{3.5}$$

and $S(\lambda, T)$ is the blackbody spectral distribution function.

For comparison of individual detectors these figures of merit are suitable. However, readout noise rather than detector performance can in some instances be the limiting factor for the sensor as a whole, and therefore *sensor* figures of merit conventionally include readout noise as well as detector noise. Furthermore, in staring IR sensors using two-dimensional detector arrays, uniformity of the array can be, and often is, more important than single detector performance. A figure of merit for staring sensors must therefore include nonuniformity effects. Because our theme is application-dependent QWIP device optimization and a major attraction of QWIPs is the possibility of large staring arrays, we need a staring-sensor figure of merit. Our approach is to generalize familiar sensor figures of merit, the *noise equivalent temperature difference* (NETD) and the *noise equivalent flux density* (NEFD), to include nonuniformity effects. This we do by treating the residual nonuniformity remaining after sensor calibration as a contribution to the total noise. In imaging applications this "noise" is usually called *fixed-pattern noise* or *correlated noise* to emphasize the fact that it changes only slowly, if at all, over time.

The standard figure of merit for thermal imagers—the most widespread application of IR sensors—is the noise equivalent temperature difference. When the sensor spectral bandwidth is not too broad, it can be expressed as

$$\mathrm{NETD} = \left[\frac{\partial(\ln N_{\mathrm{op}})}{\partial T_B}\right]^{-1} \frac{n_T}{N_{\mathrm{op}}} \approx \left[\frac{\Phi}{\Phi'}\right] \frac{n_T}{N_{\mathrm{op}}} \tag{3.6}$$

where

Φ = scene radiance (ph cm^{-2} sr^{-1} s^{-1} μm^{-1}),
Φ' = $\partial\Phi/\partial T_B$ (ph cm^{-2} sr^{-1} s^{-1} μm^{-1} K^{-1}),
T_B = scene (background) temperature (K),

N_{op} = optically generated carriers,
n_T = total noise electrons.

For a photoconductive IR detector array we can expand eq. (3.6) as

$$\text{NETD} = \left[\frac{\Phi}{\Phi'} \right] \frac{4n_T}{\pi \eta \eta_o \Phi (\theta_R D)^2 g \tau_i \Delta\lambda} \qquad (3.7)$$

where

$\Delta\lambda$ = sensor spectral bandwidth (μm), discussed later,
τ_i = integration time (sec) over which the sensor accumulates photons,
η = absorptance, discussed in Section 3.4.2,
g = photoconductive gain, discussed in Section 3.4.3,
D = optics aperture diameter (cm),
η_o = optics throughput, that is, the fraction of scene photons incident on the aperture that are focused onto the correct pixel of the focal plane,
θ_R = system IFOV (rad).

The *internal field of view* (IFOV) and aperture diameter are related to the pixel area, A_p (cm^2), and the optics f number, f_n, by

$$(\theta_R D)^2 = \frac{A_p}{f_n^2} \qquad (3.8)$$

The detector collection efficiency, given by the product ηg, is the number of signal charges collected for each in-band photon incident on the detector. For photovoltaic detector arrays, the notation η is conventionally used for the collection efficiency, and the expression corresponding to eq. (3.7) would accordingly lack the gain factor g.

Equation (3.6) expresses NETD as a noise-to-signal ratio multiplied by a temperature; in the approximate form for a narrow spectral bandwidth, that temperature is determined from scene temperature alone via the Planck radiation law. This decomposition provides a more intuitive understanding of NETD as well as simplifying calculations, and the narrowband approximation is appropriate for QWIP detectors.

Although NETD is the suitable figure of merit for the two examples we will discuss later, some sensors are better characterized in terms of noise-equivalent flux density, which is given by

$$\text{NEFD} = \frac{4}{\pi} \frac{n_T}{\eta \eta_o (\theta_R D)^2 g \tau_i \Delta\lambda} \text{ ph cm}^{-2} \text{ sr}^{-1} \text{ s}^{-1} \text{ } \mu\text{m}^{-1} \qquad (3.9)$$

For such sensors, detector optimization would involve optimizing NEFD in a way similar to our evaluation and optimization of NETD.

For both NETD and NEFD the total noise is calculated as the root sum square of several terms, which for photoconductive detectors may be written as

$$n_T^2 = 2g(N_{op} + N_{th}) + n_{ro}^2 + \beta_{gn}^2(N_{op}\sigma_{op})^2 + \beta_{of}^2[(N_{th}\sigma_{th})^2 + (N_{op}\sigma_{op})^2 + N_{ex}^2]$$

$$(3.10)$$

The first three of these represent the usual single-detector noise currents, which are evaluated from the autocorrelation function of single-detector output at different times. They are the variances in the number of optically and thermally generated carriers, and the input-referred readout noise expressed as a number of noise charges n_{ro}. In eq. (3.10) we have expressed the variances in the optically and thermally generated carrier numbers in terms of their means N_{op} and N_{th}; the conversion factor is $2g$ for photoconductors [3, 4]. This shot-noise conversion is a slight overestimate of the true generation-recombination noise of GaAs/AlGaAs QWIPs [5–7].

The remaining four terms represent the correlated or fixed-pattern noise arising from the cross correlation of the outputs of different detectors at the same time. They are the "optical correlated noise" $N_{op}\sigma_{op}$ due to nonuniform integrated response, the "thermal correlated noise" $N_{th}\sigma_{th}$ due to nonuniform dark current, and the pattern noise arising from low-frequency excess noise ($1/f$ noise) in the detectors N_{ex}.

The factors σ_{op} and σ_{th} are the ratios of the standard deviation to mean, over the array, of the optically and thermally generated charges, respectively. The terms β_{gn} and β_{of} represent gain and offset corrections determined by sensor calibration. Each is defined as the ratio of the residual, after-calibration correlated noise from one source to the raw, uncalibrated pattern noise from the same source; for no correction, $\beta = 1$. Therefore, the remaining correlated noise terms after calibration are $\beta_{gn}N_{op}\sigma_{op}$, $\beta_{of}N_{op}\sigma_{op}$, $\beta_{of}N_{th}\sigma_{th}$, and $\beta_{of}N_{ex}$. These can be significant factors for staring systems, depending on the degree of nonuniformity present in the focal plane array and the amount of $1/f$ noise. Even for small corrected nonuniformities (i.e., small $\beta\sigma$ products), the correlated noise contributions can be quite large if N_{op} and N_{th} are large. Therefore, detector collection efficiency ηg, which affects N_{op}, and detector dark current, which arises from the thermally generated carriers N_{th}, can both significantly affect the performance of a staring sensor.

In Section 3.5 we will return to the subject of sensor NETD. There we will use the preceding formulation to discuss optimization of GaAs/AlGaAs QWIP arrays for specific infrared-sensor applications.

3.3 THE DEVELOPMENT OF GaAs/AlGaAs MULTIPLE QUANTUM WELL PHOTOCONDUCTIVE DETECTORS

The last few years have seen an explosion of published reports and data concerning III–V intersubband-transition photodetectors. For an appreciation of the scope this technology offers for creative thinking and its current state of development, it is useful to review its development in brief. We shall emphasize developments concerning dark current and collection efficiency.

3.3.1 Overview

The first proposal and experimental result for infrared detection in III–V quantum well structures were reported by Chiu et al. [8] and Smith et al. [9] in 1983. They observed photocurrent in an n-type GaAs/AlGaAs MQW device that was illuminated with a broadband IR source. The mechanism proposed for infrared absorption was phonon-assisted free-carrier absorption in the GaAs wells. They also proposed a valence-intersubband absorption QW IR detector consisting of InAlAs wells and InP barriers. Coon and Karunasiri [10] proposed in 1984 an IR detector based on photoemission from an asymmetric undoped single QW. In this design, electrons are injected into the well by an applied electric field, and the "trapped" carriers are photoemitted from the well. Goossen and Lyon [11] followed this in 1985 by proposing a grating-enhanced detector to obtain high collection efficiency with normally incident radiation.

The first experiment that attracted interest for practical applications was reported by Levine et al. [12] in 1987 from a detector based on intersubband absorption in a doped GaAs/AlGaAs MQW structure. The MQW structure was sandwiched between two heavily doped GaAs layers, to which ohmic contacts were applied, and was designed to have two confined states in the wells. Carriers were photoexcited from the ground state to the first excited state and tunneled out of the wells under the influence of an applied electric field. Choi et al. [13] improved on this design by increasing barrier widths to lower the dark current, thereby improving the detector performance. The absorption and photoresponse spectra of a similar detector [14] are shown in Figure 3.2. These photoconductive detectors were the forerunners of the MQW intersubband-transition photodetectors, which are the focus of this chapter.

Further performance improvements were achieved by Levine et al. [15] by using transitions from the ground state to "resonant" states in the continuum. The continuum states are resonant in that the optical matrix elements for these transitions are enhanced compared to those for general continuum states; this happens for energies such that the condition for resonant transmission through the well (one whole wavelength within the well) is approximately satisfied. This bound-to-continuum design provided improved absorption strength and eliminated the need for the excited carriers to tunnel through the barrier to escape the well. A further increase in barrier width resulted in substantially lower dark current [16].

The first photovoltaic QW detector was demonstrated by Goossen, Lyon, and Alavi [17], who employed an asymmetric, modulation-doped single QW with a graded

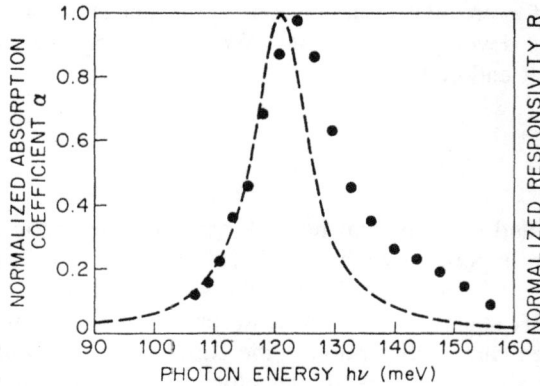

Figure 3.2 QWIP responsivity spectrum: normalized responsivity (circles) and a Lorentzian fit to the measured absorption coefficient (dashed line) for a GaAs/AlGaAs MQW detector with two confined states. (After Levine et al. [14]).

barrier on one side of the well. The intent was to lower the large dark currents experienced in the photoconductors. In this device, photoexcited electrons that escape the well are trapped in the graded barrier region producing a charge separation. Although initially thought to be a promising approach, no further work has been reported. Rosencher et al. [18] reported another photovoltaic device structure using coupled asymmetric quantum wells, where electrons are photoexcited from one well and stored in an adjacent well for a short time before tunneling back to their original ground state. These devices had virtually no dark current, but had a photoresponse much below the noise of practical amplifier circuits.

Choi et al. [19] took a different approach to lowering the dark current, building the IR-sensitive MQW into a hot electron transistor. In this three-terminal device, the GaAs/AlGaAs MQW structure is followed by a thin GaAs base region and an AlGaAs dark current blocking layer. The dark current is drained off through the base region, but the photoexcited carriers are energetic enough to pass over the dark current blocking layer. Choi et al. [20] later improved on this design by using an InGaAs base layer that provided improved photocurrent transport properties. This device has shown promising performance results but suffers from the need for two contacts to each detector, making array design and readout-circuit interconnections more complex.

Most of the recent activity in the development of LWIR detectors fabricated from GaAs/AlGaAs MQWs has focused on the bound-to-continuum transition QWIPs. A sustained effort by Levine et al. [12–15] has led to realization of these devices for practical infrared detectors. We shall focus on this detector design in this chapter.

3.3.2 Principles of Operation

A schematic diagram of the bound-to-continuum transition QWIP is shown in Figure 3.3. The device structure consists of an alternating series of thin ($\approx$40–50 Å), intentionally doped GaAs and wider ($\approx$300–500 Å), undoped $Al_xGa_{1-x}As$ layers clad by thick ($\approx$1 μm) GaAs contact layers. This structure is grown by *molecular beam epitaxy* (MBE) or *metallo-organic chemical vapor deposition* (MOCVD) on an undoped GaAs substrate to facilitate backside illumination. Typically, 50 quantum wells are grown to achieve sufficiently large total absorption. The well width is determined by the thickness of the GaAs layers, and the barrier height can be adjusted by changing the composition (x value) of the $Al_xGa_{1-x}As$. Individual detectors are delineated by mesa etching. Ohmic contacts are made to the top of each detector and to the common contact. Detectors can be wire bonded individually, or *two-dimensional* (2D) arrays can be hybridized to a fan-out structure or readout circuit with indium-bump bonds.

To operate these devices as detectors, an electric field must be applied to the structure. Under the influence of an applied electric field, the device acts as a photoconductor. Infrared radiation excites carriers from the ground state in the well to the first excited state in the well or to continuum states depending on the design of the device. The photoexcited carriers are swept out of the device by the applied electric field. For n-type GaAs/AlGaAs quantum wells, the absorption is polarization sensitive; only the component of the radiation with electric field vector perpendicular to the plane of the quantum well is absorbed. This is because the GaAs conduction band is parity invariant, so that the electromagnetic field can couple only conduction-band states of different symmetry under reflection in the plane of the well—and that requires a component of the radiation electric field perpendicular to the well. Therefore, to make a backside-illuminated n-type QWIP detector array, a

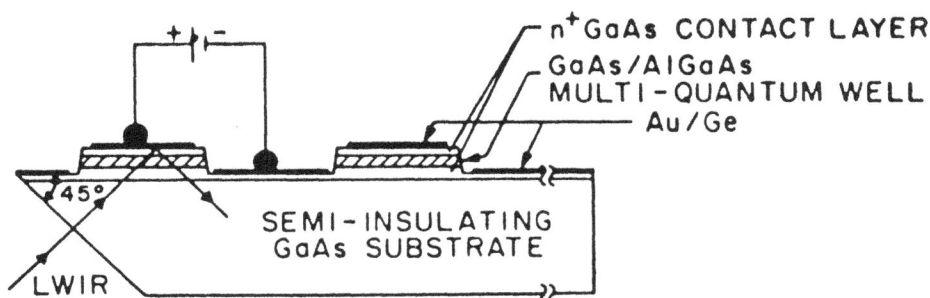

Figure 3.3 Cross-sectional view of GaAs/AlGaAs MQW detector with a 45° polished edge to facilitate absorption and responsivity measurements. (After Hasnian et al. [31]).

grating scheme must be employed to refract normally incident IR radiation so that a component of the radiation electric field is parallel to the growth direction. In p-type quantum wells, there is no such selection rule, and all polarizations of incident radiation are absorbed [21, 22].

3.3.3 Development of the Basic Device

Levine et al. [12] initial experimental results were from GaAs/AlGaAs MQW detectors with two bound states, where photoexcited carriers had to tunnel out of the well under the influence of an applied electric field. This process is illustrated in Figure 3.4. These bound-to-bound QWIPs were reported to have a responsivity in excess of 0.52 A/W at a peak wavelength of 10 μm. Infrared radiation was coupled into these detectors by polishing a 45° facet on one edge of the substrate and illuminating the detector normal to this facet with a tunable CO_2 laser as shown in Figure 3.3. By increasing the barrier width and height, the responsivity was increased to 1.9 A/W [13]. These responsivity values sparked a significant interest in MQW detectors as they began to approach values achieved in early LWIR HgCdTe detectors. However, these early reports did not address the noise arising from the large dark currents in these detectors at practical operating temperatures ($\approx$77 K).

Coon, Karunasiri, and Liu [23] showed that the maximum absorption occurred when the excited state was located just above the top of the barrier. Levine's group made use of this result in their later detectors and demonstrated a detectivity of 10^{10} cm-$\sqrt{Hz}$/W for a detector with peak wavelength at 8.3 μm for the bound-to-extended state detector [13]. A diagram depicting this type of transport in the conduction band is shown in Figure 3.5. These detectors displayed a broader spectral

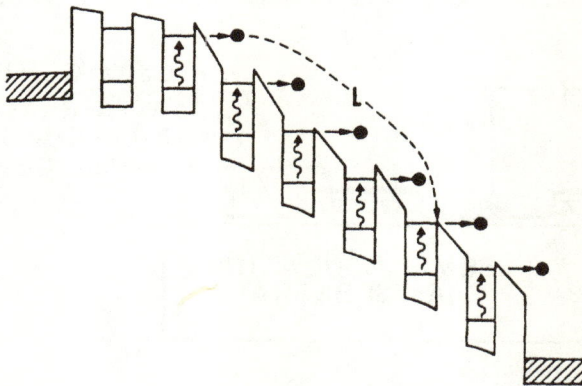

Figure 3.4 Schematic conduction band diagram of bound-to-bound transition QWIP showing photoconductivity produced by absorption of intersubband radiation followed by tunneling out of the well. (After Levine et al. [12]).

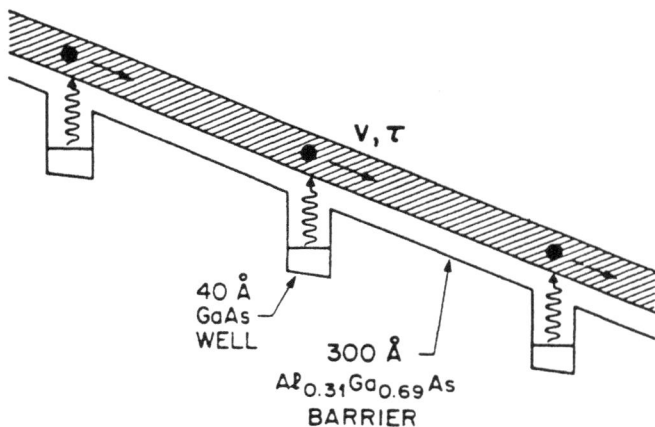

Figure 3.5 Schematic conduction band diagram of the bound to continuum transition detector. (After Levine et al. [46]).

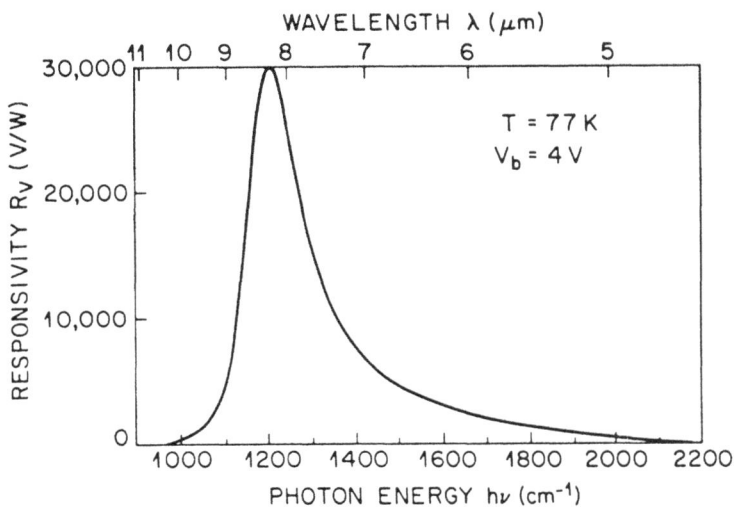

Figure 3.6 Voltage responsivity of bound-to-continuum QWIP, measured at $T = 77$ K and 4-V bias, with 100 Ω load resistance. (After Levine et al. [41]).

response than the equivalent bound-to-bound transition detectors and required a smaller applied bias voltage to achieve appreciable responsivity, as shown in Figure 3.6. The inclusion of dark current and absorptance data for these detectors made possible comparisons with existing HgCdTe detectors. Kinch and Yariv [24] pointed out that these devices would never compare favorably with equivalent ideal HgCdTe detectors, due to the intrinsically short lifetime of the excited carriers ($\approx$1–10 ps) that give rise to large thermal generation rates, and thus the observed large dark currents. Their comparison between the two detector technologies is shown in Figure 3.7. The lack of an ideal HgCdTe and the desire for large staring arrays for thermal imaging applications, where the noise from the background photocurrent would ideally dominate the detector noise, was enough justification to keep Levine, Kinch, and Yariv [25] and others pushing forward with the GaAs/AlGaAs QWIP development.

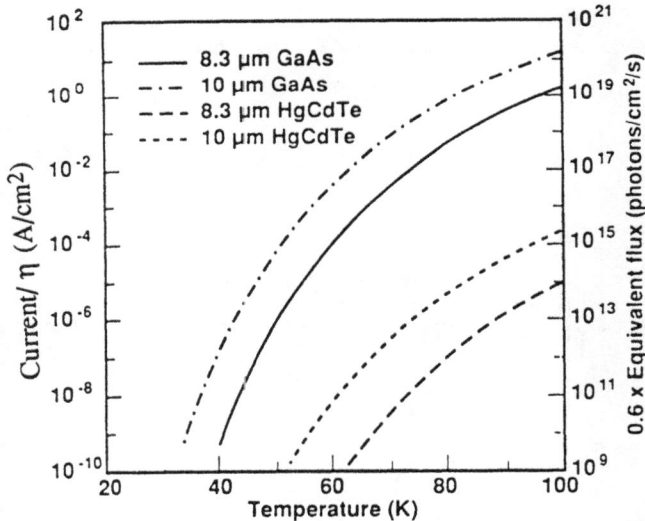

Figure 3.7 Thermal generation current versus temperature for GaAs/AlGaAs MQW detectors and HgCdTe alloys at 8.3 and 10 μm. The assumed quantum efficiencies are $\eta = 0.125$ for GaAs/AlGaAs and 0.7 for HgCdTe, respectively. (After Kinch and Yariv [24]).

With the basic GaAs/AlGaAs QWIP concept demonstrated, the development effort turned toward optimizing device parameters. This effort included demonstrating longer peak wavelength response, wider bandwidths, lower dark currents, and higher-collection-efficiency devices.

3.3.4 Long-Wavelength, Wide-Bandwidth Detectors

Following the demonstration of the bound-to-continuum GaAs/AlGaAs QWIP, Levine et al. [1] reported a wide-bandwidth QWIP, where the excited state was pushed further into the continuum, resulting in a 4 μm spectral bandwidth with peak response at 10 μm, as illustrated in Figure 3.8. This did not provide a performance improvement over narrow-bandwidth devices, because the total integrated photoresponse remains constant for a fixed doping density in the well. Later work by Zussman et al. [26] demonstrated longer wavelength detectors with cutoff wavelengths near 15 μm. In this case, two detectors were grown with nearly the same peak wavelength ($\approx$13.3 μm) but with different well widths, so that one of these detectors had the first excited state pushed further into the continuum resulting in a broader spectral bandwidth. These data are shown in Figure 3.9. As expected, the responsivities of these two devices were similar, but the detectivity of the narrow-well device was lower (Figure 3.10). This is because thermionic emission increases as well width decreases, for the ground state moves up in the well as the well is narrowed for a constant barrier height. The peak-response wavelength, however, changes much less because the resonant continuum state also rises further into in the continuum.

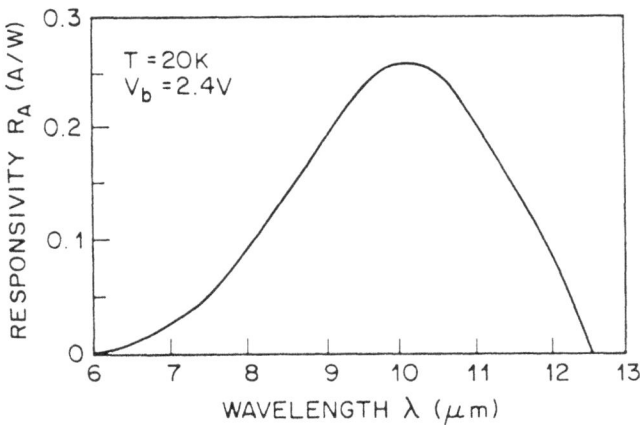

Figure 3.8 Responsivity spectrum measured at T = 20 K at bias of 2.4 V for wide bandwidth 10 μm GaAs/AlGaAs MQW detector. (After Levine et al. [1]).

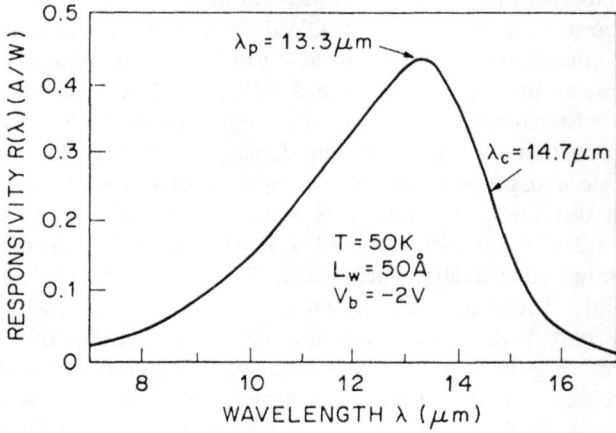

(a)

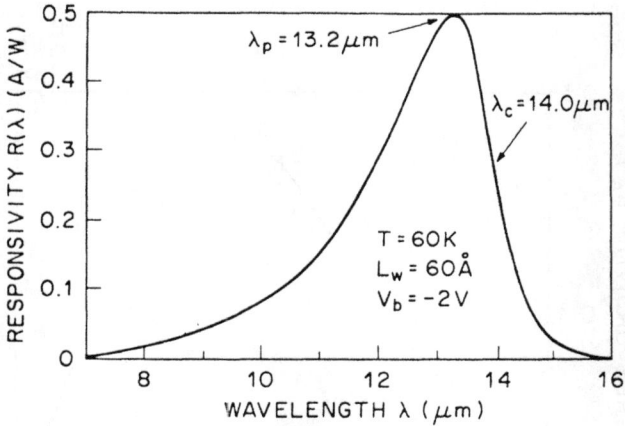

(b)

Figure 3.9 Responsivity of 13 μm bound-to-continuum transition detectors with different well widths: (a) $L_w = 5$ nm, $\lambda_p = 13.3$ μm, $\lambda_c = 14.7$ μm, $T = 50$ K; (b) $L_w = 6$ nm, $\lambda_p = 13.2$ μm, $\lambda_c = 14.0$ μm, $T = 60$ K. (After Zussman et al. [26]).

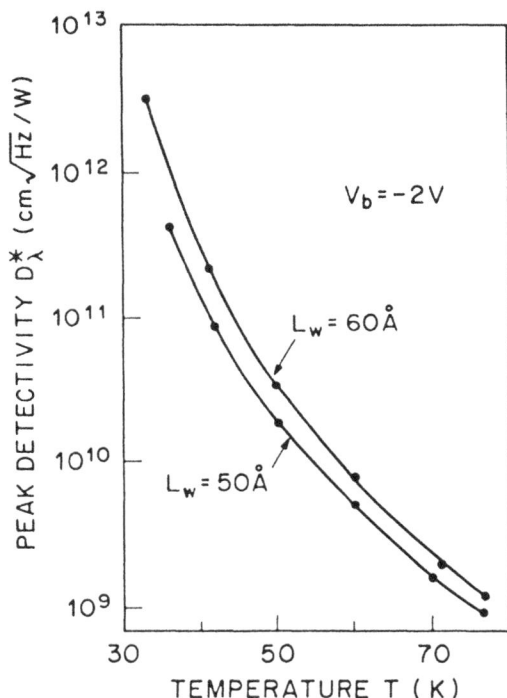

Figure 3.10 Temperature dependence of the detectivity for the two 13 μm detectors of Figure 3.9 showing the slightly reduced detectivity of the narrow-well detector. (After Zussman et al. [26]).

3.3.5 Low Dark Current Devices

Since the early results were published for the GaAs/AlGaAs MQW detectors, an ongoing effort has been made to reduce the dark current in these devices, not only to improve performance, but also to allow operation of these detectors with standard *complimentary metal oxide semiconductor* (CMOS) readout circuits. In typical readout circuits the storage capacity of the integration capacitor is limited to $\approx 10^8$ charges. In arrays with higher counts of smaller pixels, this limit will drop to $\approx 10^6$ charges. For thermal imaging applications, operating temperatures of 77 K are desired. At these temperatures, the early devices had large dark currents, mainly due to thermally assisted tunneling, that would have filled the integration well for practical integration times. One straightforward method of reducing the thermally assisted tunneling component of the dark current was to widen the barriers. Janousek et al. [5] and Levine et al. [16] showed that 400–500 Å barriers were adequate to suppress this component of the dark current for detectors with peak wavelengths out to 10 μm. These barrier widths were found to be adequate for suppressing both the thermally assisted and direct interwell tunneling down to temperatures of 30 K. A comparison of the 6 K dark current of detectors with different barrier widths is shown in Figure 3.11(a). However, as shown in Figure 3.11(b), for a small bias voltage the contribution from thermally assisted tunneling to the dark current was found to be small at 77 K, indicating that thermionic emission of carriers out of the well was the dominant contribution at that temperature. This led to the investigation of doping density effects on thermionic emission.

Analyses of the doping density effects on the dark current and detectivity were performed by several groups [27–30]. The dependence of the detectivity on doping shows a maximum value at a doping lower than initially used, as shown in Figure 3.12. The detectivity shows a broad peak near the optimum doping concentration, suggesting that doping lower than the optimum could be used while maintaining high detectivity, with dark currents reduced by orders of magnitude over the early higher doped detectors [30]. A plot of dark current versus bias voltage for three doping concentrations from Gunapala et al. [29] is shown in Figure 3.13. These results provided detectors with dark currents low enough for standard CMOS readout circuits.

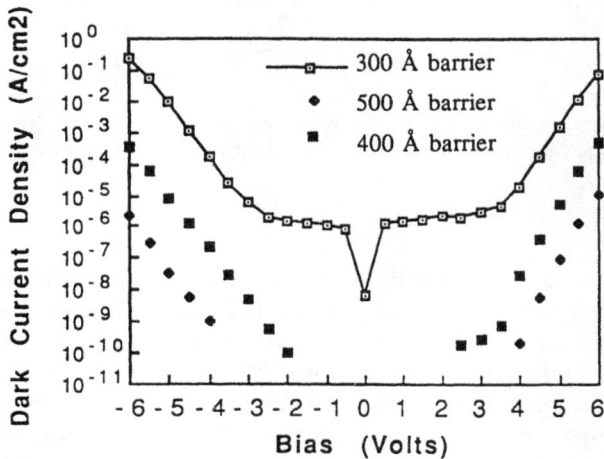

(a)

Figure 3.11 Dark current of wide-barrier GaAs/AlGaAs QWIPs. (a) Comparison of dark current of 8 μm detectors of different barrier widths, $T = 6$ K. (After [44].) (b) Mobile carrier density as a function of energy for a 10 μm detector with 50 nm barriers. (After Levine et al. [16]).

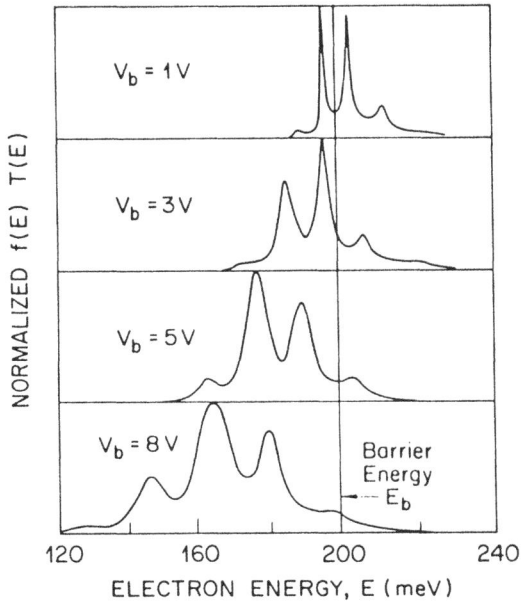

(b)

Figure 3.11 continued.

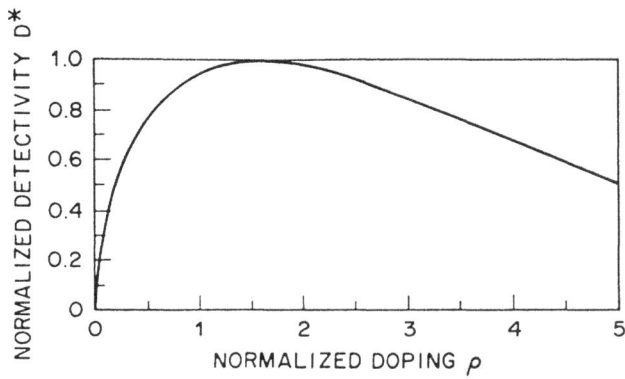

Figure 3.12 Normalized detectivity as a function of normalized doping density for GaAs/AlGaAs QWIPs showing a broad peak at the optimum doping concentration. (After Gunapala et al. [29]).

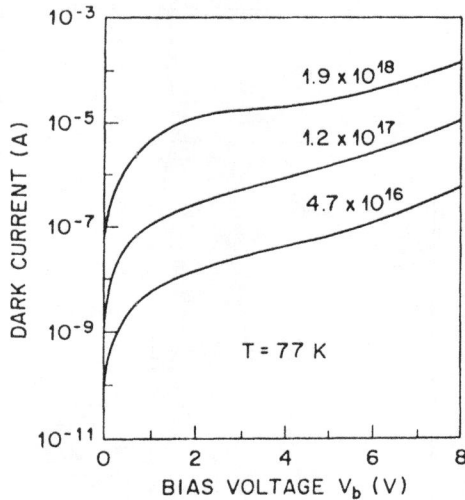

Figure 3.13 Dark current versus bias voltage for 200 μm diameter GaAs/AlGaAs MQW detectors having three different doping densities. (After Gunapala et al. [29]).

3.3.6 High-Efficiency Detectors

The major drawback in reducing the doping to minimize the dark current is that the collection efficiency is reduced as well. This fact along with the polarization sensitivity of the n-type detectors led to several attempts to increase collection efficiency. The concept of using a grating to refract the incident radiation into a more favorable angle for absorption was first proposed by Goossen and Lyon [11]. Hasnain et al. [31] demonstrated high coupling efficiency for backside-illuminated detectors with etched sawtooth gratings, as shown in Figure 3.14. Andersson, Lundqvist, and Paska [32] later demonstrated the use of a AlAs waveguide along with a lamellar reflection grating to increase the number of passes the radiation makes through the active region of the detector, resulting in an internal absorptance (corrected for reflection losses) near 60%. This design is illustrated in Figure 3.15. Later Andersson and Lundqvist [33] demonstrated 92% internal absorptance through the use of doubly periodic reflection gratings along with the waveguide structure. A plot of the measured absorptance spectrum is shown in Figure 3.16. This grating design refracts incident radiation into the transverse magnetic guided mode, which is the one absorbed through intersubband transitions in the quantum wells. This approach appears to be very promising except for the fact that the AlAs oxidizes very quickly, which means that the exposed edges of the AlAs layer must be passivated or coated to prevent oxidation and eventual disintegration.

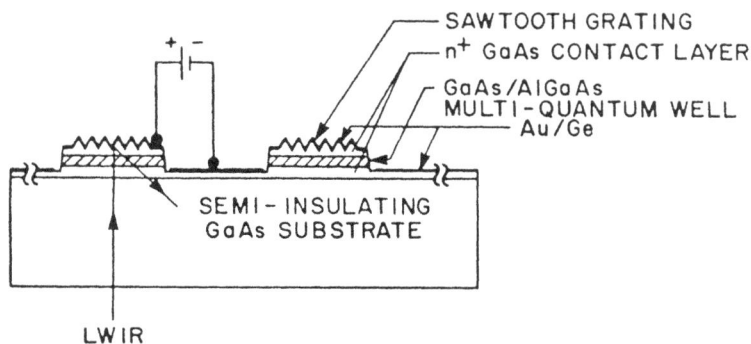

Figure 3.14 Cross-sectional view of GaAs/AlGaAs MQW detector, showing the scheme for backside illumination at normal incidence by reflection off the gold-coated grating etched into the top contact layer. (After Hasnian et al. [31]).

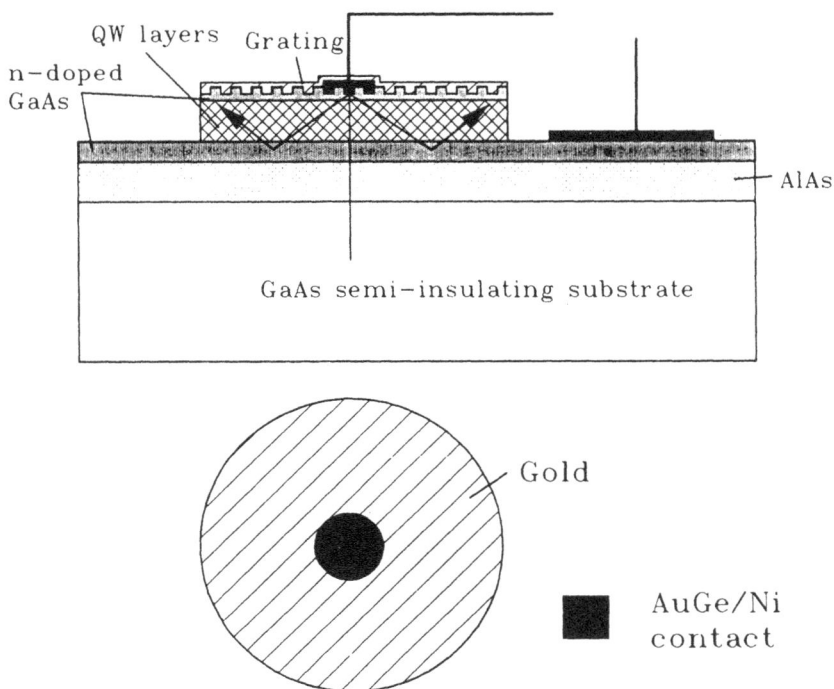

Figure 3.15 Layout of the waveguide MQW detector. Below is the mesa covered with AuGe/Ni (central part only) and pure gold, viewed from the wafer surface. (After Andersson et al. [32]).

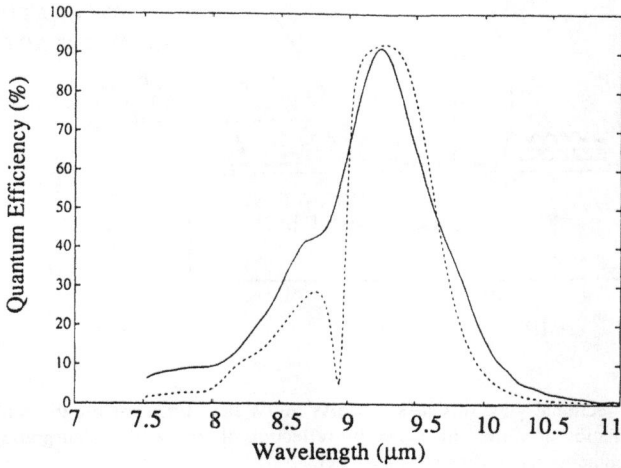

Figure 3.16 Experimental (solid curve) and theoretical (dashed curve) quantum efficiency spectra for the waveguide detector of Figure 3.15 with a doubly periodic grating. (After Andersson and Lundqvist [33]).

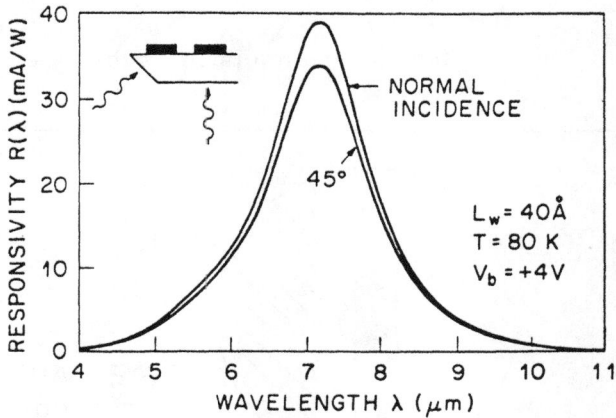

Figure 3.17 Comparison between the normal incidence and 45° responsivity spectra for a p-type GaAs/AlGaAs MQW detector. (After Levine et al. [21]).

An alternate approach to achieving high collection efficiency was demonstrated by Levine et al. [21] through the use of p-type doping in the quantum wells. In this case, the degenerate states at the valence-band maximum are not individually parity invariant, so that there is no polarization selection rule corresponding to that for n-type quantum wells. Quantum efficiencies of $\approx 30\%$ were observed in these structures without the use of gratings or waveguides. Responsivity versus wavelength data for both normal and $45°$ illumination are shown in Figure 3.17. The disadvantage of the p-type detectors is that the mean free path of the excited holes is very small, due to the large hole effective mass. This resulted in significantly smaller gain and lower detectivity than observed in equivalent n-type detectors. However, no further attempts to optimize the transport properties of these detectors have been reported. We shall show in the following sections that p-type QWIPs remain attractive alternatives to n-type devices.

3.4 PERFORMANCE MODELS FOR GaAs/AlGaAs MULTIPLE QUANTUM WELL PHOTOCONDUCTIVE DETECTORS

In this section, we shall discuss the optimization of QWIP staring arrays for several applications where they might be expected to be advantageous in some way. To clarify the impact of sensor system requirements on QWIP device design, we will attempt to relate the physical processes occurring within the detector to eventual sensor performance, as far as our current understanding permits. This will necessitate choosing specific values for material and structure properties and representing complex phenomena by simplified parametric models, sometimes in the absence of complete experimental evidence or theoretical understanding. Such assumptions and simplifications will reduce the reliability of our quantitative conclusions. However, our objective is not quantitative precision, but a qualitative understanding of the important sensor-performance issues and how they affect QWIP detector design. We believe that our assumptions and models, accepted with caution, are accurate enough for this.

The most important detector parameters for optimization of MQW detectors for IR sensor applications are dark current, collection efficiency, and correlated noise effects. The following subsections will address the important factors controlling the dark current, collection efficiency, and correlated noise. We shall develop simplified models of their dependence on material and structure parameters, which will be used later to illustrate the dependence of sensor performance on QWIP device design. We believe the models reproduce the main dependencies well enough to give an essentially correct qualitative understanding of this relationship, which is our objective. We shall try to point out the deficiencies that we are aware of, so that the reader may judge how far to trust our quantitative conclusions.

3.4.1 Dark Current

Our dark current model is based on the work of Andrews and Miller [28] and Williams et al. [34]. These authors compare experimental data with a simple theory in a way that we can use to organize large amounts of data. Their work is in turn based on that of Levine et al. [16] and Choi et al. [35]. The main points are as follows.

The dark current consists of three components: thermionic emission, thermally assisted tunneling, and temperature-independent tunneling. Thermionic emission refers to the carriers that are thermally excited into current-carrying states above the top of the well; the resulting current varies exponentially with temperature. By *temperature-independent tunneling*, we mean tunneling of carriers into a neighboring well from densely occupied states at the "bottom" of the well; that is, low in the band of two-dimensionally confined states. This current is only weakly temperature dependent, and it is observed at temperatures low enough to suppress thermionic emission.

Thermally assisted tunneling is the term usually used to refer to an intermediate case, in which carriers are thermally excited into states lying high in the confined band, but below the top of the well, and then tunnel through the triangular tip of the barrier (under an applied electric field) into current-carrying states above the top of the barrier at a different location. This process is sometimes also called *field emission*, which is consistent with the fact that it is negligible compared to thermionic emission except at relatively high electric fields. However, the reader should note that, in the present case, the term refers to a *thermally activated* process except at very high electric fields. For precision, we favor the more specific phrase *thermonic field emission*.

Another thermally activated process is possible, in which a carrier tunnels from a state high in the confined band into a confined state at the same energy in the neighboring well. For the basic QWIP photoconductor using transitions from a single confined state to the continuum, barriers are sufficiently wide that this thermally assisted well-to-well tunneling current is small compared to the sum of thermionic-emission and field-emission currents. Accordingly we shall drop it.

The remaining dark current processes are (1) the thermally activated processes of thermionic emission and thermally assisted field emission, which we shall treat together, and (2) temperature-independent tunneling of carriers low in the wells.

Thermally Activated Processes

For well-to-well tunneling, current is limited by the tunneling rate itself. We shall discuss this case later. In contrast, the rate-limiting process for thermionic-emission and field-emission currents is not the emission rate but drift of the carriers after their emission. Consequently, the thermally activated currents are conveniently modeled in a hydrodynamic picture, by analyzing [16] them into a density n_t of mobile carriers and their average drift velocity v_d:

$$I_D = eA_D v_d n_t \qquad (3.11)$$

where A_D is the detector area (which may be different than the pixel area A_p), and e is the electronic charge.

A simple model of the carrier drift velocity is

$$v_d = \mu F \left[1 + \left(\frac{\mu F}{v_s} \right)^2 \right]^{-1/2} \qquad (3.12)$$

where v_s is the saturation drift velocity, μ is the low-field carrier mobility, and F is the electric field intensity. When eq. (3.12) has been used [16, 28] to analyze QWIP dark current data, the main attention has been paid to the high-field behavior of the current, which is consistent with an approximately constant drift velocity, $v_d \approx v_s$. However, eq. (3.12) does not correctly represent the behavior of v_d observed [36] in bulk GaAs or the similar behavior predicted for AlGaAs by a Monte Carlo calculation [37]. (To our knowledge, there is as yet no experimental data for AlGaAs.) In the bulk materials, eq. (3.12) approximately represents the field dependence of the drift velocity at low fields. At electric fields greater than approximately 4×10^3 V cm^{-1}, transfer of carriers to higher effective mass valleys away from the Brillouin zone center reduces the average carrier mobility, causing drift velocity to decrease with increasing field and eventually saturate at about half the peak value. Whether v_s would show similar behavior in MBE AlGaAs layers may depend on the defect density.

An electric field of 4×10^3 V cm^{-1} corresponds to 1 V bias across a 50 period QWIP of 50 nm period. There is no gross feature at that level in reported QWIP dark current data, which has been accurately fitted to eqs. (3.11) and (3.12). We also will use eq. (3.12) for the drift velocity, while noting that the difference from bulk barrier material that it implies has not yet been explained in the literature.

As for the mobility μ and saturation velocity v_s, we would expect their values in the QWIP structure to be roughly comparable to those of the barrier material. Where mobility or saturation velocity has been determined from QWIP dark current data, this seems to be the case. The n-type mobility values of 1000[28] and 2200[16] cm^2 V^{-1} s^{-1}, quoted as representative of the dark-current data, lie at the lower end of the range of values found in MBE-grown n-type Al$_{0.26}$Ga$_{0.74}$As layers at 77 K, which may be over 4000 cm^2 V^{-1} s^{-1} [38]. More variability is seen in v_{sat}: values of v_{sat} ranging from 10^5 to 5×10^6 cm/s were reported by Andrews and Miller [28].

The uncertainty in these values somewhat degrades the quantitative predictive value of eq. (3.11). In view of two considerations, however, we consider this degree of uncertainty acceptable. In the first place, by far the greater part of dark current variation is due to the temperature and bias dependence of the density of mobile carriers. Second, we shall find that, for the two applications discussed in Section 3.5.2, the appropriate bias level for QWIP focal planes QWIPs is 0.1 to 0.2 V, corresponding to electric field intensities of 500 to 1000 V cm^{-1}. Therefore the more important of these parameters for our purposes is the mobility. We shall assume mobility values of 2000 cm^2 V^{-1} s^{-1} for n-type and 200 cm^2 V^{-1} s^{-1} for p-type GaAs/AlGaAs QWIPs. Using these values in eq. (3.12), we find that drift velocities are then sufficiently low that errors due to uncertainty in the saturation velocity are secondary.

The density of mobile carriers, n_t, is computed from the expression [16]

$$n_t = \frac{1}{L_P} \int_{E_1}^{\infty} T(E, V_1) f(E) \rho(E) \, dE \tag{3.13}$$

where $f(E)$ is the Fermi factor, $L_P = L_B + L_W$ is the period of the MQW structure, and $T(E, V_1)$ is the transmission coefficient for transport at energy E through a single barrier of width L_B biased to a potential drop of $V_1 = FL_P$. Thermionic emission is accounted for by defining $T(E, V_1)$ to be unity for energies greater than the barrier height ΔE. The two-dimensional density of states corresponding to one period of the structure is $\rho(E)$; in the case that the well binds only a single state at energy E_1, it is [24]

$$\rho_1(E) = \frac{1}{\pi \hbar^2} \left\{ m_W^* + m_B^* \, \text{Int} \left[\frac{L_P}{\pi \hbar} \sqrt{2 m_B^* (E - E_{C,B})} \right] \right\} \qquad (E > E_1) \tag{3.14}$$

where m_B^* is the effective mass in the barrier and m_W^* is that in the well, $E_{C,B}$ is the conduction-band edge in the barrier material, and Int[x] is the greatest integer in x (adopting the convention that Int[x] = 0, when x is not real). This expression partitions the continuum density of states in the QWIP structure on a per-well basis. We can verify that for large period L_P, the ratio $\rho_1(E)/L_P$ approaches the three-dimensional density of states in the barrier material. The second term in braces in eq. (3.14) affects only the thermionic-emission current, which it corrects by a factor of order unity. Andrews and Miller [28] do not include this term in their analysis. To maintain close contact with their work we shall also drop it, leaving

$$\rho_1(E) \approx \rho_1(0) = \frac{m_W^*}{\pi \hbar^2} \tag{3.15}$$

Andrews and Miller [28] have compared the model of eqs. (3.11) through (3.15) to experimental dark current data for bound-to-continuum QWIPs, using for $T(E,$

V_1) the Wentzel-Kramers-Brillouin (WKB) expression for one-dimensional tunneling [39]:[1]

$$T_{\text{WKB}}(E, V_1) = \exp\left\{ -\frac{4}{3eV_1} \sqrt{\frac{2m_B^* L_B^2}{\hbar^2}} \right.$$

$$\left. [(\Delta E - E)^{3/2} - (\Delta E - E - eV_1)^{3/2}] \right\}$$

(3.16)

where the effective mass in the barrier is m_B^*. They found quite good agreement with experiment at higher temperatures, provided that the barrier height ΔE is reduced to account for image-charge effects by an amount

$$\delta\phi = e \sqrt{\frac{eF}{4\pi\varepsilon}}$$

(3.17)

as would be expected if the emitting quantum well were a grounded conducting plane. Figures 3.18 and 3.19 illustrate the agreement they obtained by adjusting a temperature-independent saturation drift velocity v_s and well width L_W to fit the measured dark currents of QWIPs having peak spectral response at 7 and 10 μm.

The image-charge barrier-lowering of eq. (3.17) is computed by treating the quantum well as an isolated, grounded conducting plane. In the presence of an emitted charge, the charges confined in the well rearrange themselves to suppress the tangential component of the electric field in the well; the resulting potential distribution is the same as though an opposite charge were located at the reflection in the plane of the emitted charge's position [40]. Superposing a uniform electric field on this potential, one finds that the triangular tip of the barrier is rounded and thus lowered by the amount given in eq. (3.17). By straightforward calculation we can verify that the correction to eq. (3.17) due to the presence of the receiving well is unimportant.

For applied fields in the range $\hbar^2/(4m_B^* L_B^2) \ll eFL_P \ll \Delta E - E_F$, and in the range of temperatures where the dark current shows thermally activated behavior, it is possible to obtain a fairly accurate approximation to eqs. (3.13) through (3.16)

---

[1]Meshkov has pointed out that, when there is scattering in the barrier, the probability of finding a two-dimensionally confined carrier in the barrier decays with distance into the barrier at a rate determined by the carrier's total energy E, even though the Kronig-Penney model wave functions of all transverse momenta decay at the same rate. This fact was used by Levine et al. [13] to justify the use of the total energy in models such as eq. (3.16) for the transmission amplitude.

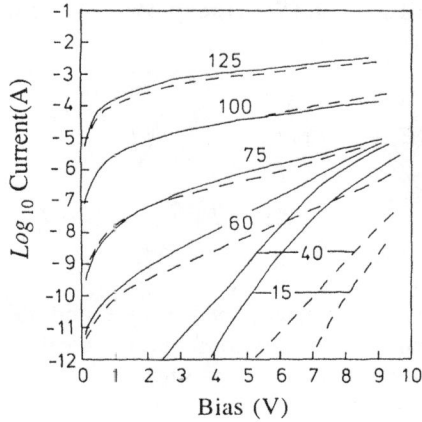

Figure 3.18 Experimental (solid curves) and theoretical (dashed curves) dark current characteristics for a GaAs/AlGaAs QWIP with a measured dark current activation energy of 99 meV. (After Andrews and Miller [28]).

by replacing the integrand with an approximation

$$T_{\text{WKB}}(E, V_1) f(E) \approx \begin{cases} \exp[-c_0 - c_1(x_1 - x)^p], & x < x_1 \\ (e^{c_0} - e^{c_1} + e^x)^{-1}, & x > x_1 \end{cases} \quad (3.18)$$

where $x = (E - E_F)/kT$, x_1 is the location of the peak in the integrand near the top of the barrier, c_0 is the magnitude of the integrand at the peak, and c_1 and p are constants. For a positive c_1, the approximate integrand is exactly integrable if the lower limit of integration is sent to $-\infty$. By judicious choice of the simplest c_1 and p that accurately match the integrand of eq. (3.16) near the top of the barrier, we obtain

$$n_t \approx \frac{\rho_0}{L_P} e^{-b} \left\{ 1 + \left[\frac{2\gamma}{1 - e^{-2\gamma}} e^\gamma - 1 \right] + 2\Gamma(5/3)(3/4\gamma)^{1/3} \left[1 + \left(\frac{3\gamma}{4a^2} \right)^{1/4} \right] e^\gamma \right\}$$
$$(3.19)$$

where Γ is Euler's gamma function, and

$$\rho_0 = kT\rho_1(0), \qquad b = \frac{\Delta E - \delta\phi - E_F}{kT},$$
$$a^2 = \frac{2m_B^* L_B^2}{\hbar^2} kT, \qquad \gamma = 1/3 \left(\frac{eV_1}{2akT} \right)^2$$
$$(3.20)$$

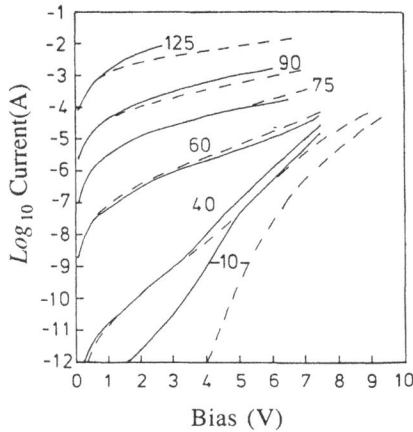

Figure 3.19 Experimental (solid curves) and theoretical (dashed curves) dark current characteristics for a GaAs/AlGaAs QWIP. (After Andrews and Miller [28]).

We have included the image-charge barrier lowering in the nondimensionalized effective barrier height b. The additional condition that eq. (3.19) accurately represent the WKB transmission given by eq. (3.16) for energies near the top of the barrier is

$$1 > 3\gamma \frac{kT}{V_1} = \frac{1}{4}\left(\frac{V_1}{kT}\right)\frac{\hbar^2}{2m_B^* L_B^2 kT} \tag{3.21}$$

Equation (3.21) is violated at low temperatures and high fields, when the thermally activated currents represented by eq. (3.19) are small compared to the temperature-independent tunneling currents.

The first term in braces of eq. (3.19) accounts for thermionic emission, whereas the remaining terms account for thermally assisted tunneling. To observe thermally assisted tunneling, the ratio γ must be large enough to noticeably correct the effective barrier height b. At low temperatures this happens at lower bias than at high temperatures, because γ varies as T^{-3}. But γ also varies as V_1^2, so by reducing the bias the thermally assisted tunneling can be made small. For this and other reasons, we shall find in Section 3.5 that QWIP focal-plane arrays are best operated at rather low bias. When γ is small, the first term in brackets vanishes linearly in γ, that is, quadratically in V_1, and the remaining term, which is proportional to $\gamma^{1/3} \sim V_1^{2/3}$, gives only a small correction to the thermionically emitted carrier density $(\rho_0/L_P) e^{-b}$. In particular, for the two examples discussed in Section 3.5.2, the optimum detector bias corresponds to a single-period bias $V_1 \sim 2$ to 4 mV. At this level, the thermally assisted tunneling current amounts to between 10 and 40% of the thermionic-emission current. In addition, because γ is inversely proportional to m_B^*, the effects are much smaller in p-type than in n-type detectors.

In Figure 3.20(a) we compare dark current computed via eq. (3.19) with that computed from the full expression, eq. (3.16), by Andrews and Miller [28]. Because eq. (3.19) represents only the thermally activated currents, we have limited the comparison to temperatures above 50 K. The solid lines are computed with the approximation, and the dots are taken from the theoretical curves in Figure 7 of Andrews and Miller [28]. Agreement is satisfactory.

Equation (3.19) can also be used to correlate published dark current data for n-type GaAs/AlGaAs QWIPs, from which we can judge the utility and limitations of the model. Summarizing eqs. (3.11), (3.12), (3.15), (3.17), and (3.19) in the expression

$$I_D = C_1 \, e^{\gamma - b} \tag{3.22}$$

we divide published dark current data by the pre-exponential factor C_1 and plot the natural logarithm of the quotient versus the argument of the exponential. To the extent that eq. (3.22) correctly represents the data, the points should lie on a line of unit slope passing through the origin. The result is shown in Figure 3.20(b): data varying over many orders of magnitude collapse into a diagonal band. The data plotted represent n-type AlGaAs/GaAs detectors having peak response near 8, [41], 10, [1, 16] and 13 [26, 28] μm, at temperatures between 50 and 125 K and applied fields between 4 and 100 mV/period.

In Figure 3.20(b), we used fixed mobility $\mu = 2000$ cm^2 V^{-1} s^{-1} and saturation drift velocity $v_s = 5 \times 10^5$ cm s^{-1} for all samples. The width of the diagonal band reflects the fact that, contradicting our use of a fixed value, v_s must vary over nearly two orders of magnitude to fit the data for individual arrays. In addition, the bowing of single curves suggests a departure from eq. (3.12) similar to what is observed in bulk GaAs.

We estimated the barrier height from the quoted cutoff wavelength λ_c (defined as the wavelength at which the photoresponse falls to half its peak value), using

$$\Delta E - E_1 = K_\lambda \frac{hc}{\lambda_c} \tag{3.23}$$

where K_λ accounts for the small (7–10 meV, or 5–10%) difference [28] between photoresponse cutoff and $\Delta E - E_1$. For Figure 3.20(b) we set $K_\lambda = 1$. We determined the Fermi energy E_F from the relation [24]:

$$\rho_s = \rho_0 \ln \left[1 + \exp\left(\frac{E_F - E_1}{kT} \right) \right] \tag{3.24}$$

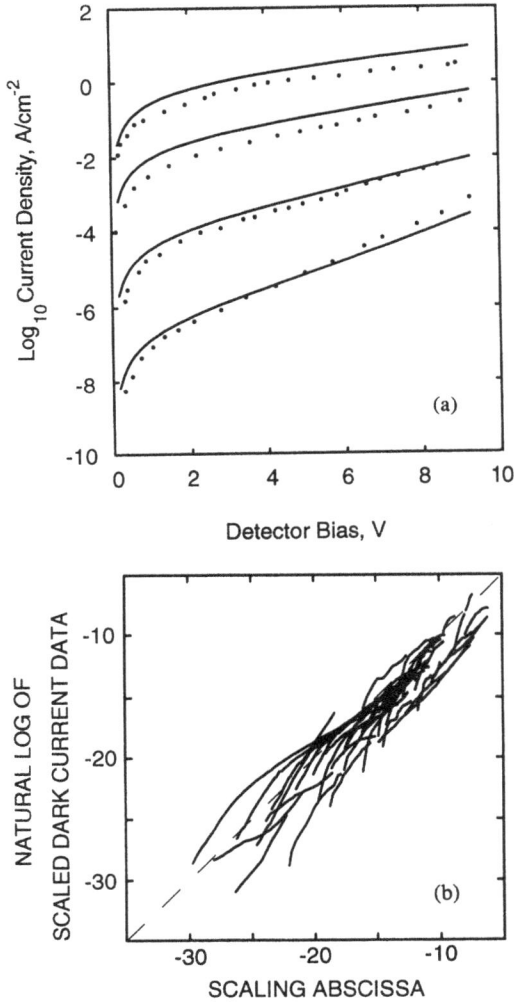

Figure 3.20 Validation of the approximate eq. (3.19) for thermally activated QWIP dark current. (a) Comparison of dark currents computed using eq. (3.19) with the values computed from the full theory of eqs. (3.11) through (3.16). Dots show the full theory, taken from Figure 7 of Andrews and Miller [28], and lines show the approximation computed from eq. (3.19), using the same parameters. From bottom to top, detector temperatures are 60, 75, 100, and 125 K. (b) Experimental dark current data from [1, 16, 26, 28, and 41] scaled by the pre-exponential factor and plotted against the argument of the exponential in eq. (3.22). The data represent dark currents of n-type GaAs/AlGaAs QWIPs having peak response near 8, 10, and 13 μm, measured at temperatures between 50 and 125 K and applied fields between 4 and 100 mV/period. Fixed values of mobility, $\mu = 2000$ cm^2 V^{-1} s^{-1}, and saturation drift velocity, $v_s = 5 \times 10^5$ cm s^{-1}, were used for all samples.

where the two-dimensional sheet charge density is expressed in terms of the three-dimensional doping density, N_D, as $\rho_s = N_D L_W$. (We have preferred to express this relationship in terms of ρ_s because it will later allow us to derive the optimum doping density independent of well width and because delta-doping, which is frequently used in the MBE growth of QWIP structures, is normally specified by ρ_s.)

Before leaving the subject of thermally activated dark currents, let us examine the behavior of its major component, the thermionically emitted carrier density n_θ. According to eq. (3.19) this is

$$n_\theta \approx \frac{\rho_0}{L_P} \exp\left(- \frac{\Delta E - \delta\phi - E_F}{kT}\right) \tag{3.25}$$

The constant ρ_0 has the numerical value

$$\rho_0 = \frac{m_W^* kT}{\pi\hbar^2} = 3.6 \times 10^{10} \left(\frac{m_W^*}{m_0}\right) T \qquad \text{cm}^{-2} \tag{3.26}$$

where m_0 is the free electron mass. Let us restate eq. (3.25) to show the dependence on detector parameters such as doping density, cutoff wavelength, and operating temperature. Using eq. (3.23) to estimate the barrier height and substituting eq. (3.24), we obtain

$$n_\theta \approx \frac{\rho_0}{L_P} \left[\exp\left(\frac{\rho_s}{\rho_0}\right) - 1 \right] \exp\left[\frac{1}{kT}\left(\delta\phi - K_\lambda \frac{hc}{\lambda_c} \right) \right] \tag{3.27}$$

To clarify the dependence of n_θ on ρ_s, we rewrite eq. (3.27) in terms of the combination

$$\hat{\rho} = \frac{\rho_s}{\rho_0} \tag{3.28}$$

obtaining

$$n_\theta \approx \frac{\rho_s}{L_P} \left\{ \frac{e^{\hat{\rho}} - 1}{\hat{\rho}} \right\} \exp\left[\frac{1}{kT}\left(\delta\phi - K_\lambda \frac{hc}{\lambda_c} \right) \right] \tag{3.29}$$

The quantity in braces in eq. (3.29) grows exponentially for $\hat{\rho} \geq 0.3$ but rapidly decreases to unity for $\hat{\rho} < 0.3$. Therefore the thermionically generated carrier density approaches

$$n_{\theta\min} \approx \frac{\rho_s}{L_P} \exp\left[\frac{1}{kT}\left(\delta\phi - K_\lambda \frac{hc}{\lambda_c} \right) \right] \tag{3.30}$$

for low doping, $\rho_s \ll \rho_0$. For $\rho_s \gtrsim \rho_0$, n_θ increases exponentially as

$$n_\theta = n_{\theta min}\left[\frac{e^{\hat{\rho}} - 1}{\hat{\rho}}\right] \qquad (3.31)$$

We will use these observations in Section 3.5, when we discuss detector design optimization. For now, we remark only that we can see from eqs. (3.7) and (3.10) that when dark current noise dominates the total noise, NETD is proportional to $(e^{\hat{\rho}} - 1)^{1/2}/\hat{\rho}$, and a minimum NETD is achieved at $\hat{\rho} \approx 1.6$.

Temperature-Independent Tunneling

At low temperatures, the temperature-independent well-bottom tunneling becomes significant, and the approximation of eq. (3.19) departs from the model of eqs. (3.13) through (3.16). More significantly, the low-temperature dark current observed by Andrews and Miller [28] is larger than that predicted by eqs. (3.11)–(3.16). They suggest that if the additional current is not surface leakage, it may be accounted for as defect-assisted tunneling or sequential resonant tunneling. The sequential resonant tunneling hypothesis has been examined by Williams et al. [34], who compare measured low-temperature dark currents to a model attributed to Choi et al. [35].

$$I_{srt} = \frac{eA_D}{\hbar L_W^2} kTD_0 \ln\left(\frac{1 + e^{u_F}}{1 + e^{u_F - u_1}}\right) \qquad (3.32)$$

where L_W is the well width, $u_F = E_F/kT$, $u_1 = V_1/kT$, and

$$D_0 = \exp\left\{-\frac{4L_B\sqrt{2m_B^*}}{3\hbar eV_1}[(\Delta E - E_1)^{3/2} - (\Delta E - E_1 - eV_1)^{3/2}]\right\} \qquad (3.33a)$$

$$\approx \exp\left\{-\frac{2\sqrt{m_B^*}L_B}{\hbar}(\Delta E - E_1)^{1/2}\left[1 - \frac{1}{4}\left(\frac{eV_1}{\Delta E - E_1}\right)\right]\right\} \qquad (3.33b)$$

Again we denote the energy of the confined state as E_1 and the voltage drop in one period as V_1. The small-bias expansion given in eq. (3.33b) is valid unless the applied field is so large that impact ionization has to be considered. (For extremely small biases, it is also necessary to include the reverse current; this may be obtained from eq. (3.33b) by the substitution $\Delta E \rightarrow \Delta E + eV_1$.) The quantity D_0 is the WKB transmission through the barrier at the energy E_1 of the confined state. The logarithm in eq. (3.32) may be obtained by integrating over energy the product $f(x)[1 - f(x + u_1)]$, where x is as used in eq. (3.19) and $f(x)$ is the Fermi factor. The additional

Fermi factor accounts for the probability of a tunneling electron finding an empty state to receive it, which could safely be omitted from eq. (3.14) for the thermally activated processes.

In their thick-barrier samples, Williams et al. [34] observed much larger dark currents (by 2.5 to 8 orders of magnitude) than predicted by eqs. (3.32) and (3.33). Measured currents nevertheless scale as predicted with well width and barrier height. Figure 3.21 shows the value of L_B required to make eqs. (3.32) and (3.33) agree with experiment for detectors of various barrier thicknesses. Good agreement with eqs. (3.32) and (3.33) for $L_B < 30$ nm combined with discrepancy and scatter for $L_B \geq 30$ nm suggests to Williams et al. that the most likely explanation is defect-assisted tunneling. Whether or not this is the correct explanation, correlation of apparent and actual barrier widths [34], reproduced in Figure 3.21, provides the information required for a simple representation of the well-bottom tunneling via eq. (3.33b), for barrier widths of up to 50 nm.

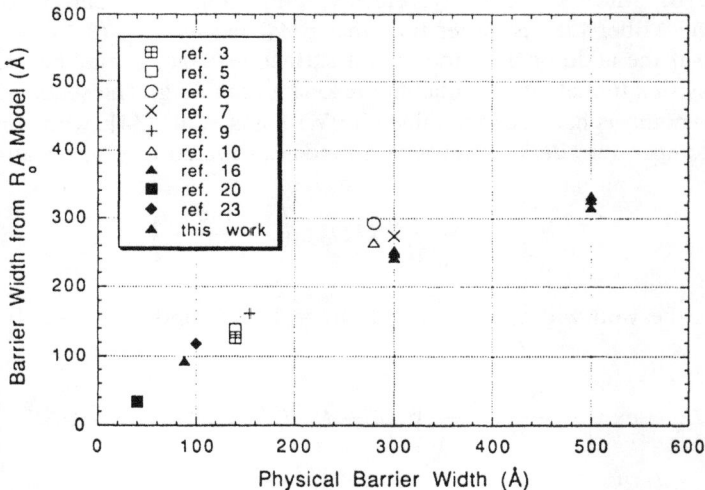

Figure 3.21 Barrier thickness required in eqs. (3.17) and (3.18) to reproduce measured dark current, plotted against physical barrier width. (After Williams et al. [34]).

When we compare the predictions of eqs. (3.32) and (3.33) to the data of Andrews and Miller [28], we find that here too the data are better represented when image charge barrier lowering is included according to eq. (3.17). This alteration has the effect of increasing the current predicted by eqs. (3.32) and (3.33). Although this correction may be significant for some low-background applications requiring low focal-plane operating temperatures, it is not important to the examples we shall consider in Section 3.5. Indeed, we shall discover that, for these examples, other factors determine the operating conditions so that temperature-independent tunneling is unimportant even for operating temperatures as low as 25 K.

Dark Current Equation

To summarize the foregoing discussion, the dark current model we shall use to optimize GaAs/AlGaAs QWIPs for specific applications is

$$I_D = eA_D \frac{\rho_s}{L_P} v_d \left(\frac{e^{\hat\rho} - 1}{\hat\rho}\right) \exp\left[\frac{1}{kT}\left(\delta\phi - K_\lambda \frac{hc}{\lambda_c}\right)\right] K_t \tag{3.34}$$

where ρ_s is the sheet-charge density with scaled value $\hat\rho$ defined in eq. (3.28) with (3.15) and (3.20), drift velocity v_d is as in eq. (3.12) with the values of low-field mobility and saturation velocity assumed to be $\mu \approx 2 \times 10^3$ cm^2 V^{-1} s^{-1} and $v_s \approx 5 \times 10^6$ cm s^{-1}, the barrier height is estimated from the cutoff wavelength λ_c with phenomenological correction factor $K_\lambda \approx 0.95$, and image-charge barrier lowering $\delta\phi$ as given in eq. (3.17), and the tunneling correction is

$$K_t = 1 + K' + K'' \tag{3.35}$$

where the thermally activated correction is

$$K' = \left[\frac{2\gamma}{1 - e^{-2\gamma}} e^\gamma - 1\right] + 2\Gamma\left(\frac{5}{3}\right)\left(\frac{3}{4}\gamma\right)^{1/3}\left[1 + \left(\frac{3\gamma}{4a^2}\right)^{1/4}\right]e^\gamma \tag{3.36}$$

with notations as in eq. (3.20), and the temperature-independent correction is

$$K'' = \frac{\pi\hbar L_p}{m_W^* L_W^2 v_d} D_0 \ln\left(\frac{1 + e^{u_F}}{1 + e^{u_F - u_1}}\right) \frac{1}{(e^{\hat\rho} - 1)} \exp\left[-\frac{1}{kT}\left(\delta\phi - K_\lambda \frac{hc}{\lambda_c}\right)\right] \tag{3.37}$$

where u_F and u_1 are as in eq. (3.32) and D_0 is given by eq. (3.33b). When using eq. (3.33b) to evaluate D_0 for GaAs/AlGaAs structures with wide barriers, we use an effective barrier width determined from the nominal one by referring to Figure 3.21.

3.4.2 Collection Efficiency

The detector collection efficiency is defined as the efficiency of converting incident radiation to carriers that are available for collection at the output of the detector. In quantum well photoconductive detectors, this involves two processes: absorption of infrared radiation by photoexcitation and transport of the photoexcited carriers to the output of the detector. In the notation of eq. (3.7), the collection efficiency is the product ηg of absorptance and photoconductive gain. We have chosen the term *absorptance* to denote what is frequently called *quantum efficiency* in the QWIP literature to avoid confusion with the conventional use of the latter term to mean the collection efficiency of a photodiode.

Absorptance

The absorptance is the ratio of photons absorbed in a detector to the number of in-band incident photons. The absorptance η can be expressed as

$$\eta = (1 - R)(1 - e^{-\kappa \alpha L})\chi \tag{3.38}$$

where R is the reflection coefficient, α is the absorption coefficient in cm^{-1}, L is the length of the multiple-quantum well structure, κ is the number of passes the infrared radiation makes through the structure,[2] and χ is a polarization correction factor. For p-type quantum wells $\chi = 1$, but because only one polarization is absorbed, $\chi = 0.5$ in n-type GaAs/AlGaAs quantum wells. The transmission coefficient $(1 - R)$ is close to 0.72 for a polished GaAs substrate.

The absorption coefficient for bound-to-bound transitions can be estimated from the expression [20]

$$\alpha = \frac{\rho_s}{L_P} \frac{e^2}{\varepsilon_0 m_W^* c} f_o \frac{\sin^2\theta}{n} \frac{1}{\Gamma} \quad cm^{-1} \tag{3.39}$$

where θ is the internal angle of the infrared radiation relative to the plane of the quantum wells, n is the refractive index, Γ is the absorption half linewidth, and f_o is the oscillator strength. Typically $f_o \approx 0.5$ for bound-to-bound GaAs/AlGaAs MQW structures with transitions corresponding to $\lambda = 10 \ \mu m$. It appears to be similar for n- and p-type GaAs/AlGaAs quantum wells. Discussion and expressions for f_o appear elsewhere [42]. The absorption linewidth is related to the detector spectral bandwidth, $\Delta\lambda_{qw}$, by

$$\frac{1}{\Gamma} = \frac{1}{\Delta\omega} = \frac{\lambda^2}{2\pi c \Delta\lambda_{qw}} \tag{3.40}$$

---

[2]The length L in eq. (3.34) is taken to be the length of the MQW structure as a matter of convention. Because the available charges are confined to the wells it would be defensible, though perhaps it would be pressing a point, to count only the widths of the actual quantum wells. If that were done in eq. (3.34), then the average volume charge density ρ_s/L_P that appears in eq. (3.35) must be replaced by the volume charge density in the well, ρ_s/L_W, so that the product αL remains invariant.

We can now express the absorption coefficient in terms of relevant detector parameters by

$$\alpha = \frac{\rho_s}{L_P} \frac{e^2}{2\pi\varepsilon_0 m_W^* c^2} f_o \frac{\sin^2\theta}{n} \frac{\lambda^2}{\Delta\lambda_{qw}} \tag{3.41a}$$

$$= 5.61 \times 10^{-17} \frac{\rho_s}{L_P} \frac{m_0}{m_W^*} f_o \sin^2\theta \frac{\lambda^2}{\Delta\lambda_{qw}} K_\alpha \quad \text{cm}^{-1} \tag{3.41b}$$

Review of reported data for bound-to-continuum transition QWIPs reveals that the measured absorption is higher than that calculated from eq. (3.41b). To account for the increased absorption as the excited state moves from a bound condition to a resonant continuum state, we have added a correction factor K_α to eq. (3.41b). The detailed calculation for the absorption is complex (see Chapter 2), and no relatively simple analytical expression is presently known to us. We might expect that the absorption goes through a maximum as the excited state moves further into the continuum, so that K_α may take on different values depending on the location of the first excited state. From a simplistic point of view, we can visualize bound-to-continuum absorption as corresponding to a transition from the bound state to N mini-subbands, each having a small spectral width. The summation of those absorptions then give rise to a stronger total absorption with a greater spectral width than a single bound-to-bound transition. Based on a review of reported data, we shall use $K_\alpha = 4$ in our examples later in this chapter.

Now for n-type GaAs/AlGaAs quantum wells, $m_W^* = 0.067 \, m_0$, assuming $\theta = 45°$ and $n = 3.27$, the absorption coefficient becomes

$$\alpha = 1.28 \times 10^{-16} \frac{\rho_s}{L_P} f_o \frac{\lambda^2}{\Delta\lambda_{qw}} K_\alpha \quad \text{cm}^{-1} \tag{3.42}$$

The assumption that $\theta = 45°$ is simplistic, because the polarization sensitivity of n-type devices requires some sort of enhancement technique applied to focal plane arrays. The most efficient approach is to use a reflection grating on each detector [11, 31, 33]. This will modify the magnitude of the absorption coefficient somewhat, as θ will be determined by the type of grating employed. Also, multiple passes may be obtained with the use of a waveguide [32], further enhancing the absorptance.

For p-type GaAs/AlGaAs quantum wells, $m_W^* = 0.48 \, m_0$. Assuming $\theta = 90°$ and $n = 3.27$, the absorption coefficient becomes

$$\alpha = 3.57 \times 10^{-17} \frac{\rho_s}{L_P} f_o \frac{\Delta\lambda_{qw}}{\lambda^2} K_\alpha \quad \text{cm}^{-1} \tag{3.43}$$

In the p-type MQW focal planes, normal-incidence absorption is a significant advantage, since no additional grating structures need be incorporated into the device structure.

Several important differences between n- and p-type MQW devices are now apparent. In n-type detectors, only one polarization of incident infrared radiation is absorbed, and the radiation must pass through the detector at an angle relative to the growth plane of the quantum well structure. In p-type detectors, both polarizations of IR radiation are absorbed for radiation passing through the detector at any angle. Assuming that the detectors are backside illuminated and have a reflecting contact and/or grating on the top of each detector, for n-type devices the incident IR radiation will not be absorbed on its first pass through. Only after it reflects back from the grating at some angle will it be absorbed; thus $\kappa = 1$ in eq. (3.38). To increase the number of passes, the addition of a waveguide layer is required. For p-type devices, radiation is absorbed during its first pass through, and after reflecting off the top contact, more is absorbed on the way back. Thus $\kappa = 2$ for p-type devices without gratings or waveguides. Further enhancements are possible for both p- and n-type detectors, but may not be required for some applications. For simplicity we shall use $\kappa = 1$ for our evaluation of n-type devices and $\kappa = 2$ for p-type devices.

Returning now to eq. (3.38), we find that for most cases of interest $\kappa \alpha L \ll 1$, so that

$$\eta \approx \kappa(1 - R) \chi \alpha L \tag{3.44}$$

Using the expression for α from eq. (3.41b) we have

$$\eta = 5.61 \times 10^{-17} \frac{\kappa(1 - R)}{n} \frac{\rho_s}{L_P} \left(\frac{m_o}{m_W^*}\right) \chi f_o \sin^2\theta K_\alpha \lambda^2 \frac{L}{\Delta\lambda_{\text{qw}}} \tag{3.45}$$

Because $\rho_s = 3.6 \times 10^{10}\, Tm_W^*\hat{\rho}$, and $L = N_W L_P$, where N_W is the number of quantum wells, eq. (3.45) can be written

$$\eta\Delta\lambda = \frac{2.02 \times 10^{-6}}{n} \kappa(1 - R)\, \lambda^2 TN(\chi f_o \sin^2\theta)(\hat{\rho})K_\alpha \frac{\Delta\lambda}{\Delta\lambda_{\text{qw}}} \tag{3.46}$$

For both n- and p-type GaAs/AlGaAs quantum wells, the absorption coefficient is proportional to the carrier density. This fact alone would drive us to dope the wells to as high a level as possible, to achieve the best absorptance. However, because detector dark current is exponentially dependent on carrier density for values of the carrier density above the level defined by $n_{t_{\text{min}}}$, the optimum doping will be determined by the operating conditions for specific applications.

Photoconductive Gain

In addition to absorptance, the other factor in the collection efficiency is photoconductive gain. This factor includes both recombination of photoexcited carriers and the injection of additional carriers from a contact to restore charge neutrality when carriers are removed at the opposite contact. Such charge injection must occur in all single-carrier devices in which a current flows out of the device. Because of this effect, a single photon can, in effect, generate more than a single excited carrier.

We will now derive an approximate relationship between bias voltage, V_B, and photoconductive gain, g, taking into account that the parameter of interest is not g itself but the product gN_W, where N_W is the number of wells in the QWIP structure. By *approximate* we mean that the relationship should be sufficiently accurate to model the gain behavior parametrically and provide an approximate result, bearing in mind that the real relationship is somewhat more complicated.

In a conventional photoconductor, $g = \tau_R/\tau_{tr}$, where τ_R is the recombination time and $\tau_{tr} = v_d/L$ is the transit time, and thus $g \propto L^{-1}$. Liu [43] has pointed out that for QWIPs, $g \approx (N_W p)^{-1}$, where p is the probability of capture of a carrier in crossing a single well; that is, $p \approx \tau_{tr,W}/\tau_R$, where $\tau_{tr,W}$ is the transit time of a carrier across a single well. Therefore we have that $g\,N_W \approx \tau_R/\tau_{tr,W}$, where $\tau_{tr,W} \approx L_W/v_d$. If the product $N_W\tau_{tr,W}$ were used for the transit time across the detector structure, then the behavior would resemble that of a conventional photoconductor. Also it should be pointed out that the relationship $g \approx (N_W p)^{-1}$ predicts that the responsivity, dark current, and photocurrent will be independent of the number of quantum wells in the detector for $\kappa\alpha L \ll 1$.

Liu showed experimentally for n-type QWIPs operated at $V_B \approx 3$ V, that $p \approx 0.05$, because

$$p = \frac{L_W}{\tau_R v_d} \tag{3.47}$$

If we take $L_W = 50$ Å, $\tau_R = 10^{-12}$ sec, and $v_d = 10^7$ cm/s at $V_B = 3$ V, we can match the experimental results with eq. (3.47). All these values are practical for the operating conditions.

Using eq. (3.12) for v_d we have

$$g\,N_W = \frac{\tau_R}{L_W}\,v_d = \frac{\tau_R \mu F}{\left[1 + \left(\dfrac{\mu F}{v_s}\right)^2\right]^{1/2}} \tag{3.48}$$

The transport of carriers in GaAs/AlGaAs QWIPs is controlled largely by the properties of the AlGaAs barrier layers, which are undoped. Even though the barriers are undoped, the type of doping in the wells will determine whether the transport is via conduction or valence band. Therefore, it is suitable to use n-type mobility values for the case of n-type doping in the wells and p-type mobility values for the case of p-type well doping. Over the temperature range of interest for operating these detectors, we will assume the appropriate low field mobilities for AlGaAs are $\mu = 2{,}000$ cm^2/Vs for n-type and $\mu = 200$ cm^2/Vs for p-type.

As remarked in Section 3.4.1, quantification of the saturation velocity v_s in n-type QWIPs is uncertain. The expected value of v_s is 5–10 $\times$ 10^6 cm/s for electric field strengths on the order of 10^4 V/cm. If we restrict ourselves to electric fields less than this, then $\mu F < v_s$. We shall find that this restriction is compatible with satisfying the constraints involving photoconductive gain that are discussed later. Under this restriction, then, we can express the drift velocity as

$$v_d = \mu F = \frac{\mu V_B}{N_W L_P} \tag{3.49}$$

and the gain as

$$g = \frac{\tau_R \mu V_B}{N_W^2 L_W L_P} \tag{3.50}$$

or equivalently, the bias voltage corresponding to gain g as

$$V_B = \frac{g N_W^2 L_W L_P}{\tau_R \mu} \tag{3.51}$$

Equation (3.51) predicts that adjusting N_W can increase gain at fixed V_B or decrease the value of V_B required to obtain a given value of g. In our discussion of device optimization we shall use this relationship to match detector gain and operating voltage to readout characteristics.

Collection Efficiency Equation

The final expression for collection efficiency combines eqs. (3.37), (3.40), and (3.46) to give

$$\eta g = \kappa(1 - R)\chi \frac{\rho_s}{L_W} \frac{V_B}{L} \frac{\tau_R \mu e^2}{2\pi\varepsilon_0 m_{W}^* c^2} f_o \frac{\sin^2\theta}{n} \frac{\lambda^2}{\Delta\lambda_{qw}} K_\alpha \tag{3.52}$$

This expression represents the model we shall use for our examples in Section 3.5.

3.4.3 Correlated Noise

Correlated noise, also called *pattern noise*, can be the dominant noise source in staring sensors. We have described in Section 3.2 the contribution of correlated noise to the total noise in terms of the number of optically and thermally generated carriers

(N_{op} and N_{th}), the standard deviation to mean ratios over the array of the optically and thermally generated charges (σ_{op} and σ_{th}), and the correlated noise reduction factors representing gain and offset correction (β_{gn} and β_{of}). The contribution of the correlated noise to the total noise depends greatly on the level of calibration (correlated noise reduction) that can be achieved and the stability of a particular sensor system.

Standard gain and offset calibration of a staring infrared sensor is accomplished by illuminating the sensor with a highly uniform, calibrated infrared source at one or more illumination intensities. For gain calibration, the residual correlated noise after calibration is limited by the uniformity of the illumination source as well as the noise of the calibration measurement. In most practical situations this limit is $\approx 0.1\%$. Therefore, a practical limit for the product $\beta_{gn}\sigma_{op}$ is 1×10^{-3}. In the offset term, the practical calibration limits are determined by the noise floor of the sensor system. Offset correction is accomplished by subtraction of the measured offset from each focal plane pixel. As a simple first approximation, we can assume that a one-frame subtraction provides correction down to one bit of the *analog-to-digital convertor* (ADC). For a 12-bit ADC, this translates to values for $\beta_{of}\sigma_{op}$ and $\beta_{of}\sigma_{th}$ of approximately 3×10^{-4}. With some frame averaging and careful ADC design, this can easily be reduced to 1×10^{-4}, and with substantially more averaging (≈ 400 frames), a value of 1×10^{-5} can be achieved. It may be possible to reduce both gain and offset limits further by using alternate approaches such as optical dithering. The ultimate calibration limit is set by the single pixel noise. For the two example applications described in Section 3.5, we will use the values 1×10^{-3} for the product $\beta_{gn}\sigma_{op}$, and 10^{-4} to 10^{-5} for both $\beta_{of}\sigma_{op}$ and $\beta_{of}\sigma_{th}$.

3.5 DEVICE OPTIMIZATION FOR APPLICATIONS

Photoconductive QWIPs made in GaAs/AlGaAs material have matured to the point where they can now be used in real system applications. Early assessment of the potential for these devices was somewhat negative because of the polarization-dependent absorption and the high thermal generation rates compared to HgCdTe IR materials [24]. However, the pessimistic outlook considered only ultimate single detector performance, not the requirements of any particular IR sensor system. Many sensor systems do not require maximum performance from the detector itself, because performance is limited by background, readout, or correlated noise. Also, for some systems, the focal-plane operating temperature is not a major system cost and weight driver. When these considerations are included, we find several areas where the MQW detectors are very competitive and possibly advantageous compared to existing technology. To properly evaluate the potential of these detectors, their noise and signal response characteristics must be well understood. Based on that knowledge, the detectors can be optimized for specific applications.

We begin in Section 3.5.1 by showing the dependence of NETD on GaAs/AlGaAs quantum well detector parameters. This is followed by a discussion of two example infrared sensor systems for which NETD is optimized.

3.5.1 Dependence of NETD on Quantum Well Detector Parameters

To illustrate the performance potential of GaAs/AlGaAs MQW detectors and to demonstrate how the detectors can be optimized with respect to an infrared sensor system, we will study two applications. We begin by expanding on the discussion in Section 3.2 to include the pertinent detector dark current and collection efficiency terms in the expression for NETD.

The Background Limit

The lowest possible NETD for a given sensor is the background-limited or *BLIP* (background-limited infrared photodetector) value, which occurs when the number of noise carriers n_T is dominated by the shot noise of the optically generated carriers:

$$n_T = n_{\mathrm{op}} = \sqrt{2gN_{\mathrm{op}}} \qquad (3.53)$$

where

$$N_{\mathrm{op}} = \frac{\pi \eta g \eta_o \Delta\lambda \Phi (\theta_R D)^2 \tau_i}{4} \qquad (3.54)$$

Putting eqs. (3.53) and (3.54) into eq. (3.7), the BLIP value becomes

$$\mathrm{NETD_{BLIP}} = \frac{\Phi}{\Phi'} \frac{2\sqrt{2}}{\theta_R D} \left(\pi \eta \eta_o \Delta\lambda \Phi \tau_i \right)^{-1/2} \qquad (3.55)$$

Note that photoconductive gain is not present in eq. (3.55) as it affects signal and noise carriers equally.

The parameters in eq. (3.55) are determined by the system design and the application, except for the absorptance of the detector. In some system designs, the system wavelength band will be defined by the detector spectral bandwidth.

Degradation of NETD by Detector Array Noise

In this subsection we will describe the factors such as dark current, readout noise, and correlated noise that can degrade sensor performance from the ideal BLIP condition. The number of thermally generated carriers can be determined from the detector dark current I_D, by

$$N_{\text{th}} = \frac{I_D \tau_i}{e} \tag{3.56}$$

and the resulting shot noise is

$$n_{\text{th}} = \sqrt{\frac{2g I_D \tau_i}{e}} \tag{3.57}$$

This is a slight overestimate of the true generation-recombination noise [5, 6, 7, 44], but is sufficient for the discussion here. The total uncorrelated noise can now be expressed by

$$n_{T_{\text{uncorr}}}^2 = \frac{\pi \eta g^2 \eta_o \Phi \Delta \lambda (\theta_R D)^2 \tau_i}{2} + \frac{2g I_D \tau_i}{e} + n_{\text{ro}}^2 \tag{3.58}$$

Similarly for the correlated noise:

$$n_{T_{\text{corr}}}^2 = \beta_{\text{gn}}^2 \left(\sigma_{\text{op}} \frac{\pi \eta g \eta_o \Phi \Delta \lambda (\theta_R D)^2 \tau_i}{4} \right)^2$$

$$+ \beta_{\text{of}}^2 \left[\left(\sigma_{\text{th}} \frac{I_D \tau_i}{e} \right)^2 + \left(\sigma_{\text{op}} \frac{\pi \eta g \eta_o \Phi \Delta \lambda (\theta_R D)^2 \tau_i}{4} \right)^2 \right] \tag{3.59}$$

We have neglected the term N_{ex} because, for GaAs/AlGaAs QWIPs, this term is insignificant.[3] By normalizing to the background shot noise we can express eq. (3.7)

[3]The low-frequency excess noise in the detectors ($1/f$ noise) can be expressed as

$$N_{\text{ex}} = \frac{i_{\text{exo}}}{e} \tau_i \left(\ln \frac{\tau_R}{\tau_i} \right)^{1/2} \approx 10^{19} \tau_i i_{\text{exo}}$$

where i_{exo} is the excess noise current at 1 Hz in A/$\sqrt{\text{Hz}}$. Experimentally [44], the state of the art observed is a corner frequency of 0.1 Hz against a total noise current of $i_n = 10^{-13}$ A/$\sqrt{\text{Hz}}$ represented by $D^* \approx 3 \times 10^{10}$ cm $\sqrt{\text{Hz}}$/W. Therefore, $i_{\text{exo}} \approx 3 \times 10^{-14}$ A/$\sqrt{\text{Hz}}$, and for $\tau_i = 0.033$ sec, $N_{\text{ex}} \approx 10^4$ noise electrons. This will be small in most cases compared to $N_{\text{th}}\sigma_{\text{th}}$ and $N_{\text{op}}\sigma_{\text{op}}$, and thus $\beta_{\text{of}}N_{\text{ex}}$ in almost all cases will be smaller than n_{ro} or thermal and optical shot noises.

as

$$\text{NETD} = \frac{\Phi}{\Phi'} \frac{2\sqrt{2}}{\theta_R D} (\pi \eta \eta_o \Delta \lambda \Phi \tau_i)^{-1/2} \left\{ 1 + \frac{4I_D}{e\pi(\theta_R D)^2 \eta g \eta_o \Phi \Delta \lambda} \right.$$

$$+ \frac{2n_{ro}^2}{\pi(\theta_R D)^2 \eta g^2 \eta_o \Phi \Delta \lambda \tau_i} + \frac{\pi(\theta_R D)^2 \eta \eta_o \Phi \Delta \lambda \tau_i}{8} (\beta_{gn} \sigma_{op})^2$$

$$\left. \left(1 + \left(\frac{\beta_{of}}{\beta_{gn}} \right)^2 \left[1 + \left(\frac{\sigma_{th}}{\sigma_{op}} \right)^2 \left(\frac{4I_D}{e\pi(\theta_R D)^2 \eta g \eta_o \Phi \Delta \lambda} \right)^2 \right] \right) \right\}^{1/2} \quad (3.60)$$

We see that this expression is simply the background-limited NETD (the first term) multiplied by a degradation factor that arises from thermal generation (I_D), readout noise (N_{ro}), and correlated noise effects. The maximum sensor performance is attained when the degradation factor is minimized.

In designing a sensor system, it is desirable to optimize the detector parameters so that readout noise is only a small fraction of total noise. For quantum well detectors, this involves a combination of detector design and bias-voltage adjustment, so that the photoconductive gain is large enough to make the thermal and optical noise contributions larger than the readout noise. To reflect this condition accurately, we will rewrite eq. (3.60) as

$$\text{NETD} = \text{NETD}_{\text{BLIP}} \left[\left(1 + H_2 + \frac{H_1 H_3}{2} \right)(1 + \Omega) \right]^{1/2} \quad (3.61)$$

where, substituting eq. (3.37) for the dark current and eq. (3.46) for the absorptance to obtain the dependence on detector parameters, the H_i are given by

$$\Omega \equiv \frac{n_{ro}^2}{n_T^2 - n_{ro}^2} = \frac{n_{ro}^2}{2g^2 N_W H_1 \left(1 + H_2 + \dfrac{N_W H_1 H_3}{2} \right)} \quad (3.62)$$

$$H_1 = \frac{N_{op}}{gN_W}$$

$$= \frac{1.59 \times 10^{-6}}{n} (\theta_R D)^2 \eta_o \Phi \kappa (1 - R) \lambda^2 (\chi f_o \sin^2 \theta)(T\hat{\rho}) \tau_i K_\alpha \frac{\Delta \lambda}{\Delta \lambda_{qw}} \quad (3.63)$$

$$H_2 = \frac{N_{th}}{N_{op}} = \frac{2.27 \times 10^{16} n A_D K_t \Delta \lambda_{qw} (e^\delta - 1) \, e^{-1.44 \times 10^4 / \lambda_c K_\lambda T} m_W^*}{\kappa (1 - R) \tau_R (\theta_R D)^2 \eta_o \Phi \hat{\rho} \lambda^2 (\chi f_o \sin^2 \theta) \Delta \lambda K_\alpha m_o} \quad (3.64)$$

$$H_3 = (\beta_{gn}\sigma_{op})^2 \left\{ 1 + \left(\frac{\beta_{of}}{\beta_{gn}}\right)^2 \left[1 + \left(\frac{\sigma_{th}}{\sigma_{op}}\right)^2 H_2^2 \right] \right\}$$ (3.65)

The quantities H_i are so defined as to be independent of photoconductive gain g. Thus, in eq. (3.61) the dependence on g has been confined to the quantity Ω, defined as the ratio of readout noise to total noise. Typical values for Ω are 0.1–0.4, depending on the bias voltage required to meet constraints imposed by the readout amplifier. We describe these constraints and develop an expression for Ω in the next subsection.

Readout Amplifier Constraints

Before discussing our example applications, we must consider the important constraints placed on the operation of a QWIP focal plane due to the interaction of the detector with the readout circuit. These constraints arise from limitations of the readout's charge-storage capacity, from the requirement for minimizing readout noise with respect to total noise, and from the requirement for high injection efficiency. For simplicity we shall assume the use of a direct-injection readout amplifier, a type commonly employed in infrared focal plane arrays.

We first consider the effect of readout-circuit charge-storage capacity on minimizing NETD. When detector current is integrated for too long a time, the charge accumulated by the readout circuit may exceed the storage capacity available at that pixel. Under given conditions of operating temperature and photon flux, this eventuality will limit integration time to a value $\tau_{i_{ST}}$ determined by readout charge-storage capacity. To obtain a longer *system* integration time or "frame time," $\tau_i = \tau_F$, it is then necessary to sum these "subframe" samples over the system frame time τ_F. The readout device itself must, of course, have the capability to operate at the shorter integration time $\tau_{i_{ST}}$. The ratio $\tau_{i_{ST}}/\tau_F$ must be less than the ratio of readout storage capacity to total charge collected in time τ_P. Setting $\tau_i \approx \tau_F$ in eqs. (3.63–3.64) to compute the total charge collected in time τ_F, we can express $\tau_{i_{ST}}$ in terms of τ_F as

$$\tau_{i_{ST}} = \frac{\beta N_{ST}\tau_F}{g N_w H_1 (1 + H_2)}$$ (3.66)

where N_{ST} is the storage capacity in carriers per pixel, and β is the maximum fraction of the storage capacity allowed by operational considerations. The appearance of the gain g in eq. (3.66) requires us to distinguish two separate cases for the constraints on gain, as follows:

Case 1: In this case we suppose that the gain can be set sufficiently low that signal can be integrated for a full frame time without exceeding the pixel charge-storage capacity. At the same time, this low value of gain must not cause the readout noise to dominate the total focal plane noise. The satellite atmospheric sounder discussed in Section 5.2 belongs to this case.

Case 2: If the conditions for case 1 cannot be satisfied, a shorter subframe integration time, $\tau_{i_{ST}}$, is required. Under these conditions we add $\tau_F/\tau_{i_{ST}}$ subframes to reach the desired system integration time τ_F. The required value of g is then limited by pixel charge-storage capacity and the readout noise level. From eq. (3.66) and the definition of Ω as the ratio of readout noise to total noise, we find that the minimum permissible gain is

$$g = \frac{n_{\text{ro}}^2}{2\Omega\beta N_{\text{st}}\left[1 + \dfrac{N_W H_1 H_3}{2(1 + H_2)}\right]} \tag{3.67}$$

This case applies mostly to high background ($>10^{15}$ ph cm^{-2} s^{-1}) applications and those dominated by large detector dark currents, such as the terrestrial thermal imager discussed in Section 3.5.2.

Minimizing readout noise relative to total noise involves optimizing the relationship between gain, bias voltage, and number of wells. Referring to eq. (3.50) in Section 3.4.2, we see that, for practical readout devices, N_W and V_B can be adjusted to optimize gain for minimizing readout noise relative to total noise. When adjusting N_W it also has to be kept in mind that as we approach BLIP or dark-current dominated NETD, the limiting NETD $\propto N_W^{-1/2}$, whereas for correlated-noise dominated systems, NETD is independent of N_W. Using eq. (3.50) for the photoconductive gain with eqs. (3.61) and (3.67), we can express the readout noise to total noise ratio, Ω, as

$$\Omega = \frac{N_W^2 L_W L_P n_{\text{ro}}^2}{\mu\tau_R 2\beta N_{\text{st}}\left[1 + \dfrac{N_W H_1 H_3}{2(1 + H_2)}\right]} \tag{3.68}$$

From eq. (3.66) for the subframe integration time we have

$$\tau_{i_{ST}} = \frac{\beta N_{ST} N_W L_W L_P \tau_F}{V_B \mu \tau_R H_1(1 + H_2)} \tag{3.69}$$

The injection efficiency, which is the ratio of the current through the readout device divided by the current generated by the detector, is given by[4]

$$n_{\text{inj}} = \frac{1}{1 + 1/g_m R_D} = \frac{1}{1 + kT/eV_B} \tag{3.70}$$

---

[4]The injection efficiency described here represents the low frequency injection efficiency, which is appropriate for practical values of integration time that might be used with these devices.

where g_m is the transconductance of the input amplifier, given by $g_m = I/(kT/e)$; I is the total current; and $R_D = V_B/I$ is the detector resistance. When $V_B < 10 \ kT/e$, injection efficiency and therefore the effective collection efficiency of the focal plane begins to degrade. Over the temperature range we will use in our examples (25–80 K), V_B, the minimum bias voltage allowed under these constraints, would range from 0.02 to 0.07 V. For other types of readout circuits, the injection efficiency may not be a significant problem, but we include it here for consistency with our example system applications.

A number of factors are involved in defining the optimum bias voltage values for quantum well detector arrays. We want to reduce the bias voltage to minimize the image charge barrier lowering effects (and the corresponding increase in dark current) and also to minimize tunneling effects. Maintaining a low bias (≈ 0.1 V) will also minimize readout storage capacity limitations, allowing fewer subframe additions. At the same time, we must minimize the readout noise contribution to the total focal plane noise. Also to maintain high injection efficiency, we want $V_B > 0.05$ V. In the following example systems, we find that, for an n-type detector array, a practical value of $V_B = 0.1$ V and $N_W = 50$ is appropriate for both examples, with the corresponding value for $\Omega = 0.14$. For p-type detector arrays, the optimum occurs at somewhat higher voltages; and for both examples we take $V_B = 0.3$ V and $N_W = 50$, with a corresponding $\Omega = 0.4$.

3.5.2 Two Optimization Examples

To illustrate the performance potential of MQW detectors, we will evaluate the performance of two specific sensor systems. The first is a terrestrial thermal imager that looks at background and scene temperatures around 300 K; and the second is a spaceborne atmospheric sounder that measures CO_2 and other gaseous absorption in the earth's atmosphere. We will assume the parameters listed in Table 3.1 for the thermal imager and the sounder. The parameters chosen are typical values for MQW properties and imager parameters. We will use them to illustrate the parametrics involved in MQW focal plane array applications. Included in these examples is the comparison of n- and p-type detectors, the parameters for which are listed in Table 3.2. For these examples, we also assume the following values for the calibration parameters, which are consistent with $\beta_{gn}\sigma_{op} \geq 10^{-3}$.

Offset Related	Gain Related
$\sigma_{th} = 0.1$	$\sigma_{op} = 0.01$
$\sigma_{op} = 0.01$	$\beta_{gn} = 0.1$
$\beta_{of} = 10^{-3}$ and 10^{-4}	

For the readout noise n_{ro}, pixel storage capacity N_{ST}, the maximum percentage of

Table 3.1
System Parameters Used for Terrestrial Imager and Atmospheric Sounder Examples

Parameters	Terrestrial Imager	Atmospheric Sounder	Units
Φ'	8.0×10^{14}	2.2×10^{14}	ph cm^{-2} (sr s μm K)$^{-1}$
Φ	4.3×10^{16}	9.7×10^{15}	ph cm^{-2} s^{-1} sr^{-1} μm^{-1}
$A_\pi = A_D$	2.5×10^{-5}	2.5×10^{-5}	cm^2
D	10	30	cm
θ_R	3.0×10^{-4}	1.67×10^{-4}	rad
η_o	.5	0.1	
τ_i	0.033	0.30	sec
$\Delta\lambda$	1.5	0.01	μm
λ_c	9.5	15.5	μm
n	3.27	3.27	
τ_R	10^{-12}	10^{-12}	sec
$1 - R$	0.72	0.72	

Table 3.2
QWIP Detector Parameters Used for Terrestrial Imager and Atmospheric Sounder Examples

Parameter		n-type	p-type
χ		0.5	1
$\sin^2\theta$		0.5	1
f_o		0.5	0.5
m_W^*		$0.067m_o$	$0.48m_o$
m_B^*		$0.088m_o$	$0.21m_o$
V_B		0.1V	0.3V
N_W		50	50
L_B		400Å	400Å
ρ		1.6	1.6
K_α		4	4
K_t	Imager	1.15	1.40
	Sounder	1.10	1.20
K_λ	Imager	1.05	1.05
	Sounder	1.10	1.10

storage capacity allowed, and number of quantum wells N_W we take

$$n_{ro} = 500 \qquad N_{ST} = 4 \times 10^7$$

$$N_W \approx 50 \qquad \beta \approx 0.5$$

In the terrestrial imager example, we have chosen a waveband placed at the lower end of the 8–14 μm spectral window to minimize the dark current. For the atmospheric sounder, the spectral band is chosen to measure several narrow CO_2 absorption lines in the 15 μm wavelength band in the upper atmosphere. These two cases have been chosen for illustration not only because they represent important applications for QWIP focal plane arrays, but also because they graphically illustrate the different parametrics for high- and low-background imaging sensors. NETD is plotted versus focal plane temperature for n- and p-type QWIP focal plane arrays for terrestrial thermal imagers in Figure 3.22 and for a spaceborne atmospheric sounder in Figure 3.23.

In these figures we have used typical parameters that apply to QWIP detectors to provide semi-quantitative models for the principal purpose of outlining the multiple factors that must be included and also to provide an example of the first-order parametric analysis required to design QWIP focal planes for various applications.

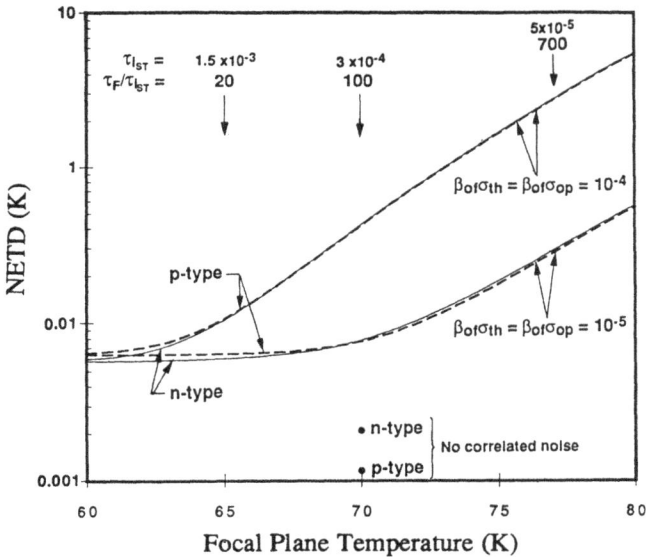

Figure 3.22 Calculated NETD versus focal-plane temperature for n- and p-type GaAs/AlGaAs MQW detectors: terrestrial thermal imager.

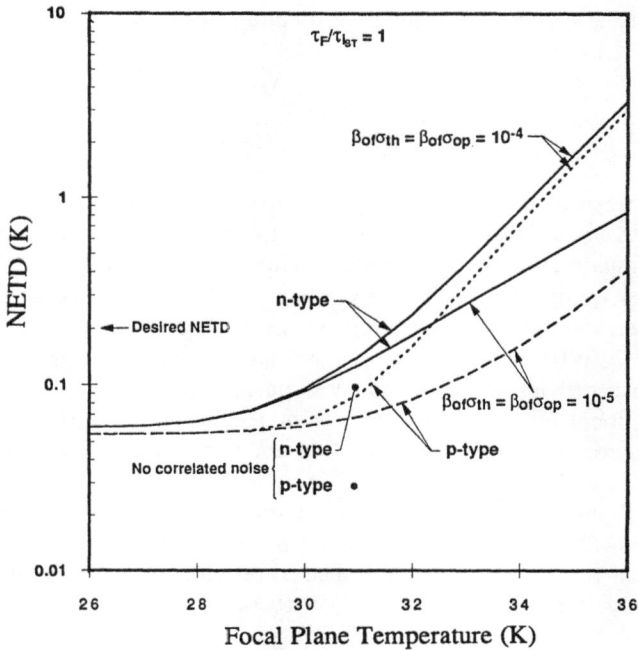

Figure 3.23 Calculated NETD versus focal-plane temperature for n- and p-type GaAs/AlGaAs MQW detectors: space-borne atmospheric sounder.

Both cases include themionic emission effects, bias voltage and gain considerations, and tunneling contributions. We have also shown the number of subframes, $\tau_F/\tau_{i_{ST}}$, required to satisfy storage capacity limitations. In addition, we have assumed that the absorption constant, which was modeled for bound-to-bound intersubband transitions, is about a factor of 4 larger for bound-to-continuum transitions where the continuum state is at or near resonance (i.e., $K_\alpha = 4$), because this approximates much of the reported data. For the curves illustrated in Figures 3.22 and 3.23, where thermal generation dominates, we have taken $\rho = 1.6$ because this minimizes the NETD as a function of doping level. Also, when thermal effects dominate (as they do in most cases of interest), we have assumed $N_W = 50$ because NETD $\propto N_W^{-1/2}$ in this case.

The figures also include the effects of various choices for pattern noise mitigation. As stated in Section 3.4.3, we will assume in these sensor examples that $\beta_{gn}\sigma_{op} = 10^{-3}$, and practical values of β_{of} range from 10^{-3} to 10^{-4}. When pattern noise dominates, NETD $= \Phi/\Phi'\beta_{gn}\sigma_{op} (1 + \Omega)^{1/2}$, independent of all other factors. This limit is shown in each of the figures.

The terrestrial thermal imager, Figure 3.22, shows NETD for $\beta_{gn}\sigma_{op} = 10^{-3}$, and for two values of β_{of}, 10^{-3} and 10^{-4}. At temperatures below 65 K the results are about the same for n- and p-type detectors. However, at and above 70 K, the results are quite different: both types can benefit considerably from an improvement in β_{of}. The modeled n-type and p-type results show similar values of NETD, and approximately the same number of subframes are required for either type. Tunneling contributions to the dark current, at the bias levels chosen, are less than 15% for this application. For practical applications, most single stage mechanical coolers designed for 77 K operation can usually do quite well down to 65–70 K. As shown in Figure 3.22, achieving residual offset variations of 10^{-5} would provide an NETD ≈ 0.07 at 70 K for both n- and p-type cases, with acceptable bias voltage and subframe addition requirements.

An NETD evaluation of an n-type QWIP focal plane [45] with properties similar to those in this example provided an average NETD $= 0.01$ K at an operating temperature of 60 K viewing a 300 K scene. The average NETD was determined from the distribution of individual pixel NETDs for the array and therefore did not accurately reflect the effects of correlated pattern noise. In the modeled results here, we obtain a single detector NETD $= 0.007$ K at 60 K. We also show the single pixel values of NETD at 70 K of 0.03 K for n-type detectors and 0.01 K for p-type, which represent the performance that would be obtained for an entire focal plane if correlated noise effects were eliminated.

For the atmospheric sounder shown in Figure 3.23, which is a relatively low-background case, we show NETD versus focal-plane temperature over the range 26–36 K for n- and p-type QWIP focal plane arrays. In this case we see that p-type focal plane arrays are superior to n-type ones for temperatures greater than 30 K. For this application thermally assisted tunneling contributions to the dark current are less than 40% for n-type and 20% for p-type arrays, with an insignificant contribution from sequential resonant tunneling. These atmospheric sounders generally want NETD ≈ 0.2 K as shown in the figure. We see that this can be provided by n-type QWIP focal plane arrays operating at 31 K, whereas p-type QWIP focal planes could operate up to about 32–34 K, depending on the level of offset calibration achieved. These temperatures should be attainable by practical two-stage coolers. For this example, the number of subframes required to obtain $\tau_F = 0.3$ is essentially one for all temperatures less that 34 K. We have also shown in the figure that, in the absence of correlated noise, NETDs of 0.03 K for the p-type case and 0.08 K for the n-type case would be obtained at a temperature of 30 K.

In both examples, the dominating effects of correlated noise on the performance of staring sensors is very clear. These effects will have a strong influence on the design of the QWIP structure, as well as on the entire sensor design due to the need for accurate calibration. Having a clear understanding of the origin and magnitude of these effects is paramount to optimizing sensor performance.

3.6 CONCLUSION

In this chapter we have shown that the specifics of infrared sensor applications can significantly affect the design and operation of QWIP focal plane arrays. Background flux, desired operating temperature, sensor wavelength band, readout considerations, and calibration accuracy will strongly affect design and optimization of the quantum well structure.

In Section 3.2, we discussed the basic framework for evaluating the performance of QWIP focal plane arrays in staring infrared sensor systems. The fundamental concepts presented there are applicable to any type of focal plane array. We used these basic concepts to analyze the performance GaAs/AlGaAs multiple-quantum well photoconductive arrays in staring infrared sensors. For the two examples that we discussed in detail, we showed that the correlated noise of the GaAs/AlGaAs detector array dominates sensor performance over much of the temperature range of interest. This result is also likely to hold for other types of focal plane, including those using HgCdTe detectors.

We have developed simplified models of the dark current and collection efficiency of GaAs/AlGaAs multiple-quantum well detectors as functions of detector design parameters such as temperature, cutoff wavelength, well sheet-charge density, operating bias, and barrier width. Although the expressions presented here represent approximations and are not highly accurate, they give a summary of complex detector behavior that varies over many orders of magnitude. This summary is simple enough to optimize performance of GaAs/AlGaAs QWIPs for specific applications; and we believe it is reliable enough, within its limits, to determine whether they are a suitable choice.

The two cases we have illustrated here represent only a small fraction of the applications for which these focal plane arrays might be considered. The general expressions presented here can be adapted to other infrared-sensor applications, using our examples as a guide.

REFERENCES

[1]   Levine, B.F., G. Hasnain, C.G. Bethea, and N. Chand, *Appl. Phys. Lett.*, Vol. 54, 1989, p. 2704.

[2]   Capasso, F., in *Semiconductors and Semimetals*, ed. R.K. Willardson and A.C. Beer, Vol. 24, Applications of Multiquantum Wells, Selective Doping, and Superlattices, Academic Press, San Diego, CA, 1987, Ch. 6.

[3]   Keyes, R.J., ed., *Optical and Infrared Detectors*, 2d ed., Springer-Verlag, New York, 1980.

[4]   Kruse, P.W., in *Semiconductors and Semimetals*, ed. R.K. Willardson and A.C. Beer, Vol. 5, Infrared Detectors, Academic Press, New York, 1970, pp. 15ff.

[5]   Janousek, B.K., M.J. Daugherty, W.L. Bloss, M.L. Rosenbluth, and M.J. O'Loughlin, *J. Appl. Phys.*, Vol. 67, 1990, p. 7608.

[6]   Beck, W.A., *J. Appl. Phys.*, Vol. 69, 1991, p. 4129.
[7]   Janousek, B.K., M.J. Daugherty, W.L. Bloss, R. Lacoe, M.J. O'Loughlin, H. Kanter, F.J. de Luccia, and L.E. Perry, *J. Appl. Phys.*, Vol. 69, 1991, p. 4130.
[8]   Chiu, L., J.S. Smith, S. Margalit, and A. Yariv, *Appl. Phys. Lett.*, Vol. 43, 1983, p. 331.
[9]   Smith, J.S., L.C. Chiu, S. Margalit, A. Yariv, and A.Y. Cho, *J. Vac. Sci. Technol. B*, Vol. 1, 1983, p. 376.
[10]   Coon, D.D., and R.P.G. Karunasiri, *Appl. Phys. Lett.*, Vol. 45, 1984, p. 649.
[11]   Goossen, K.W., and S.A. Lyon, *Appl. Phys. Lett.*, Vol. 47, 1985, p. 1257.
[12]   Levine, B.F., K.K. Choi, C.G. Bethea, J. Walker, and R.J. Malik, *Appl. Phys. Lett.*, Vol. 50, 1987, p. 1092.
[13]   Choi, K.K., B.F. Levine, C.G. Bethea, J. Walker, and R.J. Malik, *Appl. Phys. Lett.*, Vol. 50, 1987, p. 1814.
[14]   Levine, B.F., C.G. Bethea, K.K. Choi, J. Walker, and R.J. Malik, *Appl. Phys. Lett.*, Vol. 53, 1988, p. 231.
[15]   Levine, B.F., C.G. Bethea, K.K. Choi, J. Walker, and R.J. Malik, *J. Appl. Phys.*, Vol. 64, 1988, p. 1591.
[16]   Levine, B.F., C.G. Bethea, G. Hasnain, V.O. Shen, E. Pelvé, R.R. Abbott, and S.J. Hsieh, *Appl. Phys. Lett.*, Vol. 56, 1990, p. 851.
[17]   Goossen, K.W., S.A. Lyon, and K. Alavi, *Appl. Phys. Lett.*, Vol. 52, 1988, p. 1701.
[18]   Rosencher, E., P. Bois, B. Vinter, J. Nagle, and D. Kaplan, *Appl. Phys. Lett.*, Vol. 56, 1990, p. 1822.
[19]   Choi, K.K., M. Dutta, P.G. Newman, M.L. Saunders, and G.J. Iafrate, *Appl. Phys. Lett.*, Vol. 57, 1990, p. 1348.
[20]   Choi, K.K., L. Fotiadis, M. Taysing-Lara, W. Chang, and G.J. Iafrate, *Appl. Phys. Lett.*, Vol. 59, 1991, p. 3303.
[21]   Levine, B.F., S.D. Gunapala, J.M. Juo, S.S. Pei, and S. Hui, *Appl. Phys. Lett.*, Vol. 59, 1991, p. 1864.
[22]   Chang, Y.C., and R.B. James, *Phys. Rev. B*, Vol. 39, 1989, p. 12672.
[23]   Coon, D.D., R.P.G. Karunasiri, and L.Z. Liu, *Appl. Phys. Lett.*, Vol. 47, 1985, p. 289.
[24]   Kinch, M.A., and A. Yariv, *Appl. Phys. Lett.*, Vol. 55, 1989, p. 2093.
[25]   Levine, B.F., M.A. Kinch, and A. Yariv, *Appl. Phys. Lett.*, Vol. 56, 1990, p. 2354.
[26]   Zussman, A., B.F. Levine, J.M. Kuo, and de Jong, *J. Appl. Phys.*, Vol. 70, 1991, p. 5101.
[27]   Kane, M.J., M.T. Emeny, D. Lee, and C.R. Whitehouse, in *Gallium Arsenide and Related Compounds 1990. Proc. 17th Int. Symp.*, ed. K.E. Singer; Inst. Phys. Conf. Ser. Vol. 112, Bristol, England, 1990, pp. 597–602.
[28]   Andrews, S.R., and B.A. Miller, *J. Appl. Phys.*, Vol. 70, 1991, p. 993.
[29]   Gunapala, S.D., B.F. Levine, L. Pfeiffer, and K.West. *J. Appl. Phys.*, Vol. 69, 1991, p. 6517.
[30]   Lyon, S.A. *Surf. Sci.*, Vol. 228, 1990, p. 508.
[31]   Hasnain, G., B.F. Levine, C.G. Bethea, R.A. Logan, and J. Walker, *Appl. Phys. Lett.*, Vol. 54, 1989, p. 2515.
[32]   Andersson, J.Y., L. Lundqvist, and Z.F. Paska, *Appl. Phys. Lett.*, Vol. 58, 1991, p. 2264.
[33]   Andersson, J.Y., and L. Lundqvist, *Appl. Phys. Lett.*, Vol. 59, 1991, p. 857.
[34]   Williams, G.M., R.E. DeWames, C.W. Farley, and R.J. Anderson, *Appl. Phys. Lett.*, Vol. 60, 1992, p. 1324.
[35]   Choi, K.K., B.F. Levine, R.J. Malik, J. Walker, and C.G. Bethea, *Phys. Rev. B*, Vol. 35, 1987, p. 4172.
[36]   Ruch, J.G., and G.S. Kino, *Phys. Rev.*, Vol. 174, 1968, p. 921.
[37]   Brennan, K.F., D.H. Park, K. Hess, and M.A. Littlejohn, *J. Appl. Phys.*, Vol. 63, 1988, p. 5004.

[38]   Liu, W.C., *J. Materials Science*, Vol. 25, 1990, p. 1765.

[39]   Meshkov, S.V., *Zh. Eksp. Teor. Fiz.*, Vol. 91, 1986, p. 2252 *Trans. in Sov. Phys. JETP*, Vol. 64, 1986, p. 1337.

[40]   See, for example, Jackson, J.D., *Classical Electrodynamics*, John Wiley and Sons, New York, 1962, Ch. 2.

[41]   Levine, B.F., C.G. Bethea, G. Hasnain, J. Walker, and R.J. Malik, *Appl. Phys. Lett.*, Vol. 53, 1988, p. 296.

[42]   West, L.C., and S.J. Eglash, *Appl. Phys. Lett.*, Vol. 46, 1985, p. 1156.

[43]   Liu, H.C., *Appl. Phys. Lett.*, Vol. 60, 1992, p. 1507.

[44]   Rosenbluth, M., M. O'Loughlin, W. Bloss, F. De Luccia, H. Kanter, B. Janousek, E. Perry, and M. Daugherty, *SPIE Proc. 1283: Quantum-Well and Superlattice Physics III*, 1990, p. 82.

[45]   Levine, B.F., in *Proc. NATO Advanced Research Workshop on Intersubband Transitions in Quantum Wells*, ed. E. Rosencher, B. Vinter, and B. Levine, Cargese, September 10–14, 1991, Plenum Press, New York, 1992.

[46]   Levine, B.F., C.G. Bethea, G. Hasnain, J. Walker, and R.J. Malik, *SPIE Proc. 930: Infrared Detectors and Arrays*, 1988, p. 114.

Chapter 4

Far-Infrared Materials Based on InAs/GaInSb Type II, Strained-Layer Superlattices[1]

Christian Mailhiot

Physics Department
Lawrence Livermore National Laboratory
University of California

4.1 INTRODUCTION

Semiconductor superlattices are synthetically modulated nanostructures consisting of alternating thin layers of different semiconductors. Such artificial structures are currently of great technological interest because they afford the possibility of tailoring the electronic structure by controlled modifications of the growth parameters; for example, layer thickness, alloy compositions, strain, growth orientation [1, 2]. The flexibility of tuning the electronic structure of semiconductor superlattices currently serves as the basis for the design of semiconductor diode lasers (see [3–5]), electrooptical modulators (see [6–11]), and nonlinear optical devices (see [12–18]). The purpose of the present chapter is to demonstrate the usefulness of III–V semiconductor superlattices as long-wavelength infrared-sensitive materials characterized by cutoff wavelength $\lambda_c \equiv (hc/E_g)$ (where h is Planck's constant, c is the speed of light, and E_g is the energy bandgap) near or exceeding 10–12 μm.

The emergence of epitaxial growth techniques has allowed the synthesization of a wide range of superlattice materials systems. In particular, lattice-mismatched heterostructures can be grown with essentially no misfit dislocations if the layers are sufficiently thin [19, 20], in which case the lattice-constant mismatch is accommodated by coherent strain in the individual layers. This effect was first explored in the context of semiconductor superlattices by Matthews and Blakeslee [21–23].

[1]Work was performed at Lawrence Livermore National Laboratory under the auspices of the U.S. Department of Energy under contract no. W-7405-ENG-48.

Removal of the constraints imposed by close lattice-constant matching greatly extends the number of superlattice materials systems that can be synthesized. In addition, Osbourn [24–28] demonstrated that lattice-mismatch induced strains lead to interesting new behavior in such *strained-layer superlattices* (SLS) by deformation-potential effects. Strain effects modify the bandgaps of the constituent materials and split the degeneracy of the heavy- and light-hole bands at the center of the bulk Brillouin zone. In Figure 4.1, we indicate a schematic diagram of the band-edge lineup for type I and type II superlattices for three lattice-constant matching conditions: equal lattice constant ($a_A = a_B$); material A in compression with the lattice constant of material A greater than that of material B ($a_A > a_B$); and material A in tension with the lattice constant of material A smaller than that of material B ($a_A < a_B$). Inspection of Figure 4.1 reveals that a wide range of possibilities are available for the growth of superlattice systems covering a broad spectrum of electronic and optical properties. The effect of lattice-mismatch induced strains on the band struc-

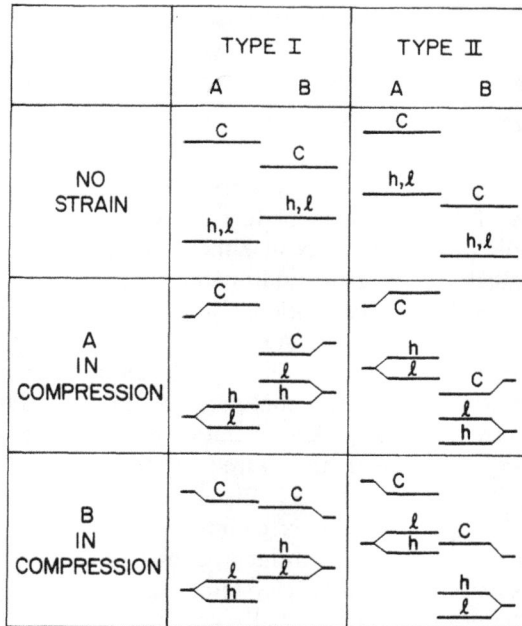

	TYPE I A	TYPE I B	TYPE II A	TYPE II B
NO STRAIN	c / h,ℓ	c / h,ℓ	c / h,ℓ	c / h,ℓ
A IN COMPRESSION	c / h ℓ	c / ℓ h	c / h ℓ	c / ℓ h
B IN COMPRESSION	c / ℓ h	c / h ℓ	c / ℓ h	c / h ℓ

Figure 4.1 Schematic illustration of the band-edge lineup in type I and type II superlattices. Band lineups for three strain conditions are indicated—equal lattice constant ($a_A = a_B$); material A in compression with the lattice constant of material A greater than that of material B ($a_A > a_B$); and material A in tension with the lattice constant of material A smaller than that of material B ($a_A < a_B$).

ture of the constituent materials is illustrated by Figure 4.1. In the material under biaxial compression (tension), the heavy-hole states are split up (down) from the light-hole states at the center of the bulk Brillouin zone. Strain-induced splitting in the bulk constituent materials can reach several tens of meV and, in some cases, be comparable to the magnitude of the valence band offset. The epitaxial growth of high-quality SLS has been conclusively demonstrated a decade ago [29, 30].

The different types of band-edge lineups indicated in Figure 4.1 lead to very distinct features in the electronic structure of the superlattice. In superlattices characterized by a type I band lineup (left column in Figure 4.1), electrons and holes are quantum-confined in the small bandgap material. On the other hand, for superlattices characterized by a type II band lineup (right column in Figure 4.1), electrons and holes are quantum-confined in adjacent layers. The essential difference in quantum confinement between type I and type II superlattices is illustrated in Figure 4.2, where the squared, symmetry-resolved, envelope functions corresponding to type I and type II superlattices are plotted along the superlattice growth axis ($\hat{z}$). The first superlattice conduction state (C_1) and the first superlattice heavy-hole (HH$_1$) and light-hole (LH$_1$) states are shown. The type I superlattice is lattice matched and consists of 70 Å of Ga$_{0.47}$In$_{0.53}$As (quantum wells) alternating with 30 Å of Al$_{0.48}$In$_{0.52}$As (barriers) grown along the (001) orientation. The type II superlattice consists of 24 Å of InAs alternating with 24 Å of GaSb grown along the (001) orientation. Inspection of Figure 4.2 reveals that, in type I superlattices, electrons and holes are quantum confined within the same layer, whereas in type II superlattices, electrons and holes are mostly quantum confined in adjacent layers. Consequently, the electron-hole wave-function overlap is large in type I superlattices, but tends to be small in type II superlattices and decreases exponentially with increasing the layer thickness of the constituent materials forming the type II superlattice. However, a unique and interesting feature of type II superlattices is that the superlattice bandgap can be smaller than that of either of the constituent materials. In fact, the superlattice can be metallic [31]. This property forms the basis for the utilization of type II superlattices for far-infrared sensitive materials, and it will be discussed in detail in Section 4.3.

Numerous superlattice systems have been grown and studied. Because of different band-edge lineups, strain conditions, and growth orientations, the various materials systems can exhibit qualitatively different physical behavior. In this chapter, we indicate the usefulness of type II SLS grown from III–V zinc-blende structure compound semiconductors as long-wavelength infrared-sensitive materials. In particular, we consider InAs/Ga$_{1-x}$In$_x$Sb type II SLS and demonstrate that the electronic and optical properties of this superlattice compare favorably with those of conventional Hg$_{1-x}$Cd$_x$Te alloys for far-infrared cutoff wavelength λ_c exceeding 10 μm.

The chapter is organized as follows. Following this Introduction, a review of the theoretical methods used in the calculations of the electronic structure of semiconductor superlattices is presented. Special attention is given to an interface-adapted

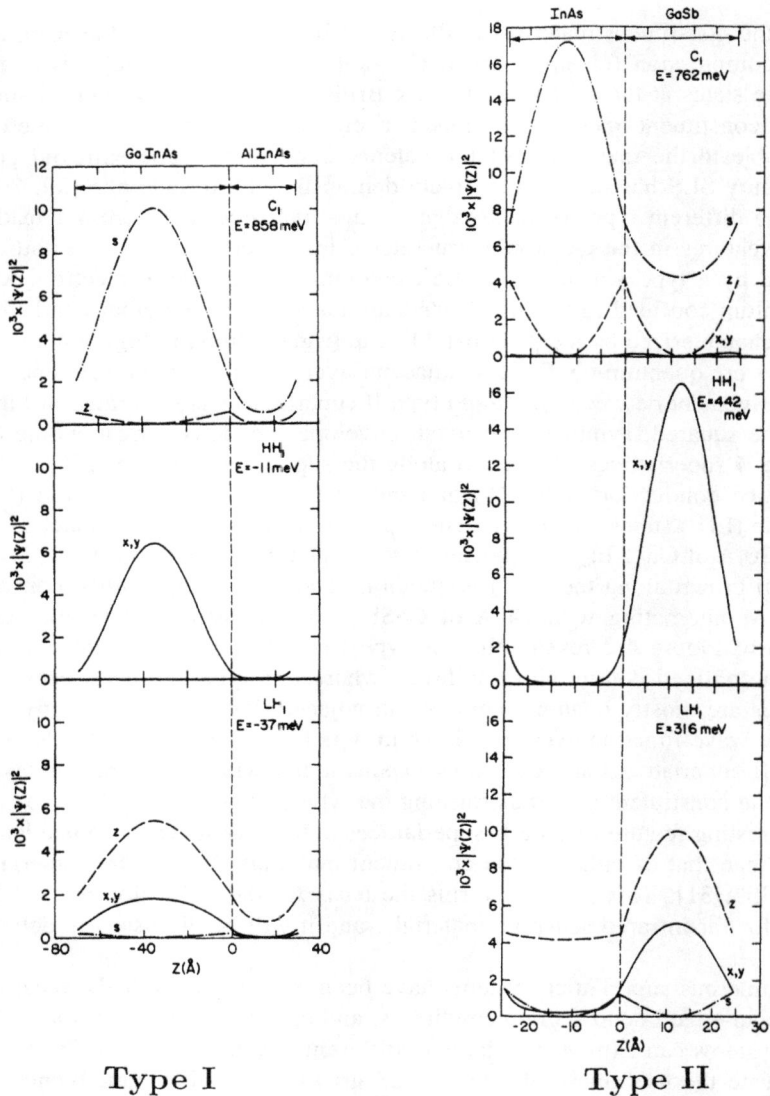

Figure 4.2 Squared, symmetry-resolved, envelope functions corresponding to type I and type II superlattices plotted along the superlattice growth axis ($\hat{z}$). The first superlattice conduction state (C_1) and the first superlattice heavy-hole (HH$_1$) and light-hole (LH$_1$) states are shown. The type I superlattice is lattice-matched and consists of 70 Å of Ga$_{0.47}$In$_{0.53}$As (quantum wells) alternating with 30 Å of Al$_{0.48}$In$_{0.52}$As (barriers) grown along the (001) orientation. The type II superlattice consists of 24 Å of InAs alternating with 24 Å of GaSb grown along the (001) orientation.

$\mathbf{k} \cdot \mathbf{p}$ method, which serves as the basis for the numerical results presented in Section 4.3. In Section 4.3, we indicate theoretical results on the electronic and optical properties of $InAs/Ga_{1-x}In_xSb$ thin-layer SLS. The experimental situation is reviewed in Section 4.4, and a summary is provided in Section 4.5.

4.2 ELECTRONIC STRUCTURE THEORY OF SEMICONDUCTOR SUPERLATTICES

4.2.1 Overview of Theoretical Electronic Structure Methods

Electronic structure theory has a truly unique role to play in superlattice research because of the wide variety of materials systems that can be synthesized. Theory serves as the basis for the design of superlattice structures exhibiting novel physical phenomena without analogs in bulk. In most cases, these novel phenomena have technological applications. Theory is also used to tailor the electronic structure of superlattices, either to enhance the observation of a particular effect or to optimize a particular superlattice system for a specific technological purpose. There is an essential difference between the role of electronic structure theory in superlattice research and its role in the area of conventional bulk semiconductors. For conventional bulk semiconductors, theory is often used to describe the electronic structure and interpret experimental observations that depend on the details of the electronic structure. However, few parameters are available to design new materials. On the other hand, the large number of possible superlattice systems allows great freedom in materials design. Consequently, electronic structure theory plays an important role in the *design process* characteristic of the field of semiconductor superlattices.

The energy offsets (that is, the band-edge lineups) between corresponding bands in the constituent materials forming a superlattice are typically on the order of few hundred meV. Modifications of the electronic structure due to the periodic superlattice potential occur on this energy scale. Consequently, a theoretical description of the electronic structure of superlattices must treat effects on this energy scale. Those electronic states whose mean free path is comparable to, or longer than, the thickness of the constituent material layers are significantly perturbed by the spatial modulation of the superlattice potential. States whose mean free path is much less than the thickness of the constituent material layers essentially are kinetically confined within a particular material. Therefore, such states are little affected by the periodic superlattice potential modulation. As a result, electronic states close in energy to the band edges, which have relatively long lifetimes and long mean free paths, are of greatest interest in describing the electronic structure of superlattice materials. Consequently, we are interested in an accurate description of the energy band edges because band-edge states are those most perturbed by the superlattice potential. For this reason, the most useful theories of the electronic structure of superlattices are essentially *band-edge* electronic structure theories, although secondary energy minima can be important as well. Perturbations such as spin-orbit interactions, strain, and electric and magnetic fields are also important on the energy scale of interest.

Over the last decade, many theoretical formalisms have been developed to describe the electronic structure of semiconductor superlattices and other synthetically modulated quantum-confined structures [32]. These theoretical formalisms can generally be divided into two classes: supercell methods and boundary-condition (or wave-function matching) methods. In the supercell methods, the superlattice is viewed essentially as a new material, whose unit cell is enlarged to the dimension of the superlattice period. Superlattice energies and eigenstates are obtained by diagonalization of the Hamiltonian describing the interactions within the supercell. Application of the supercell method yields a description of the superlattice band structure over a large range of energies throughout the Brillouin zone. However, the computational effort involved in this method increases rapidly with the dimension of the supercell, and consequently, this approach is impractical for the thick-layer superlattices of interest in most photonics applications. In the boundary-condition method, on the other hand, superlattice states are interpreted in terms of the bulk eigenstates associated with the constituent materials. At a given energy, superlattice states are expanded in terms of the bulk states of the constituent materials. The superlattice energies and eigenstates are determined by imposing a set of boundary conditions on the superlattice state at the interface. In the boundary-condition methods, the energy is an input to the problem and therefore can be chosen to lie within the range of most interest, usually near the energy band edges of the constituent materials. The computational effort associated with boundary-conditions methods is independent of the superlattice layer thickness, and consequently, boundary-condition methods are well-suited to treat thick-layer superlattices. This approach can also yield the superlattice band structure over a large range of energies and throughout the Brillouin zone. However, boundary-condition methods have been applied mostly in a context where a description of the bulk band structure of the constituent materials near the center of the Brillouin zone was adequate. This is the case for superlattices containing direct bandgap materials, but not for superlattices containing indirect bandgap materials.

4.2.2 $k \cdot p$ Electronic Structure Theory of Semiconductor Superlattices

In this subsection, we describe a boundary-condition approach to the calculation of the electronic structure of semiconductor superlattices based on a $k \cdot p$ theory adapted to hetero-interfaces [33, 34]. We first present a description of the electronic structure of the individual materials forming the superlattice. We then indicate how the superlattice energies and wave functions are obtained from the imposition of suitable boundary conditions.

k · p *Electronic Structure of Bulk Constituent Materials*

We first describe the electronic structure of the individual constituent materials. For the calculation of the electronic structure of the constituent materials, we want to describe the states in both materials in terms of a *single* zone-center basis set. Consider a superlattice consisting of the alternate sequencing of two constituent materials *a* and *b*. We define a reference Hamiltonian constructed by averaging the potentials of the constituent materials,

$$H_R = \frac{\mathbf{p}^2}{2m} + \frac{1}{2}[V_a(\mathbf{x}) + V_b(\mathbf{x})] \tag{4.1}$$

where $V_i(\mathbf{x})$ is the pseudopotential of material *i* described in terms of pseudopotential form factors [35]. The lattice constant in the reference Hamiltonian is the average of the lattice constants of the constituent materials *a* and *b*. Alloys are treated in the virtual crystal approximation.

The reference Hamiltonian is diagonalized at $\mathbf{k} = 0$ to yield the zone-center energies ε_β and the zone-center cell-periodic functions in terms of a plane-wave expansion:

$$u_\beta(\mathbf{x}) = \frac{1}{\sqrt{\Omega}} \sum_{\mathbf{G}} \exp(i\mathbf{G} \cdot \mathbf{x}) \, U(\mathbf{G}, \beta) \tag{4.2}$$

where $\mathbf{G}$ is a reciprocal lattice vector, Ω is the volume of the unit cell, and β labels the zone-center functions. The expansion coefficients $U(\mathbf{G}, \beta)$ form the unitary transformation matrix between the plane-wave $\exp(i\mathbf{G} \cdot \mathbf{x}) \sim |\mathbf{G}\rangle$ and the zone-center $u_\beta(\mathbf{x}) \sim |\beta\rangle$ representations. Phases are chosen such that the zone-center basis functions $u_\beta(\mathbf{x})$ are real.

In terms of the reference Hamiltonian, H_R, the Hamiltonian describing the bulk constituent material *i* can be written as

$$H^{(i)} = H_R + \Delta V^{(i)}(\mathbf{x}) + H^{(i)}_{\text{s.o.}} + H^{(i)}_{\text{strain}} \tag{4.3}$$

where the potential difference $\Delta V^{(i)}(\mathbf{x})$ is

$$\Delta V^{(i)}(\mathbf{x}) \equiv V^{(i)}(\mathbf{x}) - \frac{1}{2}[V_a(\mathbf{x}) + V_b(\mathbf{x})] \tag{4.4}$$

The term $H^{(i)}_{\text{s.o.}}$ represents the spin-orbit interactions in material *i* and $H^{(i)}_{\text{strain}}$ represents the stress interactions due to the lattice mismatch in material *i*.

We use the zone-center states $u_\beta(\mathbf{x})$ as a basis set to describe the cell-periodic functions in each of the constituent materials. We divide the set of states $u_\beta(\mathbf{x})$ into two groups. One group consists of the threefold Γ_{15}^v valence-band states ($|x\rangle$, $|y\rangle$, $|z\rangle$) and of the Γ_1^c conduction-band state ($|s\rangle$) combined with a spinor. These eight states are treated explicitly and are denoted by the label d as $u_d(\mathbf{x})$. The other spatial states are also combined with a spinor and form the second group of states. These states, labeled $u_\delta(\mathbf{x})$, are included to first order for the wave function in Löwdin perturbation theory [36]. The perturbation Hamiltonian is

$$H_{\text{pert}}^{(i)} = \Delta V^{(i)}(\mathbf{x}) + \frac{\hbar}{m}\, \mathbf{k}^{(i)} \cdot \mathbf{p} \tag{4.5}$$

Therefore, in each material i, the cell-periodic functions are expanded as

$$u_d^{(i)}(\mathbf{k}_j^{(i)}, \mathbf{x}) = u_d(\mathbf{x}) + \sum_\delta u_\delta(\mathbf{x}) \frac{\langle u_\delta | \Delta V^{(i)}(\mathbf{x}) + (\hbar/m)\mathbf{k}_j^{(i)} \cdot \mathbf{p} | u_d \rangle}{(\varepsilon_d - \varepsilon_\delta)} \tag{4.6}$$

where the summation over the states δ does not include the explicitly treated states labeled by d.

The Bloch and evanescent states in material i are expanded as

$$\psi^{(i)}(\mathbf{k}_j^{(i)}, \mathbf{x}) \equiv |\psi_j^{(i)}\rangle = \frac{1}{\sqrt{N}} \exp(i\mathbf{k}_j^{(i)} \cdot \mathbf{x}) \sum_d C^{(i)}(d, \mathbf{k}_j^{(i)})\, u_d^{(i)}(\mathbf{k}_j^{(i)}, \mathbf{x}) \tag{4.7}$$

where N is the number of bulk primitive cells in the superlattice. The expansion coefficients $C^{(i)}(d, \mathbf{k}_j^{(i)})$ are obtained by solving the eigenvalue equation:

$$[H^{(i)}(d, d';\ \mathbf{k}_j^{(i)}) - E\delta_{d,d'}]\, C^{(i)}(d', \mathbf{k}_j^{(i)}) = 0 \tag{4.8}$$

where E is the energy of the state. The Hamiltonian matrix elements $H^{(i)}(d, d';\ \mathbf{k}_j^{(i)})$ are defined as

$$H^{(i)}(d, d';\ \mathbf{k}_j^{(i)}) \equiv \langle u_d | \exp(-i\mathbf{k}_j^{(i)} \cdot \mathbf{x})\, H^{(i)} \exp(i\mathbf{k}_j^{(i)} \cdot \mathbf{x}) | u_{d'}^{(i)}(\mathbf{k}_j^{(i)})\rangle \tag{4.9}$$

In the interface problem, the wave vector $\mathbf{k}_j^{(i)}$ is allowed to take on complex values, and a complex-k band structure calculation must be performed. We denote $\mathbf{k}_\parallel$ and $k_j^{(i)}$ the components of $\mathbf{k}_j^{(i)}$ parallel and perpendicular to the interface plane, respectively. In the complex-k band structure problem, we fix $\mathbf{k}_\parallel$ (real) and the energy E and determine the values of $k_j^{(i)}$ (complex), which are solutions of Schrödinger's equation. The complex solutions $k_j^{(i)}$ are most conveniently found when the Hamiltonian is factored to explicitly display the dependence on $k_j^{(i)}$:

$$\hat{H}(d, d';\ \mathbf{k}_j) = H^{(2)}(d, d')k_j^{(i)2} + H^{(1)}(d, d';\ \mathbf{k}_\parallel)k_j^{(i)} + H^{(0)}(d, d';\ \mathbf{k}_\parallel) \tag{4.10}$$

Explicit expressions for the matrix elements $H^{(i)}(d, d'; \mathbf{k}_j^{(i)})$ have appeared in the literature [33] and will not be repeated here. The complex solutions k for fixed $\mathbf{k}_\parallel$ and E can be found by transforming the original eigenvalue problem for the energies into an eigenvalue problem for k [33]. Because the Hamiltonian is hermitian, if k is a solution, so is k^*.

Description of a Hetero-Interface

We consider an abrupt interface between materials a and b grown along the $\hat{\mathbf{z}}$ axis. The interface plane is defined by $z = z_0$ and the Hamiltonian is

$$H = H^{(a)}\theta(-z + z_0) + H^{(b)}\theta(z - z_0) \tag{4.11}$$

where $H^{(a,b)}$ is the Hamiltonian in material $a(b)$ and $\theta(z)$ is a unit step function. The wave function has the form

$$|\Psi\rangle = |\Psi^{(a)}\rangle\theta(-z + z_0) + |\Psi^{(b)}\rangle\theta(z - z_0) \tag{4.12}$$

where $|\Psi^{a(b)}\Psi\rangle$ is the wave function in material $a(b)$. The wave function $|\Psi^{a(b)}\Psi\rangle$ is expanded in terms of the bulk states of each material with a fixed $\mathbf{k}_\parallel$ and energy E:

$$|\Psi^{(a)}\rangle = \sum_j |\psi_j^{(a)}\rangle A_j \qquad \text{in material } A \tag{4.13a}$$

$$|\Psi^{(b)}\rangle = \sum_i |\psi_i^{(b)}\rangle B_i \qquad \text{in material } B \tag{4.13b}$$

where $|\psi_j^{(i)}\rangle \equiv \psi^{(i)}(\mathbf{k}_j^{(i)}, \mathbf{x})$ is the bulk solution associated with the complex wave vector $\mathbf{k}_j^{(i)} \equiv \mathbf{k}_\parallel + \hat{\mathbf{z}}k_j^{(i)}$. The coefficients A_j and B_i determine the wave function in material a and b, respectively.

The expansion coefficients A_j and B_i are determined by appropriate boundary conditions. Because of the form of the Hamiltonian, the wave function and its derivative must be continuous across the interface. Although these conditions are certainly correct, they are inconvenient to use directly for the interface problem. A much more convenient approach is to formulate the problem in terms of the $\hat{\mathbf{z}}$ component of the current density operator evaluated at the point $\mathbf{R}$:

$$
\begin{aligned}
J_z(\mathbf{R}) &= \frac{1}{2m}[\delta(\mathbf{x} - \mathbf{R})p_z + p_z\delta(\mathbf{x} - \mathbf{R})] \\
&= \frac{1}{2m}\left[2\delta(\mathbf{x} - \mathbf{R})p_z + \left\{\frac{\hbar}{i}\frac{\partial}{\partial z}\delta(\mathbf{x} - \mathbf{R})\right\}\right]
\end{aligned}
\tag{4.14}
$$

Because $\delta(\mathbf{x} - \mathbf{R})$ and $\partial\delta(\mathbf{x} - \mathbf{R})/\partial z$ are linearly independent, the interface boundary condition can be written, with $\mathbf{R} = x\hat{\mathbf{x}} + y\hat{\mathbf{y}} + z_0\hat{\mathbf{z}}$, as

$$J_z(x, y, z_0)|\Psi^{(a)}\rangle = J_z(x, y, z_0)|\Psi^{(b)}\rangle \qquad (4.15)$$

To convert eq. (4.15) into a matrix equation relating the expansion coefficients A_j and B_i, we substitute eq. (4.13) into eq. (4.15), overlap with $\langle\psi_{j*}^{(a)}|$ and average over the x-y plane:

$$\sum_{j'} \frac{1}{S} \int dx \, dy \, \langle\psi_{j*}^{(a)}|J_z(x, y, z_0)|\psi_{j'}^{(a)}\rangle A_{j'} = \sum_i \frac{1}{S} \int dx \, dy \, \langle\psi_{j*}^{(a)}|J_z(x, y, z_0)|\psi_i^{(b)}\rangle B_i$$

$$(4.16)$$

where S is the area in the x-y plane.

We now substitute the form eq. (4.13) for the bulk states to write the left-hand side of eq. (4.16) as,

$$\frac{1}{S} \int dx \, dy \, \langle\psi_{j*}^{(a)}|J_z(x, y, z_0)|\psi_{j'}^{(a)}\rangle = \exp[i(k_{j'}^{(a)} - k_j^{(a)}) z_0]$$

$$\times \frac{1}{\hbar\Omega} [\mathbf{C}(k_j^*)]^\dagger \{\mathbf{H}^{(2)} (k_j^{(a)} + k_{j'}^{(a)})^2 + \mathbf{H}^{(1)}\} [\mathbf{C}(k_{j'}^{(a)})] \equiv \delta_{jj'} J_{j*j}^{(a)} \qquad (4.17a)$$

and similarly with the right-hand side of eq. (4.16),

$$\frac{1}{S} \int dx \, dy \, \langle\psi_{j*}^{(a)}|J_z(x, y, z_0)|\psi_i^{(b)}\rangle = \exp[i(k_i^{(b)} - k_j^{(a)})z_0]$$

$$\times \frac{1}{\hbar\Omega} [\mathbf{C}(k_j^*)]^\dagger \{\mathbf{H}^{(2)} (k_j^{(a)} + k_i^{(b)})^2 + \mathbf{H}^{(1)}\} [\mathbf{C}(k_i^{(b)})] \qquad (4.17b)$$

$$\equiv \exp[i(k_i^{(b)} - k_j^{(a)}) z_0] J_{j*i}^{(ab)} (z_0)$$

where the matrices $\mathbf{H}^{(1)}$ and $\mathbf{H}^{(2)}$ (defined in eq. (4.10)) and the expansion coefficients $\mathbf{C}$ (defined in eq. (4.7)) are written in the $u_d(\mathbf{x})$ basis.

Therefore, the matching condition eq. (4.16) becomes

$$\exp(ik_j^{(a)}z_0) A_j = \frac{1}{J_{j*j}^{(a)}} \sum_i J_{j*i}^{(ab)}(z_0) \exp(ik_i^{(b)}z_0)B_i \qquad (4.18)$$

The calculation of non-bulk matrix elements of the current density operator, $J_{j*i}^{(ab)}(z_0)$, is unambiguous because delta functions in the current density operator restrict the integration to the interface where both wave functions are defined and expanded on the same basis.

Equation (4.18) describes the interface matching condition. Superlattice periodicity can be used, together with the interface matching condition, to formulate an eigenvalue equation for the superlattice wave vector and wave function. A derivation of this eigenvalue has appeared in the literature [33] and need not be repeated here. Let us simply state that the eigenvalue problem yielding the superlattice solutions is

$$\sum_{j'} M_{jj'} A_{j'\alpha} = \exp[iq_\alpha(a + b)] A_{j\alpha} \qquad (4.19a)$$

where

$$M_{jj'} \equiv \exp(ik_j^{(a)}a) \frac{1}{J_{j*j}^{(a)}} \sum_i J_{j*i}^{(ab)} \exp(ik_i^{(b)}b) \frac{1}{J_{i*i}^{(b)}} J_{i*j'}^{(ba)} \qquad (4.19b)$$

In eqs. (4.19), $a(b)$ is the thickness of semiconductor $a(b)$ within the superlattice unit cell, q_α is the superlattice wave vector, and α is a quantum number labeling the superlattice solutions. Once the expansion coefficients A_j in material a are determined by the solution of eqs. (4.19), the expansion coefficients B_i in material b are found by the matching condition eq. (4.18).

Extensive applications of the $\mathbf{k} \cdot \mathbf{p}$ formalism presented here to the calculation of the electronic structure and optical properties of semiconductor superlattices have been documented in the literature [11, 13, 34, 37–43].

Strained-Layer Superlattices

Because a central topic of this chapter is the electronic structure of *strained-layer superlattices* (SLS), we indicate how lattice-mismatch induced strains are incorporated in the description of the bulk electronic structure of the constituent materials.

The lattice-mismatch induced strains in the constituent materials forming the superlattice are calculated by minimizing the strain energy subject to the constraints imposed by the pseudomorphic character of the interfaces. We define the set of orthonormal vectors $\{\hat{\mathbf{n}}_i\}$ so that $\hat{\mathbf{n}}_1$ and $\hat{\mathbf{n}}_2$ lie in the superlattice plane and $\hat{\mathbf{n}}_3$ is orthogonal to this plane. We define the vectors

$$\begin{bmatrix} \mathbf{x}_i \\ \mathbf{y}_i \\ \mathbf{z}_i \end{bmatrix} = \begin{bmatrix} (1 + \varepsilon_{xx}^i) & \varepsilon_{xy}^i & \varepsilon_{xz}^i \\ \varepsilon_{yx}^i & (1 + \varepsilon_{yy}^i) & \varepsilon_{yz}^i \\ \varepsilon_{zx}^i & \varepsilon_{zy}^i & (1 + \varepsilon_{zz}^i) \end{bmatrix} \begin{bmatrix} \hat{\mathbf{x}} \\ \hat{\mathbf{y}} \\ \hat{\mathbf{z}} \end{bmatrix} \qquad (4.20)$$

where i labels the two constituent materials of the superlattice. The off-diagonal strain components are diagonal ($\varepsilon_{ij} = \varepsilon_{ji}$), so that six unknown strain components are to be determined in each of the two constituent materials. Without loss of generality, we consider a superlattice grown from two zinc-blende structure compound semiconductors. The zinc-blende structure primitive lattice translation vectors are

$$\mathbf{a}_i = \frac{a_i}{2}(\hat{\mathbf{y}} + \hat{\mathbf{z}}) \tag{4.21a}$$

$$\mathbf{b}_i = \frac{a_i}{2}(\hat{\mathbf{z}} + \hat{\mathbf{x}}) \tag{4.21b}$$

$$\mathbf{c}_i = \frac{a_i}{2}(\hat{\mathbf{x}} + \hat{\mathbf{y}}) \tag{4.21c}$$

where a is the length of the cube edge.

When the crystal is strained, the primitive translation vectors become

$$\mathbf{a}'_i = \frac{a_i}{2}(\mathbf{y}_i + \mathbf{z}_i) \tag{4.21d}$$

$$\mathbf{b}'_i = \frac{a_i}{2}(\mathbf{z}_i + \mathbf{x}_i) \tag{4.21e}$$

$$\mathbf{c}'_i = \frac{a_i}{2}(\mathbf{x}_i + \mathbf{y}_i) \tag{4.21f}$$

The conditions for pseudomorphic interfaces are

$$\mathbf{a}'_i \cdot \hat{\mathbf{n}}_1 = \mathbf{a}'_j \cdot \hat{\mathbf{n}}_1 \tag{4.22a}$$

$$\mathbf{a}'_i \cdot \hat{\mathbf{n}}_2 = \mathbf{a}'_j \cdot \hat{\mathbf{n}}_2 \tag{4.22b}$$

$$\mathbf{b}'_i \cdot \hat{\mathbf{n}}_1 = \mathbf{b}'_j \cdot \hat{\mathbf{n}}_1 \tag{4.22c}$$

$$\mathbf{b}'_i \cdot \hat{\mathbf{n}}_2 = \mathbf{b}'_j \cdot \hat{\mathbf{n}}_2 \tag{4.22d}$$

$$\mathbf{c}'_i \cdot \hat{\mathbf{n}}_1 = \mathbf{c}'_j \cdot \hat{\mathbf{n}}_1 \tag{4.22e}$$

$$\mathbf{c}'_i \cdot \hat{\mathbf{n}}_2 = \mathbf{c}'_j \cdot \hat{\mathbf{n}}_2 \tag{4.22f}$$

These constraint conditions are linearized in the lattice constant difference, $\Delta a_0 = a_0^{(a)} - a_0^{(b)}$. The constraint equations reduce the number of unknowns (from 12 to 7), which are determined by minimizing the superlattice strain energy density.

The strain energy density of a zinc-blende structure material is

$$U_i = \frac{1}{2} c_{11}^i \left[(\varepsilon_{xx}^i)^2 + (\varepsilon_{yy}^i)^2 + (\varepsilon_{zz}^i)^2 \right] + 2c_{44}^i \left[(\varepsilon_{xy}^i)^2 + (\varepsilon_{yz}^i)^2 + (\varepsilon_{zx}^i)^2 \right]$$
$$+ c_{12}^i \left[\varepsilon_{yy}^i \varepsilon_{zz}^i + \varepsilon_{xx}^i \varepsilon_{zz}^i + \varepsilon_{xx}^i \varepsilon_{yy}^i \right] \tag{4.23}$$

where the coefficients c_{11}, c_{12}, and c_{44} are the elastic constants. For a strained-layer superlattice synthesized from two zinc-blende structure materials, the strain energy density is

$$U_{ab} = \frac{U_a h_a + U_b h_b}{h_a + h_b} \tag{4.24}$$

where h_a (h_b) is the layer thickness of material a (b). The strain components in the constituent materials are determined by minimizing the strain energy density U_{ab} subject to the pseudomorphic constraints, eq. (4.22).

For a superlattice grown along the (001) axis from two zinc-blende structure materials, only diagonal strain components are generated (e.g., $\varepsilon_{xy} = 0$) by the lattice mismatch. The diagonal components are given by [24]

$$\varepsilon_{zz}^i = - \left(\frac{2c_{12}^i}{c_{11}^i} \right) \varepsilon_{xx}^i \tag{4.25a}$$

$$\varepsilon_{xx}^a = \varepsilon_{yy}^a = - \left(\frac{\Delta a_0}{a_0} \right) \left(\frac{G^b h_b}{G^a h_a + G^b h_b} \right) \tag{4.25b}$$

$$\varepsilon_{xx}^b = \varepsilon_{yy}^b = \left(\frac{\Delta a_0}{a_0} \right) \left(\frac{G^a h_a}{G^a h_a + G^b h_b} \right) \tag{4.25c}$$

where

$$\Delta a_0 = a_0^{(a)} - a_0^{(b)} \tag{4.26a}$$

$$a_0 = \frac{1}{2} \left[a_0^{(a)} + a_0^{(b)} \right] \tag{4.26b}$$

$$G^i = 2 \left[c_{11}^i + c_{12}^i - \frac{2(c_{12}^i)^2}{c_{11}^i} \right] \tag{4.26c}$$

Neglecting the spin dependence of the stress interactions, the strain Hamiltonian for a (001)-oriented SLS is independent of spin and has nonzero matrix elements [44]:

$$\langle s|H^{(i)}_{\text{strain}}|s\rangle = c^i(2\varepsilon^i_{xx} + \varepsilon^i_{zz}) \tag{4.27a}$$

$$\langle x|H^{(i)}_{\text{strain}}|x\rangle = \langle y|H^{(i)}_{\text{strain}}|y\rangle = (l^i + m^i)\varepsilon^i_{xx} + m^i\varepsilon^i_{zz} \tag{4.27b}$$

$$\langle z|H^{(i)}_{\text{strain}}|z\rangle = 2m^i\varepsilon^i_{xx} + l^i\varepsilon^i_{zz} \tag{4.27c}$$

where c, l, and m are deformation-potential constants. In eqs. (4.27), the valence-band functions $|x\rangle$, $|y\rangle$, and $|z\rangle$ belong to the irreducible representation Γ^v_{15} and the conduction-band function $|s\rangle$ belongs to the irreducible representation Γ^c_1.

4.3 InAs/Ga$_{1-x}$In$_x$Sb TYPE II STRAINED-LAYER SUPERLATTICES

The purpose of this section is to use the electronic structure method described in the preceding section to demonstrate the usefulness of III–V SLS with a type II band alignment as novel materials for long-wavelength infrared detector applications. In particular, we consider InAs/Ga$_{1-x}$In$_x$Sb SLS and show that the favorable electronic and optical properties exhibited by this system make it a very promising candidate for a far-infrared sensitive material. In fact, our results indicate that thin-layer InAs/Ga$_{1-x}$In$_x$Sb type II SLS are expected to exhibit materials properties for long-wavelength infrared detection applications superior overall to conventional bulk Hg$_{1-x}$Cd$_x$Te alloys. Recent experimental observations have confirmed these expectations and are summarized in Section 4.4.

4.3.1 Introduction

Major efforts have been devoted to the fabrication of two-dimensional arrays of photovoltaic detectors for the purpose of infrared imaging [45]. Such detectors are usually fabricated from the Hg$_{1-x}$Cd$_x$Te alloy. Major, well-known difficulties are associated with the use of Hg$_{1-x}$Cd$_x$Te alloys in the fabrication of such detector arrays: for infrared wavelengths exceeding 10 μm, large tunneling dark currents, the requirement of extremely precise composition control to accurately determine the bandgap, and Auger recombination rates that exceed radiative recombination rates are serious obstacles in making high-performance infrared arrays from the Hg$_{1-x}$Cd$_x$Te alloy. These problems are particularly severe for long-wavelength applications. Small-bandgap superlattices have been proposed as alternatives to Hg$_{1-x}$Cd$_x$Te for long-wavelength photovoltaic detectors. The most promising small-bandgap superlattices proposed so far are the HgTe-CdTe superlattice [46, 47], in which one of the constituent material (HgTe) is a semimetal, and the InAs$_{0.4}$Sb$_{0.6}$/InAs$_{1-x}$Sb$_x$ [27], in which internal strain effects are used to reduce the bandgap of the $InAs_{0.4}Sb_{0.6}$ quantum wells. However, the epitaxial growth of the HgTe-CdTe [48] and $InAs_{0.4}Sb_{0.6}$/ $InAs_{1-x}Sb_x$ [49] has proven to be difficult.

The purpose of this section is to demonstrate that thin-layer III–V SLS with a type II band alignment show great promise as novel materials for long-wavelength infrared detector applications with cutoff wavelength λ_c exceeding 10 μm. This proposal is based on the use of internal strains in type II superlattices to achieve small-bandgap materials with sufficiently thin alternating layers to obtain large optical absorption coefficients. Moreover, because the superlattice layers are thin, the proposed superlattice will exhibit favorable transport properties due to the large values of electron-effective masses.

The InAs/GaSb superlattice has a type II band alignment in which the conduction band minimum of InAs is lower in energy than the valence band maximum of GaSb [31, 50–53]. Because of this type II alignment, the superlattice can have a bandgap smaller than that of either of the constituent materials. In fact, the superlattice becomes metallic beyond a critical layer thickness [50]. However, because of this type II energy band lineup, electrons and holes tend to be localized in adjacent layers: electrons are localized within the InAs layers, whereas holes are localized within the GaSb layers. Consequently, the electron-hole wave function overlap— and the resulting optical matrix elements—decreases exponentially with increasing the InAs and GaSb layer thickness in the superlattice. As a result, the optical absorption coefficient of InAs/GaSb superlattices is much too small for thick-layer superlattices having small bandgaps in the far-infrared region. Indeed, for superlattice layer thicknesses required to reach long-wavelength sensitivity, that is, individual material layer thickness d exceeding 50 Å, the optical absorption properties of the InAs/GaSb type II superlattice are inadequate for infrared detector applications [54]. Therefore, the *simultaneous* requirements of small bandgaps and large optical matrix elements appear to be two mutually exclusive demands in InAs/GaSb type II superlattices.

We now demonstrate that by alloying GaSb and InAs, so that the system becomes an $InAs/Ga_{1-x}In_xSb$ type II SLS, it is possible to *simultaneously* achieve small-bandgap superlattices with cutoff wavelengths exceeding 10 μm and large optical absorption [13, 37–39, 41, 43]. The large optical absorption exhibited by the InAs/ $Ga_{1-x}In_xSb$ type II SLS is possible because the thickness of the InAs and $Ga_{1-x}In_xSb$ layers necessary to achieve small superlattice bandgaps is small, and consequently, the electron-hole optical matrix elements are large because the wave-function overlap is sizeable. Because the alternating layers are thin, $InAs/Ga_{1-x}In_xSb$ type II superlattices exhibit good optical absorption properties at a cutoff wavelength of $\lambda_c \approx 12$ μm. In fact, we indicate later that the optical absorption properties of $InAs/Ga_{1-x}In_xSb$ type II SLS compare favorably with those of conventional bulk $Hg_{1-x}Cd_xTe$ alloys. Moreover, on the basis of the theoretical band structure results presented later, the transport properties of $InAs/Ga_{1-x}In_xSb$ type II SLS are expected to be superior to those of $Hg_{1-x}Cd_xTe$ alloys. Overall, $InAs/Ga_{1-x}In_xSb$ thin-layer SLS are expected to exhibit materials properties for long-wavelength infrared detection applications superior to the $Hg_{1-x}Cd_xTe$ alloy. Recent experimental observations have quantitatively confirmed these expectations and will be summarized in Section 4.4.

4.3.2 Optical Properties of $InAs/Ga_{1-x}In_xSb$ Type II Superlattices

In this subsection, we present results of calculations of the electronic structure and optical properties of $InAs/Ga_{1-x}In_xSb$ type II SLS. Our calculations are based on the $\mathbf{k} \cdot \mathbf{p}$ formalism presented in Section 4.2.2. Momentum matrix elements are calculated from a plane-wave expansion of the zone-center basis functions. We explicitly consider the eight zone-center states belonging to the irreducible representations Γ_6, Γ_7, and Γ_8 of the T_d double group. Higher lying zone-center states are included in Löwdin perturbation theory. Treatments of lattice-mismatch induced strains have been described in Section 4.2.2. The empirical parameters entered into these calculations are the pseudopotential form factors, the deformation potential constants, and the valence band offset. Relevant quantities are indicated in Table 4.1.

Table 4.1
Empirical Parameters

Materials Parameter	InAs	GaSb	InSb
Δ_0 (eV)	0.38	0.77	0.81
a_0 (Å)	6.0584	6.0954	6.4788
c_{11} (10^{11} dyn.cm^{-2})	8.33	8.84	6.67
c_{12} (10^{11} dyn.cm^{-2})	4.53	4.03	3.65
c_{44} (10^{11} dyn.cm^{-2})	3.96	4.32	3.02
c (eV)	−5.80	−8.28	−7.92
b (eV)	−1.70	−1.80	−1.80
d (eV)	−4.55	−4.20	−4.20

Materials parameters for InAs, GaSb, and InSb used for the calculation of the electronic structure of $InAs/Ga_{1-x}In_xSb$ SLS: Δ_0 is the spin-orbit splitting; a_0 is the bulk lattice constant; c_{11}, c_{12}, and c_{44} are elastic constants; c, b, and d are deformation potential constants. Parameters for the $Ga_{1-x}In_xSb$ alloy are obtained from the corresponding average of the quantities in GaSb and InSb.

We now describe the procedure to obtain the valence-band offsets for the $InAs/Ga_{1-x}In_xSb$ SLS. The InAs/GaSb system is a type II superlattice. The conduction-band minimum of InAs lies 0.1 eV below the valence-band maximum of GaSb, resulting in a valence band offset of $\Delta E_v = 0.51$ eV [51, 55]. However, the valence-band offset of the $InAs/Ga_{1-x}In_xSb$ superlattice is unknown. In Figure 4.3, we indicate the prescription used to obtain the valence-band offset at the $InAs/Ga_{1-x}In_xSb$

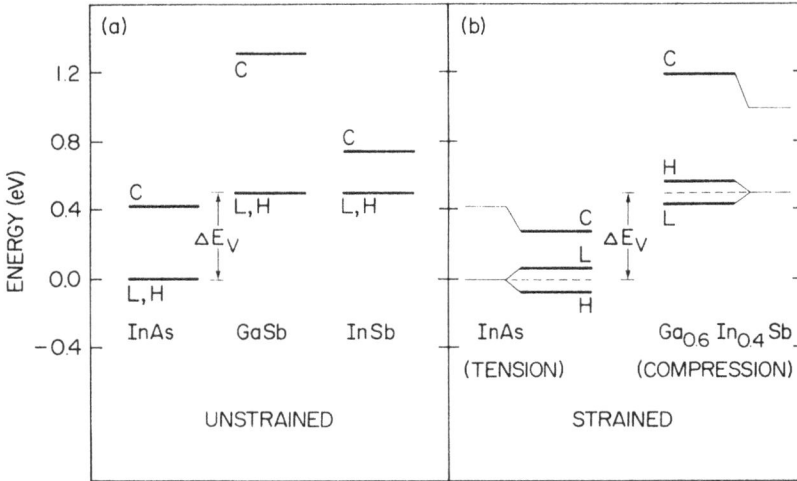

Figure 4.3 (a) Assumed relative energy positions for unstrained InAs, GaSb and InSb. (b) Effect of lattice-mismatch induced internal strain on the energy band offset for an InAs/Ga$_{0.60}$In$_{0.40}$Sb.

heterointerface (Figure 4.3(b)) from a knowledge of the known valence-band offset at the InAs/GaSb interface (Figure 4.3(a)). For illustrative purposes, the numerical results shown in Figure 4.3(b) correspond to an In composition of $x = 0.40$ in the Ga$_{1-x}$In$_x$Sb alloy. We assume that the valence-band maximum of GaSb and InSb line up (see Figure 4.3(a)). This assumption is consistent with photoemission threshold experiments [56] and with various theoretical results [57]. The consequence of this choice of offset on our results is the following: if the valence-band maximum of InSb is actually at a higher (lower) energy than that of GaSb, the predicted electronic and optical properties of the InAs/Ga$_{1-x}$In$_x$Sb SLS discussed later would be better (worse) than the ones obtained from the prescription indicated schematically in Figure 4.3. We further assume that the band edges of the Ga$_{1-x}$In$_x$Sb alloy interpolate linearly with composition x between those of InAs and GaSb.

The presence of lattice-mismatch induced internal strain splits the otherwise fourfold degenerate Γ_8 valence-band maximum states at the center of the bulk Brillouin zone. In the material under biaxial compression (the Ga$_{1-x}$In$_x$Sb alloy), the heavy-hole states—as defined by the energy band dispersion along the superlattice growth axis—split up from the light-hole states at the center of the Brillouin zone. In the material under biaxial tension (InAs), the heavy-hole states split down from the light-hole states at the center of the Brillouin zone. These splittings are indicated in Figure 4.3(b) for the InAs/Ga$_{0.60}$In$_{0.40}$Sb SLS. The effects of internal strain on the valence-band offsets are incorporated by assuming that the "center of mass" of the strain-split valence bands coincides with the unstrained valence bands [34, 58].

We now indicate how the presence of internal strains is responsible for the favorable properties exhibited by InAs/Ga$_{1-x}$In$_x$Sb thin-layer SLS as useful materials for far-infrared detection applications. The internal strain lowers the conduction-band minimum of InAs and raises the strain-split heavy-hole band in Ga$_{1-x}$In$_x$Sb (see Figure 4.3(b)) by deformation potential effects. The superlattice bandgap derives from electron states quantum confined up from the InAs conduction-band minimum and heavy-hole states quantum confined down from the Ga$_{1-x}$In$_x$Sb valence-band maximum. Because deformation potential effects simultaneously lower the InAs conduction band and raise the Ga$_{1-x}$In$_x$Sb heavy-hole band, the presence of internal strain *reduces the superlattice bandgap at a given layer thickness*. Consequently, deformation potential effects increase the superlattice cutoff wavelength λ_c for a fixed layer thickness. This strain-induced bandgap reduction is illustrated in Figure 4.4(a), where cutoff wavelengths of InAs/Ga$_{1-x}$In$_x$Sb type II SLS are calculated as a function of layer thickness for alloy compositions in the range $0 \leq x \leq 0.40$. The superlattice consists of M_a layers of InAs alternating with N_b layers of Ga$_{1-x}$In$_x$Sb grown along the (001) axis. Without any loss of generality, we consider the special case $M_a = N_b$ and assume that the superlattice is grown in a free-standing configuration; that is, both the InAs and Ga$_{1-x}$In$_x$Sb layers are strained, with the InAs layers under biaxial tension and the Ga$_{1-x}$In$_x$Sb layers under biaxial compression. For fixed layer thickness, the superlattice cutoff wavelength λ_c increases (the superlattice bandgap decreases) significantly with increasing the In composition x in the Ga$_{1-x}$In$_x$Sb layers. This increase in the superlattice cutoff wavelength is almost entirely due to deformation potential effects resulting from the increase in the internal strain.

For a type II superlattice to be a useful material for long-wavelength infrared detection, it is extremely important that the superlattice layers be thin because the electron-hole optical matrix elements decrease exponentially with increasing layer thickness in type II superlattices. Inspection of Figure 4.4(a) reveals that the layer thickness required to obtain a given cutoff wavelength decreases with increasing the In composition x in the Ga$_{1-x}$In$_x$Sb layers, leading to improved optical absorption properties. Therefore, to achieve a given value of cutoff wavelength λ_c, InAs/Ga$_{1-x}$In$_x$Sb SLS with large values of x (large internal strain) are more advantageous because the layers are thinner and, consequently, the electron-hole optical matrix elements are larger.

Calculations of optical matrix elements as a function of cutoff wavelength for (001)-oriented InAs/Ga$_{1-x}$In$_x$Sb type II SLS are indicated in Figure 4.4(b) and reveal that the optical matrix elements are very rapidly decreasing functions of the superlattice cutoff wavelength λ_c. The superlattice cutoff wavelength is determined by the superlattice layer thickness, as illustrated in Figure 4.4(a).

Because small-bandgap superlattices can be achieved for InAs/Ga$_{1-x}$In$_x$Sb thin-layer SLS, reasonably large optical matrix elements can be obtained (see Figure 4.4(b)). Consequently, this materials system exhibits desirable optical properties at long cutoff wavelengths. In Figure 4.5, we compare the calculated optical absorption

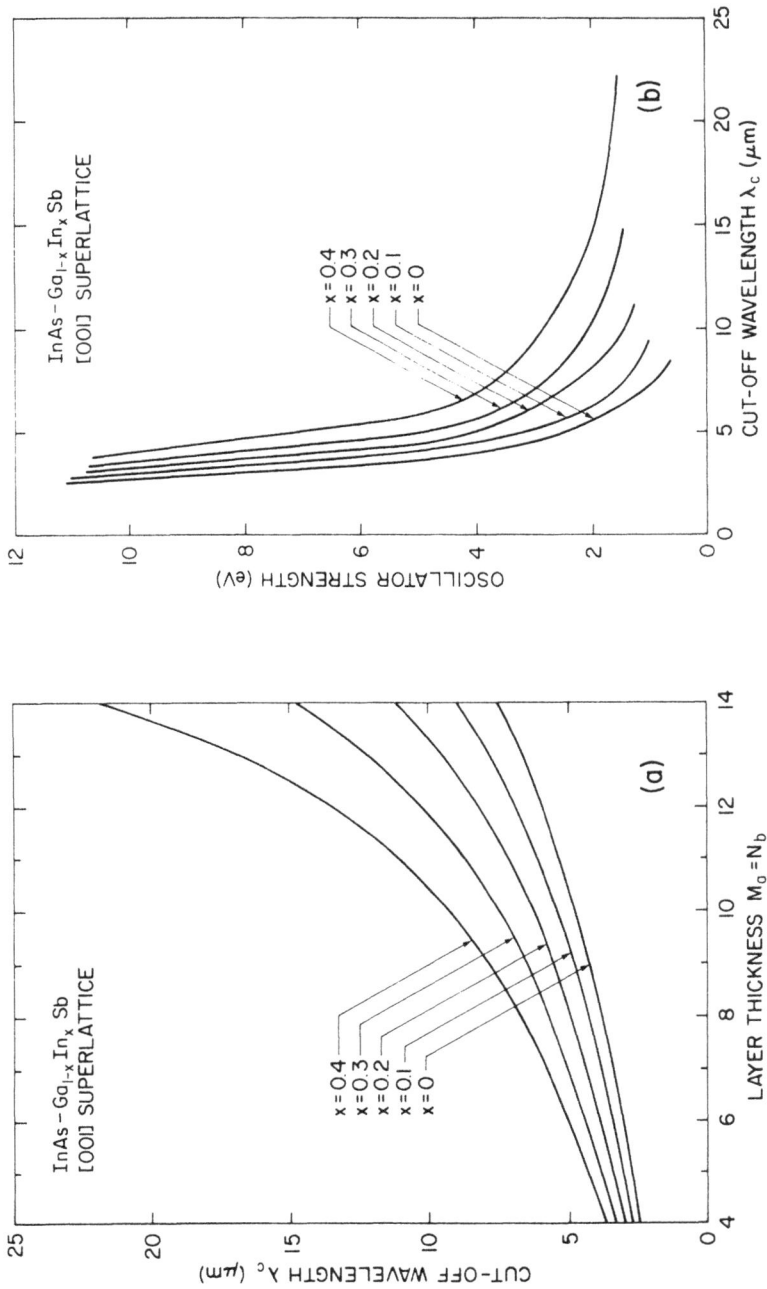

Figure 4.4 (a) Cutoff wavelength λ_c calculated as a function of InAs/Ga$_{1-x}$In$_x$Sb superlattice layer thickness for various In compositions x. (b) Optical matrix elements calculated as a function of the InAs/Ga$_{1-x}$Sb cutoff wavelength λ_c for various In compositions x. The superlattice consists of M_a layers of InAs alternating with $N_b = M_a$ layers of Ga$_{1-x}$In$_x$Sb grown along the (001) axis.

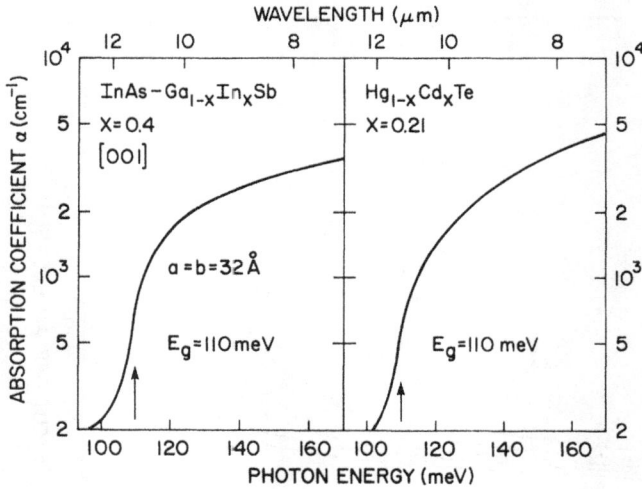

Figure 4.5 Calculated absorption coefficient as a function of photon energy for a InAs/Ga$_{0.60}$In$_{0.40}$Sb SLS and for a bulk Hg$_{1-x}$Cd$_x$Te alloy with $x = 0.21$. Both materials have a bandgap of 0.11 eV ($\lambda_c = 11.2 \, \mu$m). The superlattice consists of 11 layers of InAs alternating with 11 layers of Ga$_{1-x}$In$_x$Sb grown along the (001) axis.

coefficient for the InAs/Ga$_{0.60}$In$_{0.40}$Sb SLS with the absorption coefficient for a bulk Hg$_{1-x}$Cd$_x$Te alloy with $x = 0.21$. Both materials have a bandgap of 110 meV ($\lambda_c = 11.2 \, \mu$m). The free-standing superlattice consists of 11 layers of InAs alternating with 11 layers of Ga$_{1-x}$In$_x$Sb grown along the (001) axis. Because the superlattice is free-standing, the lattice mismatch is approximately 1.5% in each material. The calculations shown in Figure 4.5 reveal that, close to the threshold, the optical absorption properties of thin-layer InAs/Ga$_{0.60}$In$_{0.40}$Sb SLS are comparable to those of the Hg$_{1-x}$Cd$_x$Te alloy. The optical matrix element of the superlattice is smaller than that of the Hg$_{1-x}$Cd$_x$Te alloy. However, the parallel (in-plane) and perpendicular (along the growth axis) electron effective masses of the superlattice ($m_{c,\parallel}^{*\text{super}}/m_0 = 0.054$, $m_{c,\perp}^{*\text{super}}/m_0 = 0.042$, see later) are much larger than that of the Hg$_{1-x}$Cd$_x$Te alloy ($m_{c,\text{HgCdTe}}^* = 0.0088$). Consequently, the larger density of state in the InAs/Ga$_{0.60}$In$_{0.40}$Sb SLS overcomes the larger optical matrix elements in the Hg$_{1-x}$Cd$_x$Te alloy, and therefore the optical absorption properties of the two systems are comparable. The density of states in neither the superlattice nor the alloy is completely characterized by effective masses because of nonparabolicity effects, however. Only the first heavy-hole–to–first-conduction-band transitions contribute to the optical absorption spectrum over the small energy range shown in Figure 4.5 (the calculated energy band structure of the superlattice is shown in Figure 4.6). Exciton effects

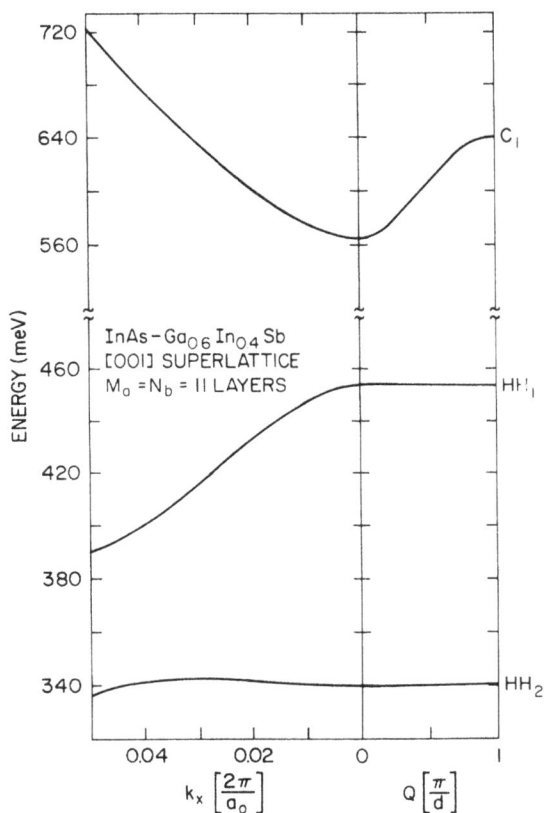

Figure 4.6 Calculated energy band structure for a free-standing superlattice consisting of 11 layers of
InAs alternating with 11 layers of $Ga_{1-x}In_xSb$ (≈ 32 Å of each material) grown along the
(001) axis. The optical absorption spectrum of this superlattice is shown in Figure 4.5.
Superlattice energy subband dispersions are calculated for wavevector parallel (Q) and perpendicular (k_x) to the (001) growth axis of the superlattice. The bandgap of this superlattice
corresponds to a cutoff wavelength of $\lambda_c = 11.2$ μm. The large energy dispersion of the
electron subband leads to favorable transport properties along the growth axis of the superlattice.

were not included either for the superlattice or the alloy as they are quite small in these small-bandgap materials.

4.3.3 Electronic and Transport Properties of InAs/Ga$_{1-x}$In$_x$Sb Type II Superlattices

We now discuss the electronic structure of thin-layer InAs/Ga$_{1-x}$In$_x$Sb type II SLS. Calculated superlattice energy subband dispersions for a (001)-oriented InAs/Ga$_{0.60}$In$_{0.40}$Sb SLS are indicated in Figure 4.6. The free-standing superlattice consists of 11 layers of InAs alternating with 11 layers of Ga$_{1-x}$In$_x$Sb ($\approx$32 Å of each material) grown along the (001) axis. The calculated optical absorption spectrum of this superlattice is illustrated in Figure 4.5. Superlattice energy subband dispersions are calculated for wave vectors parallel (Q) and perpendicular (k_x) to the (001) growth axis of the superlattice. The bandgap of this superlattice corresponds to a cutoff wavelength of $\lambda_c = 11.2$ μm. A particularly important, and advantageous, feature of the energy band structure shown in Figure 4.6 is the large energy dispersion of the electron subband along the growth direction. This large energy dispersion occurs because the superlattice layers are thin, and as a result, the electrons are not strongly quantum confined. This large conduction subband energy dispersion leads to good electron transport properties along the growth axis. The electron effective masses can be calculated from the energy dispersion relations shown in Figure 4.6 and are found to be nearly isotropic: $m_{c,\parallel}^{*\text{super}}/m_0 = 0.054$ and $m_{c,\perp}^{*\text{super}}/m_0 = 0.042$. These numerical values of electron effective masses are very convenient for the purpose of device design: they are large enough to lead to small diode tunneling currents and slow Auger recombination rates compared with existing Hg$_{1-x}$Cd$_x$Te-based diodes at the same cutoff wavelength [46] yet they are small enough to lead to large mobilities and diffusivity.

The energy band structure of InAs/Ga$_{0.60}$In$_{0.40}$Sb type II SLS shown in Figure 4.6 illustrates an important feature of SLS not afforded by conventional bulk semiconductors; namely, that it is possible to independently vary the superlattice bandgap and the superlattice effective mass. In particular, it is possible to *simultaneously* achieve *small energy gaps and large effective masses* in SLS. For example, in the InAs/Ga$_{0.60}$In$_{0.40}$Sb type II SLS whose calculated band structure is shown in Figure 4.6, the superlattice bandgap is small ($E_g^{\text{super}} \approx 0.1$ eV), yet the electron effective mass is relatively large ($m_{c,\parallel}^{*\text{super}} \approx m_{c,\perp}^{*\text{super}} \approx 0.05m_0$). This ability to independently tailor the superlattice bandgap and effective mass is absent in bulk semiconductors. In bulk semiconductors, the effective mass is related to the bulk bandgap by the relation

$$\frac{m_0}{m_{\text{bulk}}^*} \sim \frac{1}{E_g^{\text{bulk}}} \left[\frac{4}{3} \frac{P^2}{m_0} \right] \tag{4.28}$$

where P is the electron-hole momentum matrix element. Consequently, in bulk semi-conductor materials, a small bandgap leads to a small effective mass. For example, in $Hg_{1-x}Cd_xTe$ alloys used for far-infrared applications ($x = 0.21$), the electron effective mass corresponding to a bandgap of $E_g^{HgCdTe} \approx 0.11$ eV is $m_{c,HgCdTe}^* = 0.0088$.

4.3.4 Growth Considerations for $InAs/Ga_{1-x}In_xSb$ Type II Superlattices

We close our discussion of the properties of $InAs/Ga_{1-x}In_xSb$ SLS by considering some aspects of the hetero-epitaxial growth of this system. The substrate GaSb is a convenient material on which to grow $InAs/Ga_{1-x}In_xSb$ superlattices. For an In composition of $x = 0.40$, GaSb is lattice matched to the free-standing $InAs/Ga_{0.60}In_{0.40}Sb$ superlattice (with alternating layers of equal thicknesses) to better than 1%. Moreover, the lattice constant of GaSb is slightly smaller than that of the free-standing $InAs/Ga_{0.60}In_{0.40}Sb$ SLS. Consequently, the buffer layer would be under biaxial compression when epitaxially grown onto the GaSb substrate. This strain configuration is very desirable as it eliminates problems originating from microcrack formation [49].

4.4 EXPERIMENTAL SITUATION

Recently, thin-layer $InAs/Ga_{1-x}In_xSb$ SLS have been successfully grown by various groups [59–63]. Transmission electron microscopy results indicate that a much lower dislocation density is generated when GaSb substrates are used rather than GaAs substrates. Many physical characterization techniques have been used to verify the high degree of crystalline quality of $InAs/Ga_{1-x}In_xSb$ SLS grown onto GaSb substrates. In this section, we describe selected experimental observations that have confirmed the favorable optical properties of thin-layer $InAs/Ga_{1-x}In_xSb$ SLS for far-infrared detection applications. We present far-infrared ($\lambda_c \approx 10$ μm) photoluminescence, photoabsorption, and photoconductivity spectra that are in excellent quantitative agreement with the theoretical predictions presented in the preceding section.

Measured photoluminescence spectra and spectrally resolved photoconductive responses from three $InAs/Ga_{1-x}In_xSb$ SLS ($x = 0$ and $x = 0.25$) are shown in Figure 4.7. The luminescence peaks are associated with heavy-hole-to-conduction transitions across the $InAs/Ga_{1-x}In_xSb$ SLS bandgap. Inspection of the photoluminescence and photoconductivity spectra shown in Figure 4.7 reveals the systematic decrease of the superlattice bandgap as the In composition x is increased from 0 to 0.25 and as the InAs layer thickness is increased. These observations are in quantitative agreement with the theoretical predictions discussed in Section 4.3. A comparison between measured superlattice bandgaps as extracted from photoluminescence and photoconductivity and theoretically predicted bandgaps is provided in Table 4.2. A column

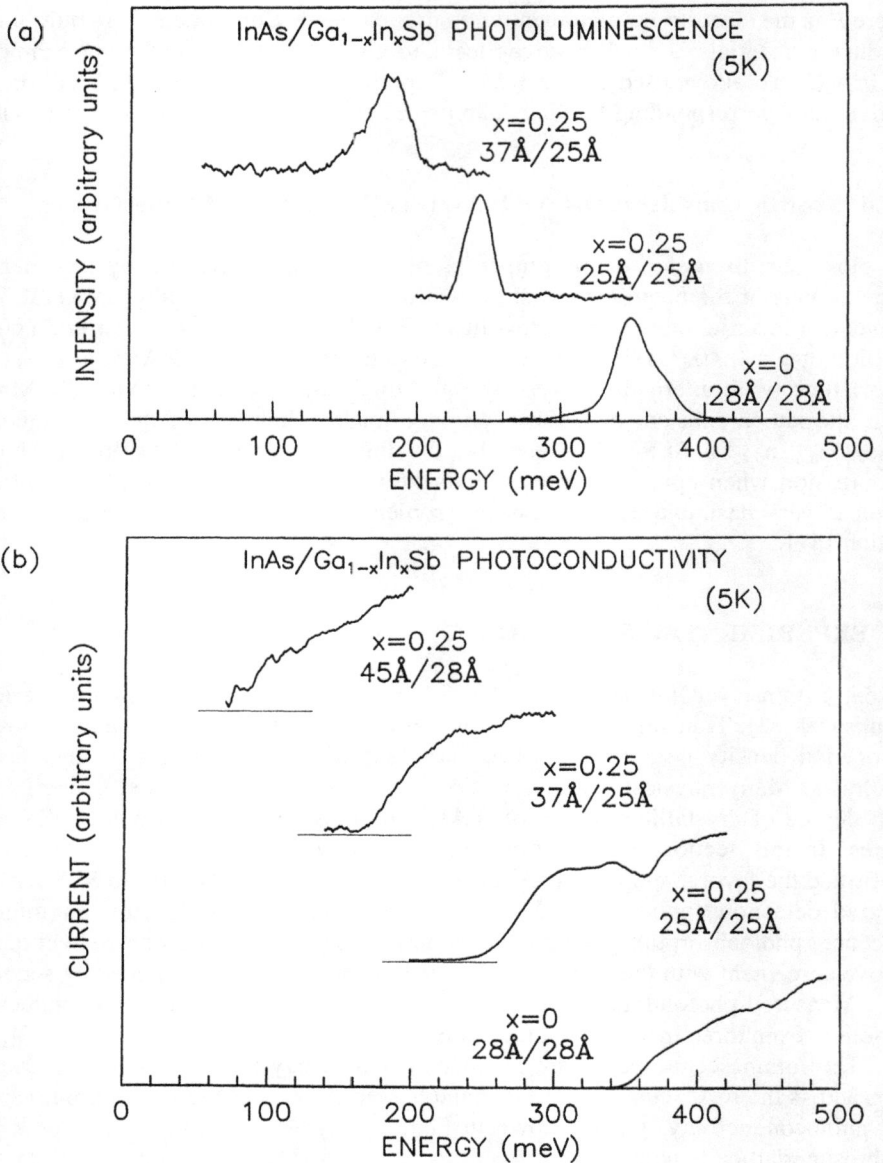

Figure 4.7 (a) Measured photoluminescence spectra from three InAs/Ga$_{1-x}$In$_x$Sb superlattices: 28 Å/ 28 Å InAs/GaSb, 25 Å/25 Å InAs/Ga$_{0.75}$In$_{0.25}$Sb, and 37 Å/25 Å InAs/Ga$_{0.75}$In$_{0.25}$Sb. (b) Measured photoconductivity spectra from the same superlattice samples and an additional 45 Å/28 Å InAs/Ga$_{0.75}$In$_{0.25}$Sb superlattice. All data were taken at 5 K. (Reproduced from Miles et al. [60]).

Table 4.2
Theoretical and Experimental Energy Bandgaps

| Layer Thickness (Å) | | | | Energy Bandgap (meV) | | |
InAs	$Ga_{1-x}In_xSb_{1-y}As_y$	x	y	PL	PC	Theory
28	28	0	0.07	330 ± 10	350 ± 10	320
25	25	0.25	0.08	240 ± 10	250 ± 10	280
37	25	0.25	0.05	150 ± 10	170 ± 10	180
41	25	0.25	0	—	110 ± 10	110
45	28	0.25	0	—	80 ± 10	100

Comparison between energy bandgaps extracted from photoluminescence and photoconductivity measurement and theoretically calculated for $InAs/Ga_{1-x}In_xSb$ SLS examined in Figures 4.7 and 4.8 (reproduced from Miles et al. [60]).

has been added to Table 4.2 to identify the extent of arsenic incorporation within the antimonide layers. Inspection of Table 4.2 reveals the excellent agreement between theoretically predicted superlattice bandgaps and experimental observations.

Experimental and theoretical photoabsorption spectra for a 41 Å/25 Å $InAs/Ga_{0.75}In_{0.25}Sb$ SLS are illustrated in Figure 4.8. These results clearly indicate that both the observed superlattice energy bandgap of 108 meV and the measured absorption coefficient are in excellent agreement with the theoretical predictions presented in the preceding section. The superlattice absorption edge is abrupt, reaching a value of $\alpha \approx 2 \times 10^3$ cm^{-1} at 10 μm. This value is comparable to that of bulk $Hg_{1-x}Cd_xTe$ alloy with the same gap and supports the use of $InAs/Ga_{1-x}In_xSb$ as far-infrared detector materials. Moreover, the magnitude of the absorption coefficient for the superlattice is consistent with the prediction that the perpendicular (along the growth axis) electron effective mass in the superlattice, $m_{c,\perp}^{*\,super}$, is greater than that of the $Hg_{1-x}Cd_xTe$ alloy. As mentioned in the preceding section, the larger density of states in the $InAs/Ga_{0.60}In_{0.40}Sb$ SLS overcomes the larger optical matrix elements in the $Hg_{1-x}Cd_xTe$ alloy, and therefore the optical absorption properties of the two systems are comparable. The larger superlattice effective masses present potential for improved performance over conventional $Hg_{1-x}Cd_xTe$, as they will reduce tunneling dark currents and Auger recombination rates.

In Figure 4.9, we show the measured optical absorption coefficient of a $InAs/Ga_{1-x}In_xSb$ SLS at $T = 300$ K and 5 K as a function of photon energy. The superlattice consists of 37 Å of InAs alternating with 19 Å of $Ga_{0.65}In_{0.35}Sb$ grown onto a (001)-oriented p-type GaSb substrate. At 5 K, a clear absorption threshold is observed at approximately 121 meV. The absorption threshold is found to shift to lower energy with increasing temperature, which is typical of III–V bulk semiconductors

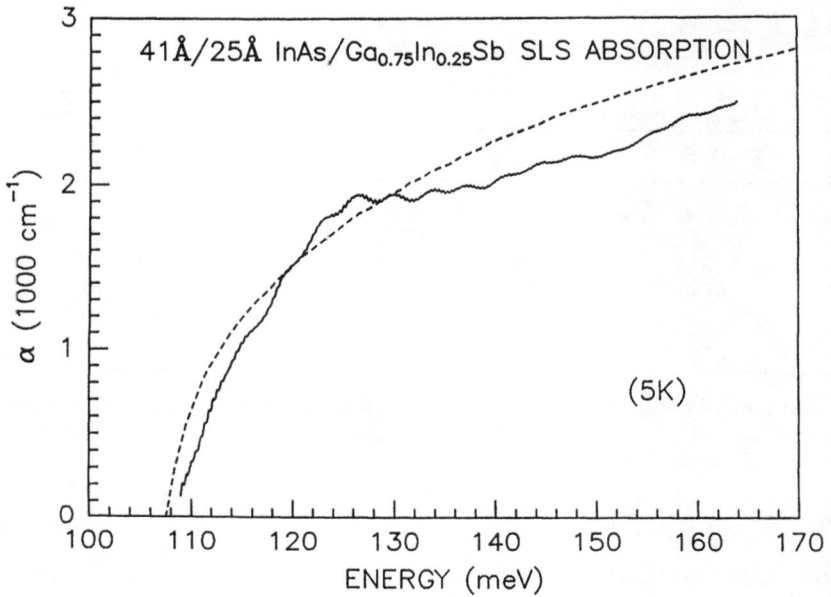

Figure 4.8 Experimental (solid line) and theoretical (dashed line) absorption coefficients for a 41 Å/ 25 Å InAs/Ga$_{0.75}$In$_{0.25}$Sb superlattice. The absorption coefficient at 10 μm is comparable to that of bulk Hg$_{1-x}$Cd$_x$Te with the same bandgap. (Reproduced from Miles et al. [60]).

[61]. A theoretical calculation for the absorption coefficient of a superlattice consisting of 12 molecular layers (36.1 Å) of InAs alternating with 6 molecular layers (19.1 Å) of Ga$_{0.65}$In$_{0.35}$Sb at $T = 0$ K is also indicated in Figure 4.9. The measured absorption coefficient at 5 K is in good agreement with the theoretical result. There are no adjustable parameters in the calculations. The lateral photoconductive response at 5 K measured as a function of photon energy is shown in Figure 4.10 for the InAs/Ga$_{0.65}$In$_{0.35}$Sb SLS. A clear photoconductive threshold is observed at about 124 meV. This photoconductive threshold is in good agreement with the absorption threshold (121 meV) and with the calculated bandgap (126 meV).

The selected experimental results discussed in this section demonstrate that, thus far, the measured properties of thin-layer InAs/Ga$_{1-x}$In$_x$Sb SLS indicate that this superlattice system shows great promise for long-wavelength infrared detection. In particular, it has been experimentally demonstrated that the InAs/Ga$_{1-x}$In$_x$Sb superlattice exhibits absorption coefficients comparable to those of the Hg$_{1-x}$Cd$_x$Te alloy in the $\lambda_c = 10$ μm region. All experimental observations reported so far are in excellent quantitative agreement with the theoretical predictions presented in Section 4.3.

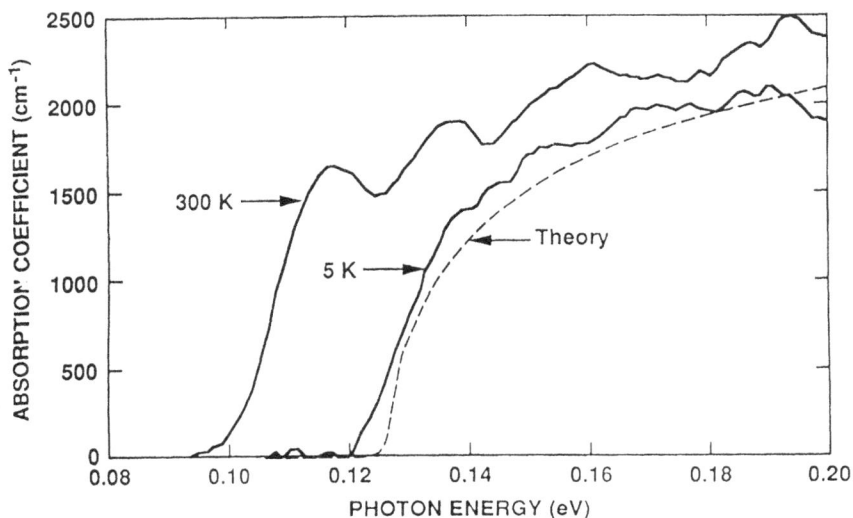

Figure 4.9 Measured superlattice optical absorption coefficient of a InAs/Ga$_{0.65}$In$_{0.35}$Sb SLS at T = 300 K and 5 K as a function of photon energy. Calculation of the superlattice absorption coefficient at 0 K as a function of photon energy is also indicated (dashed line). (Reproduced from Campbell et al. [61]).

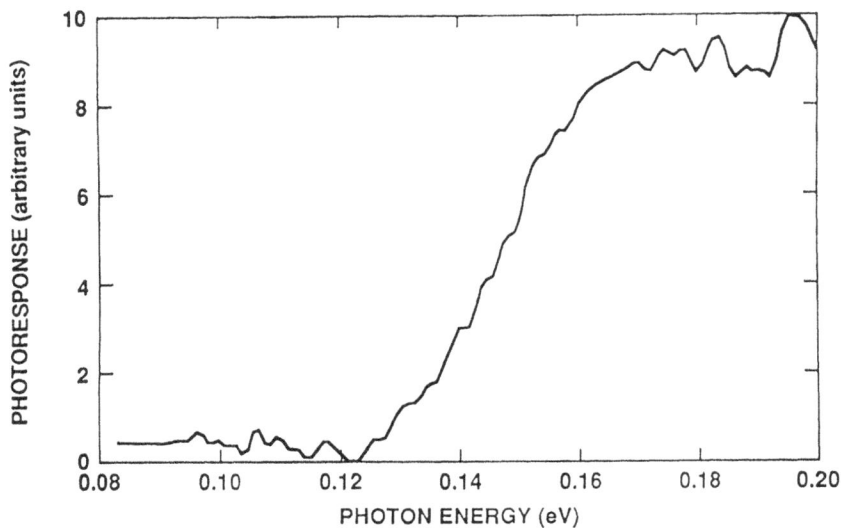

Figure 4.10 Measured superlattice photoconductive response as a function of photon energy at 5 K for a InAs/Ga$_{0.65}$In$_{0.35}$Sb SLS. (Reproduced from Campbell et al. [61]).

4.5 CONCLUSION

An overview of the modern theoretical methods that serve as the basis for the calculation of the electronic structure of strained-layer semiconductor superlattices was reviewed. These electronic structure methods were used to predict that the electronic and optical properties of $InAs/Ga_{1-x}In_xSb$ type II SLS are very favorable for long-wavelength λ_c exceeding 10 μm infrared detection. In particular, the optical absorption coefficient exhibited by $InAs/Ga_{1-x}In_xSb$ SLS is found to be as large as that of conventional $Hg_{1-x}Cd_xTe$ alloys in the $\lambda_c = 10$–12 μm region. An additional desirable feature of the $InAs/Ga_{1-x}In_xSb$ SLS is the nearly isotropic electron effective mass in the superlattice: $m_{c,\parallel}^{*\,\mathrm{super}}/m_0 = 0.054$ and $m_{c,\perp}^{*\,\mathrm{super}}/m_0 = 0.042$. These numerical values of electron effective masses are very convenient for the purpose of device design: they are large enough to lead to small diode tunneling currents and slow Auger recombination rates compared with existing $Hg_{1-x}Cd_xTe$-based diodes at the same cutoff wavelength, yet they are small enough to lead to large mobilities and diffusivity. Available experimental observations of the optical properties of $InAs/Ga_{1-x}In_xSb$ SLS are in excellent quantitative agreement with these theoretical predictions. Overall, $InAs/Ga_{1-x}In_xSb$ thin-layer SLS are expected to exhibit materials properties for long-wavelength infrared detection superior to conventional $Hg_{1-x}Cd_xTe$ alloys.

REFERENCES

[1]   Esaki, L., in *Proc. 17th Int. Conf. Physics of Semiconductors*, ed. D.J. Chadi and W.A. Harrison, Springer-Verlag, New York, 1984, p. 473.

[2]   Chang, L.L., and K. Ploog, eds., *Proc. NATO Advanced Study Institute on Molecular Beam Epitaxy and Heterostructures*, Martinus Nijhoff, Leyden, 1985.

[3]   Van der Ziel, J.P., R. Dingle, R.C. Miller, W. Wiegmann, and W.A. Nordland, Jr., *Appl. Phys. Lett.*, Vol. 26, 1975, p. 463.

[4]   Holonyak, N., Jr., R.M. Kolbas, R.D. Dupuis, and P.D. Dapkus, *IEEE J. Quantum Electron.*, Vol. QE-16, 1980, p. 170.

[5]   Tsang, W.T., *Appl. Phys. Lett.*, Vol. 39, 1981, p. 786.

[6]   Miller, D.A.B., D.S. Chemla, T.C. Damen, A.C. Gossard, W. Wiegman, T.H. Wood, and C.A. Burrus, *Phys. Rev. Lett.*, Vol. 53, 1984, p. 2173.

[7]   Miller, D.A.B., D.S. Chemla, T.C. Damen, A.C. Gossard, W. Wiegman, T.H. Wood, and C.A. Burrus, *Appl. Phys. Lett.*, Vol. 45, 1984, p. 13.

[8]   Miller, D.A.B., D.S. Chemla, T.C. Damen, A.C. Gossard, W. Wiegman, T.H. Wood, and C.A. Burrus, *Phys. Rev. B*, Vol. 32, 1985, p. 1043.

[9]   Chang, Y.C., J.N. Schulman, and U.J. Efron, *Appl. Phys.*, Vol. 62, 1987, p. 453.

[10]   Wood, T.H., R.W. Tkach, and A.R. Chraplyvy, *Appl. Phys. Lett.*, Vol. 50, 1987, p. 798.

[11]   Mailhiot, C., and D.L. Smith, *Phys. Rev. B*, Vol. 37, 1988, p. 10415.

[12]   Chemla, D.S., D.A.B. Miller, P.W. Smith, A.C. Gossard, and W. Wiegmann, *IEEE J. Quantum Electron.*, Vol. QE-20, 1984, p. 265.

[13]   Smith, D.L., and C. Mailhiot, *J. Vac. Sci. Technol. A*, Vol. 5, 1987, p. 2060.

[14]   Smith, D.L., and C. Mailhiot, *Phys. Rev. Lett.*, Vol. 58, 1987, p. 1264.

[15]   Bloss, W.L., and L.J. Friedman, *J. Appl. Phys.*, Vol. 41, 1982, p. 1023.
[16]   Bloss, W.L., and L.J. Friedman, *J. Vac. Sci Technol. B*, Vol. 1, 1983, p. 150.
[17]   Chang, Y.C., *J. Appl. Phys.*, Vol. 58, 1985, p. 499.
[18]   Chang, Y.C., *Appl. Phys. Lett.*, Vol. 46, 1985, p. 710.
[19]   Frank, F.C., and J.H. van der Merwe, *Proc. R. Soc. London*, Ser. A, Vol. 198, 1949, p. 216.
[20]   Frank, F.C., *J. Appl. Phys.*, Vol. 34, 1963, p. 117.
[21]   Matthews, J.W., and A.E. Blakeslee, *J. Cryst. Growth*, Vol. 27, 1974, p. 118.
[22]   Matthews, J.W., and A.E. Blakeslee, *J. Cryst. Growth*, Vol. 29, 1975, p. 273.
[23]   Matthews, J.W., and A.E. Blakeslee, *J. Cryst. Growth*, Vol. 32, 1976, p. 265.
[24]   Osbourn, G.C., *J. Appl. Phys.*, Vol. 53, 1982, p. 1586.
[25]   Osbourn, G.C., *J. Vac. Sci. Technol.*, Vol. 21, 1982, p. 469.
[26]   Osbourn, G.C., *Phys. Rev. B*, Vol. 27, 1983, p. 5126.
[27]   Osbourn, G.C., *J. Vac. Sci. Technol. B*, Vol. 2, 1984, p. 176.
[28]   Osbourn, G.C., *J. Vac. Sci. Technol. A*, Vol. 3, 1985, p. 826.
[29]   Osbourn, G.C., H.M. Biefeld, and P.L. Gourley, *Appl. Phys. Lett.*, Vol. 41, 1982, p. 172.
[30]   Fritz, I.J., L.R. Dawson, and T.E. Zipperian, *Appl. Phys. Lett.*, Vol. 43, 1983, p. 846.
[31]   Sai-Halasz, G.A., L. Esaki, and W.A. Harrison, *Phys. Rev. B*, Vol. 18, 1978, p. 2812.
[32]   Smith, D.L., and C. Mailhiot, *Rev. Mod. Phys.*, Vol. 62, 1990, p. 173.
[33]   Smith, D.L., and C. Mailhiot, *Phys. Rev. B*, Vol. 33, 1986, p. 8345.
[34]   Mailhiot, C., and D.L. Smith, *Phys. Rev. B*, Vol. 33, 1986, p. 8360.
[35]   Cohen, M.L., and T.K. Bergstresser, *Phys. Rev.*, Vol. 141, 1966, p. 789.
[36]   Löwdin, P.O., *J. Chem. Phys.*, Vol. 19, 1951, p. 1396.
[37]   Mailhiot, C., and D.L. Smith, *Phys. Rev. B*, Vol. 35, 1987, p. 1242.
[38]   Mailhiot, C., and D.L. Smith, *Phys. Rev. B*, Vol. 36, 1987, p. 2942.
[39]   Mailhiot, C., and D.L. Smith, *J. Vac. Sci. Technol. B*, Vol. 5, 1987, p. 1268.
[40]   Mailhiot, C., and D.L. Smith, in *Crystal Properties and Preparation*, ed. R.M. Biefeld, Vol. 21, Trans Tech Publications, Switzerland, 1989, pp. 1–28.
[41]   Mailhiot, C., and D.L. Smith, *J. Vac. Sci. Technol. A*, Vol. 7, 1989, p. 445.
[42]   Mailhiot, C., and D.L. Smith, *CRC Critical Reviews in Solid State and Materials Sciences*, ed. J.E. Greene, Vol. 16, 1990, pp. 131–160.
[43]   Smith, D.L., and C.J. Mailhiot, *J. Appl. Phys.*, Vol. 62, 1987, p. 2545.
[44]   Bir, G.L., and G.E. Pikus, *Symmetry and Strain Effects in Semiconductors*, John Wiley and Sons, New York, 1974.
[45]   Reine, M.B., A.K. Sood, and T.J. Tredwell, in *Semiconductors and Semimetals*, ed. R.K. Willardson and A.C. Beer, Academic Press, New York, 1981, Vol. 18, p. 201.
[46]   Smith, D.L., T.C. McGill, and J.N. Schulman, *Appl. Phys. Lett.*, Vol. 43, 1983, p. 180.
[47]   Schulman, J.N., and T.C. McGill, *Appl. Phys. Lett.*, Vol. 34, 1979, p. 663.
[48]   Faurie, J.P., S. Sivananthan, and J. Reno, *J. Vac. Sci. Technol. A*, Vol. 4, 1986, p. 2096.
[49]   Osbourn, G.C., L.R. Dawson, H.M. Biefeld, T.E. Zipperian, I.J. Fritz, and B.L. Doyle, *J. Vac. Sci. Technol. A*, Vol. 5, 1987, p. 3150.
[50]   Sai-Halasz, G.A., R. Tsu, and L. Esaki, *Appl. Phys. Lett.*, Vol. 30, 1977, p. 651.
[51]   Sai-Halasz, G.A., L.L. Chang, J.M. Walter, C.A. Chang, and L. Esaki, *Solid State Commun.*, Vol. 27, 1978, p. 935.
[52]   Kawai, N.J., L.L. Chang, G.A. Sai-Halasz, C.A., Chang, and L. Esaki, *Appl. Phys. Lett.*, Vol. 36, 1980, p. 369.
[53]   Schulman, J.N., and Y.C. Chang, *Phys. Rev. B*, Vol. 31, 1981, p. 2056.
[54]   Arch, D.K., G. Wicks, T., Tonaue, and J.L. Staudenman, *J. Appl. Phys.*, Vol. 58, 1985, p. 3933.
[55]   Kroemer, H., *J. Vac. Sci. Technol. B*, Vol. 2, 1984, p. 433.
[56]   Gobeli, G.W., and F.G. Allen, *Phys. Rev. A*, Vol. 137, 1965, p. 245.

[57]   Harrison, W.A., *J. Vac. Sci. Technol. B*, Vol. 3, 1985, p. 1232.

[58]   Van de Walle, C.G., and R.M. Martin, *Phys. Rev. B*, Vol. 35, 1987, p. 8154.

[59]   Chow, D.H., R.H. Miles, J.R. Söderström, and T.C. McGill, *Appl. Phys. Lett.*, Vol. 56, 1990, p. 1418.

[60]   Miles, R.H., D.H. Chow, J.N. Schulman, and T.C. McGill, *Appl. Phys. Lett.*, Vol. 57, 1990, p. 801.

[61]   Campbell, I.H., I. Sela, B.K. Laurich, D.L. Smith, C.R. Bolognesi, L.A. Samoska, A.C. Gossard, and H. Kroemer, *Appl. Phys. Lett.*, Vol. 59, 1991, p. 846.

[62]   Fashe, R., J.T. Zborowski, T.D. Golding, H.D. Shih, P.C. Chow, K. Matsuichi, B.C. Covington, A. Chi, J. Zheng, and H.F. Schaake, *J. Cryst. Growth*, Vol. 111, 1991, p. 677.

[63]   Golding, T.D., H.D. Shih, J.T. Zborowski, W.C. Fan, C.C. Horton, B.C. Covington, A. Chi, J.M. Anthony, and H.F. Schaake, *J. Vac. Sci. Technol.*, 1991.

Chapter 5

Infrared Detectors Using SiGe/Si Quantum Well Structures

Kang L. Wang and R.P.G. Karunasiri

Department of Electrical Engineering
University of California, Los Angeles

Infrared detection and imaging have been demonstrated using intersubband transitions based on III–V heterojunction quantum wells and superlattices. There are many advantages to using such structures. Among them is the convenience of energy tuning and process control as well as potential affordable cost. The major drawback of the intersubband transition, as in the case of AlGaAs/GaAs multiple quantum well structures, is the selection rule of the Γ band, forbidding the detection of normal incident light. Therefore, usually, a grating coupler is needed for coupling the normal incident light in focal plane application. The recent advent of Si-based, low-temperature epitaxy techniques gives great successes in the growth of strained SiGe layers. The strained-layer $Si_{1-x}Ge_x/Si$ heterostructures and multiple quantum well structures have created a great deal of interest due to the potential of monolithic integration with the conventional silicon very large scale integration signal processing technology. For infrared detection applications, the intersubband transition using $Si_{1-x}Ge_x/Si$ offers normal incidence detection possibilities. In this chapter, we will discuss the principle and applications of Si and SiGe quantum well structures for infrared detection using intersubband transition. We will begin with a brief introduction of the current status of the SiGe material system, followed by a brief discussion of some relevant basic properties of the strained layers. Elements of intersubband transition at the Γ point will be reviewed. The physics of intersubband transition in Si-based quantum wells and superlattices will then be discussed for both n-type and p-type materials. Several normal incidence detection mechanisms will be illustrated: nonvanishing off-diagonal elements of the effective mass tensor in n-type, free-carrier absorption, and intervalence band transition for p-type dopants. The effect of the strain in determining the occupancy of the valleys will be described. For illustration, oscillator strength is calculated for several different structures. The experimental observation of normal incident intersubband transition in SiGe/Si quantum wells and δ-doped Si layers will be discussed. The importance of many-body effects in determination of transition energy in very heavily doped structures and δ-doped layers will also be shown for tuning of a wide range of transition energy (from 2 μm to tens of μm and perhaps longer). Finally, infrared detector using SiGe/Si multiple quantum well structures will be illustrated.

5.1 INTRODUCTION

The advances of epitaxial growth techniques make possible the synthesis of artificial quantum well and superlattice structures with novel optical and electrical properties not existing in the bulk constituents. Initially, much of the effort focused on the lattice matched materials (e.g., GaAs/AlGaAs and InP/InGaAsP), and later it expanded into lattice mismatch systems as the growth of strained layers came to be understood. Among the lattice mismatch systems, $Si_{1-x}Ge_x$/Si heterostructures have created an added interest due to their potential to be integrated with the conventional silicon very large scale integration (VLSI) electronics. With the current advances in silicon molecular beam epitaxy and other low-temperature epitaxy (such as *chemical vapor deposition*, CVD) techniques, device quality $Si_{1-x}Ge_x$/Si layers with controlled strain were obtained. Many optical and electronic devices, such as *modulation doped field effect transistors* (MODFET) [1, 2], *heterojunction bipolar transistors* (HBT) [3], photodetectors and photomodulators [4, 5], quantum well *metal oxide semiconductor field effect transistors* (MOSFET) [6], and tunneling devices [7, 8] have been demonstrated with the $Si_{1-x}Ge_x$/Si system. More recently, intersubband transition was also demonstrated, and indeed potential normal incidence detectors using free carrier absorption followed by photoemission over a Schottky barrier have been realized [9, 10]. In addition, there are recent advances in the understanding of intersubband transition physics, and several additional normal incidence detection mechanisms have been established. In this chapter, we will present a brief review of the development of the SiGe growth with different techniques. Some basic properties of strained SiGe relevant to the subsequent discussions of intersubband transition will be given next. Then a general discussion of the Γ point intersubband transition for both n-type and p-type doping will be followed. The principles for normal incidence detection using both n- and p-type dopants will be discussed. For n-type doping, the use of the off-diagonal elements of the effective mass tensor to induce the intersubband transition will be presented and a (110)-oriented structure is used to illustrate the principle. The substrate orientation for maximizing the normal incidence absorption is shown for the Si-like valley (X-valley); similar results are obtained for Ge-like valley (L-valley). For p-type dopants, the use of free carrier absorption and photoemission process is discussed. Another normal incidence detection using inter-valence band transition (from the heavy hole to the light hole or to the spin-orbital band) is demonstrated. The use of δ-doping is shown to have a substantially large increase of transition energy for Si-based material in comparison with the AlGaAs/GaAs quantum wells, due to many-body effects. This large increase is unique for Si-based material owing to the high density of states, and this large effect is used to tune the transition energy range from far-IR to close to 2 μm, which was otherwise believed to be impossible in the past. Finally, detector structures and experimental demonstrations are presented. The chapter concludes with a discussion assessing the status of the use of SiGe quantum wells for infrared focal-plane technology.

5.2 GROWTH OF SiGe STRAINED LAYERS

The growth of $Si_{1-x}Ge_x$ had been attempted earlier by Kasper and Herzog [11, 12] using MBE above 750°C and also by others using high-temperature CVD techniques [13]; however, in these approaches three-dimensional growth and islanding were observed. In the early 1980s Bean et al. [14] first reported the successful low-temperature commensurate growth of $Si_{1-x}Ge_x$ on Si at 580°C. The important feature of the latter work and all other recent works [7, 15, 16] is the low-temperature virtue, which enables the films to be grown pseudomorphically or in two-dimension coherent (commensurate) growth. Successful demonstrations of high-quality epitaxial SiGe films by low-temperature epitaxy have stimulated a great deal of interest in several potential applications in advancing the speed performance and techniques in Si-based technology. Owing to the lattice mismatch of 4.17% between Ge and Si at room temperature, the epitaxial growth of $Si_{1-x}Ge_x$ on Si may take any of three forms. At sufficiently high temperatures, islanding growth often occurs. At low temperatures, the growth occurs commensurately, or in other words, the in-plane growth lattice will be strained to match that of the substrate and the vertical lattice constant will be elastically strained (although free of stress), resulting in a tetragonal distortion of the cubic lattice. The growth proceeds at a reasonably low temperature until the thickness of the film exceeds a critical value such that the energy in the strained layer becomes greater than the energy needed for dislocations to develop. In the latter case, a generation of misfit dislocations occurs at the interface, and the lattice resumes the natural lattice constant as determined by Vegard's law. For this case, a relatively high density of dislocations remain in the film. Several models are available to predict the critical thickness for various Ge contents [17, 18, 19], and the critical thickness is meaningful only for equilibrium. The fact that all the growth techniques used are in nonequilibrium conditions suggest that the critical thickness is not unique for prediction of the maximum pseudomorphic film thickness for a given Ge content. Indeed, experimentally, the so-called critical thickness is not a reproducible number, owing to the nonequilibrium nature of the growth. The pseudomorphic thickness depends on the growth temperature, pressure, and nucleation mechanisms of misfit dislocations [20, 21]. As shown in Figure 5.1, we illustrate

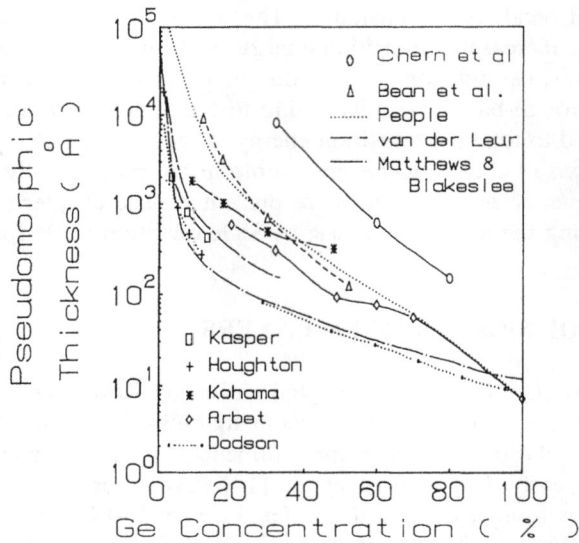

Figure 5.1 Pseudomorphic thickness (or so-called critical thickness) obtained by different investigators: Chern (et al.) [21]; Bean (et al.) [14]; People (and Bean) [18, 36], Matthews and Blakeslee [105]; Van der Leur (et al.) [109]; Dodson (and Taylor) [110]; Kohama (et al.) [111]; Arbet (et al.) [112]; and Houghton (et al.) [113].

the pseudomorphic thickness obtained from differently calculated values as well as several experimentally obtained data as a function of Ge concentration, illustrating a wide variation from various reports. One recent report from this group shows that a low temperature with a low background pressure results in an order of magnitude increase of the pseudomorphic film thickness [21]. This substantiates the nonequilibrium nature of the growth and its dependences on the growth conditions. Therefore, the so-called critical thickness can be used only as a guide. For multiple layer growth, the pseudomorphic thickness can be estimated by using the average Ge composition of a single period of the structure (i.e., the average Ge content of a period, $x_{Ge} = (x_1 d_1 + x_1 d_2)/(d_1 + d_2)$, where x and d are the Ge content and thickness of each constituent layer, respectively) [19]. If a relaxed SiGe buffer layer is used with a Ge fraction equal to the preceding x_{Ge}, strain symmetrization in the multilayered film can be accomplished [22]. In this case, the net strain of the multilayered film on top of the buffer layer vanishes and the total thickness of the film is no longer limited.

Comparing the two most successful techniques, MBE and CVD, the MBE is a physical process, and the deposited species from an electron gun or a Knudsen cell arrive at the surface in the molecular flow and are subsequently deposited onto the surface. The absorbed species, called *adatoms*, migrate to the surface step edges and are incorporated as the film growth proceeds. The doping species are also evaporated by knudsen cells and the incorporation depends on the surface migration and diffusion. For some species, a surface resulting in segregation of the species can occur. Although there is no problem in p-type doping using boron, the surface segregation of n-type dopant Sb causes difficulties in controlling the doping profile particularly at a relatively high temperature (580°C). Details of Si MBE techniques and means for improving doping can be found in several review papers and edited volumes [23, 24].

One of the best-known low-temperature CVD technique is the *ultra-high vacuum* (UHV) low-pressure CVD, which uses the clean ambient feature of MBE to prevent contamination species such as O (and C) from arriving at the sample surface. A low-water partial pressure is crucial for eliminating oxygen incorporation to obtain high-quality films, as discussed by Ghidini and Smith [25]. Under UHV, the epitaxial growth can be done at temperatures below 550°C, and the pseudomorphic growth can be accomplished without significant O and C contaminants. Normally, the UHV-CVD is done in a hot wall reactor in a gaseous ambient of 5×10^{-2} Pa (close to the laminar flow condition), and the deposition in this case is accomplished by pyrolysis of silane (disilane, or dichlorosilane) and germane [15, 26, 27]. Doping in this instance is done by using diborane or phosphine and arsine for p-type and n-type, respectively, in line with other CVD techniques.

The *limited-reaction processing* (LRP) [16] is another version of chemical vapor deposition. Specific to this process is the use of the wafer temperature rather than the gas flow switching to initiate and terminate growth in a cold wall reactor. In this technique, the cleaning is done at a very high temperature (close to 1200°C) in hydrogen. The growth is done usually at the millitorr to 10 torr range [16, 28]. Si deposition was originally done at above 740°C to reduce the oxygen and carbon concentration. For $Si_{1-x}Ge_x$ growth, the temperature range of 625–650°C is preferred to avoid strain relaxation or generation of a large density of misfit dislocations and hence island growth. The gases used are dichlorolsilane and germane and H_2, and the dopants are introduced by diborane and arsine for p- and n-types, respectively. The major problems are the removal of the unwanted species, such as O and C, at low temperatures below 900°C, for $Si_{1-x}Ge_x$ growth. There can be a very high concentration of O of about 2×10^{20} cm^{-3} (or about 1 atomic percent). With improved vacuum and reduced water partial pressure using a loadlock, the oxygen concentration in the film has been substantially reduced and the growth temperature has been reduced to below 650°C. For multilayer growth, such as bipolar transistors, the top Si layer, for example, is grown at above 850°C for a short time (e.g., 5.5 min.). The reason for using the high temperature for overlayer growth is to reduce the O concentration and increase the doping density [16, 28]. Gaseous beam molecular beam epitaxy, which combines the advantages of CVD and solid source MBE, may be a possible candidate for the next evolution of the low-temperature epitaxy system. Hirayama and Aizaki [29, 30, 31] have demonstrated the use of a SiH_4 or Si_2H_6

gaseous species in lieu of an electron gun Si source for the deposition. The p-type doping is done similarly to the CVD processes. The n-type doping problem appears to have been solved by the use of phosphine (PH_3).

Using these techniques many heterojunction devices, quantum wells, and superlattices have been fabricated. In comparison of all these techniques, MBE seems to be the best experimental tool, particularly for preparation of quantum well structures. MBE offers the best thickness control and reproducibility as it operates in the molecular flow condition. In contrast, CVD gives a higher growth rate and hence a greater throughput. Particularly, multiple wafer processing is usually done in UHV-CVD, and the throughput is further increased. However, for precise control of thickness, as needed in quantum well engineering for intersubband infrared detection, MBE still offers the most convenient and reproducible approach.

5.3 BAND STRUCTURES AND PROPERTIES OF SiGe LAYERS UNDER STRAIN

The band structures of $Si_{1-x}Ge_x$ change under strain, and this change depends on substrate orientation. The band splitting under strain can be calculated with available deformation potential data. The knowledge of the change of the band structure under strain is important for understanding intersubband transition and its engineering.

5.3.1 Conduction Band Splitting of Si and Ge Under Strain

According to deformation potential theory [32, 33], the energy shift of the conduction band at valley i for an arbitrary homogeneous deformation can be expressed as

$$\Delta E_c^i = [\Xi_d \mathbf{1} + \Xi_u \{a_i a_i\}] : \varepsilon \tag{5.1}$$

where $\mathbf{1}$ is the unit tensor, a_i is a unit vector parallel to the $\vec{k}$ vector of the valley i, ε is the strain tensor, Ξ_d and Ξ_u are deformation potentials, and $\{ \ \}$ denotes a dyadic product. The shift of the mean energy of the band extrema is

$$\Delta E_c^0 = (\Xi_d + \Xi_u/3)\mathbf{1} : \varepsilon \tag{5.2}$$

When Si or Ge are coherently grown on a substrate with the same diamond structure but different lattice constant, the strain tensor is similar to that of the biaxial stress case. The quantity $\Delta E_c^i - \Delta E_c^0$ is given in Table 5.1 for Si and Ge when the growth direction is in (001), (111), and (110).

In Table 5.1, the strain tensors for the three growth directions are

$$\varepsilon_{001} = \begin{pmatrix} \varepsilon_x' & 0 & 0 \\ 0 & \varepsilon_x' & 0 \\ 0 & 0 & \varepsilon_z' \end{pmatrix} \tag{5.3}$$

Table 5.1

Conduction Band Splittings in Strained SiGe Alloys on (100), (111), and (110) Substrates

Valleys	$\Delta E_c^i - \Delta E_c^0$
(100) substrate	
(001), (00$\bar{1}$)	$2/3\Xi_u(\varepsilon_{zz} - \varepsilon_{xx})$
(010), (0$\bar{1}$0)	$-1/3\Xi_u(\varepsilon_{zz} - \varepsilon_{xx})$
($\bar{1}$00), (100)	
(111) substrate	
[111]	$2\Xi_u\varepsilon_{xy}$
[11$\bar{1}$], [1$\bar{1}$1]	$-2/3\Xi_u\varepsilon_{xy}$
[$\bar{1}$11]	
(110) substrate	
(001), (00$\bar{1}$)	$2/3\Xi_u(\varepsilon_{zz} - \varepsilon_{xx})$
(010), (0$\bar{1}$0)	$-1/3\Xi_u(\varepsilon_{zz} - \varepsilon_{xx})$
(100), ($\bar{1}$00)	
(111), (11$\bar{1}$)	$2/3\Xi_u\varepsilon_{xy}$
($\bar{1}$11), (1$\bar{1}$1)	$-2/3\Xi_u\varepsilon_{xy}$

$$\varepsilon_{111} = \begin{pmatrix} 2\varepsilon_x' + \varepsilon_z' & -\varepsilon_x' + \varepsilon_z' & -\varepsilon_x' + \varepsilon_z' \\ -\varepsilon_x' + \varepsilon_z' & 2\varepsilon_x' + \varepsilon_z' & -\varepsilon_x' + \varepsilon_z' \\ -\varepsilon_x' + \varepsilon_z' & -\varepsilon_x' + \varepsilon_z' & 2\varepsilon_x' + \varepsilon_z' \end{pmatrix} \times \frac{1}{3} \tag{5.4}$$

$$\varepsilon_{110} = \begin{pmatrix} \varepsilon_x' + \varepsilon_z' & -\varepsilon_x' + \varepsilon_z' & 0 \\ -\varepsilon_x' + \varepsilon_z' & \varepsilon_x' + \varepsilon_z' & 0 \\ 0 & 0 & 2\varepsilon_x' \end{pmatrix} \times \frac{1}{2} \tag{5.5}$$

where

ϵ_x' = strain component perpendicular to the growth direction,

ε_z' = strain component parallel to the growth direction,

ε = strain tensor in the original crystal coordinate system.

5.3.2 Valence Band Splitting of Strained Si and Ge

Including the spin-orbit splitting, the valence band of Si and Ge contains a fourfold $p_{3/2}$ multiplet and a $p_{1/2}$ doublet with a separation of $\Lambda = 0.044$ eV in Si and $\Lambda = 0.29$ eV in Ge. In the coherent growth case, the valence bands will be shifted by the strain. For (001) and (111) growth directions the top valence bands at $\vec{k} = 0$ are split into three doubly degenerate energy levels [34]:

$$E_v(3/2, \pm 3/2) = a\mathbf{1} : \varepsilon + e_x \tag{5.6}$$

$$E_v(3/2, \pm 1/2) = a\mathbf{1} : \varepsilon - (e_x + \Lambda)/2 + \sqrt{9e_x^2 + \Lambda^2 - 2e_x\Lambda}/2 \tag{5.7}$$

$$E_v(1/2, \pm 1/2) = a\mathbf{1} : \varepsilon - (e_x + \Lambda)/2 - \sqrt{9e_x^2 + \Lambda^2 - 2e_x\Lambda}/2 \tag{5.8}$$

where for (001) growth, $e_x = 2/3D_\mu(\varepsilon_{zz} - \varepsilon_{xx})$; for (111) growth, $e_x = 2D'_\mu\varepsilon_{xy}$ [35]. The physical meaning of the valence band deformation potentials are as follows: a gives the shift of the center of gravity of the entire valence band; D_μ and D'_μ define the valence band splitting for uniaxial stress along (001) and (111), respectively.

For the (110) growth direction, the linear splitting can be calculated from the 6×6 of the strained Hamiltonian matrix [36]. However, nonlinear splitting has been reported and can be obtained by using a tight binding calculation [37].

5.3.3 Band Offsets for Strained Si/Ge Systems

Van de Walle and Martin [38] have calculated the heterojunction band lineups at the Si/Ge interface with an *ab initio* pseudopotential method. A list of values of $\Delta E_{v,av}$ are given in Table 5.2 The deformation potentials used in the estimation of band offsets are given in Table 5.3. The term $\Delta E_{v,av}$ refers to the discontinuity in the weighted average of the valence bands including the spin-orbit band at Γ'_{25}. For all orientations and different strain conditions, $\Delta E_{v,av}$ between Si and Ge is almost constant: $\Delta E_{v,av} = 0.54 \pm 0.04$ eV. This suggests that $\Delta E_{v,av}$ might qualify as a parameter characteristic of the Si/Ge heterojunction, irrespective of orientation and strain condition. It was also noted that ΔE_v varies almost linearly with composition. To determine the valence band offset, ΔE_v was used as a reference parameter to calculate relative positions between the valence bands in the adjacent layers. Assuming the alloying effect on ΔE_v is linear, we can express ΔE_v for pseudomorphic Ge_xSi_{1-x}/Si interface on $Ge_{x_s}Si_{1-x_s}$ substrate in a simple form:

$$\Delta E_{v,x} = [A - (A - B) \times x_s] \times x \tag{5.9}$$

where A and B are ΔE_v for the cases corresponding to strained Ge on the Si substrate and strained Si on the Ge substrate, respectively. The values of ΔE_v between strained Ge_xSi_{1-x} and Ge_ySi_{1-y} can be obtained by considering an insertion of a Si layer in between:

$$\Delta E_v(Ge_xSi_{1-x}/Ge_ySi_{1-y}) = \Delta E_{v,x} - \Delta E_{v,y} \tag{5.10}$$

Table 5.2

Valence Band Offsets Between Strained Si and Ge on Si, $Ge_{0.4}Si_{0.6}$, and Ge Substrates for (100), (111), and (110) growth directions [38, 39]

	Substrate	$\Delta E_v, av(eV)$	$\Delta E_v(eV)$
(001)	Si	0.54	0.84
	$Ge_{0.4}Si_{0.6}$	0.53	0.61
	Ge	0.51	0.31
(111)	Si	0.58	0.85
	Ge	0.56	0.37
(110)	Si	0.52	0.76
	Ge	0.50	0.22

Table 5.3

Deformation Potentials Used in the Estimation of Band Offsets

	X Valley	L Valley
Ξ_u	9.2 [33]	16.2 [33]
$\Xi_d + 1/3\Xi_u - a$	1.5 [40]	−4.5 [40]
	Si	Ge
$D_u(100)$	3.4 [41]	3.3 [42]
$D'_u(111)$	4.4 [41]	3.8 [42]

The relative positions of the mean energy of the conduction bands at the X and L points to the relaxed valence band top are calculated by adding the gap linearly interpolated from bulk Si to Ge at the X and L points, respectively, and the shift of the mean energy due to the strain. Then the energies of the conduction valleys at the X and L points are obtained by adding the differences, $\Delta E_c^i - \Delta E_c^0$. From these procedures, the conduction band offsets for various conduction valleys are calculated. The results for three commonly used orientations (001), (110), and (111) are given as shown in Figures 5.2 through 5.4, incorporating the band splittings and offsets. For $Si_{1-y}Ge_y$ substrate with any Ge content, the band structure may be obtained by linear interpolation of the two set of data given for Si and Ge substrates, respectively. For (110), it can be referred to elsewhere [37, 43, 36]. From the band

(100 Si Sub.)

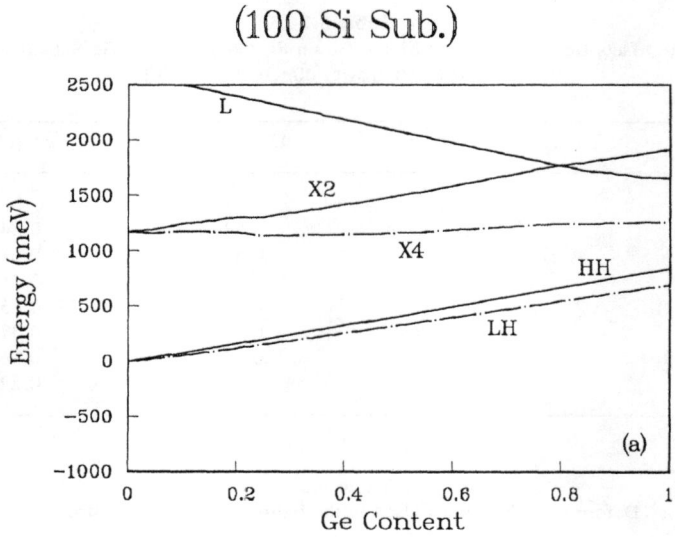

(a)

(100 40% Ge Sub.)

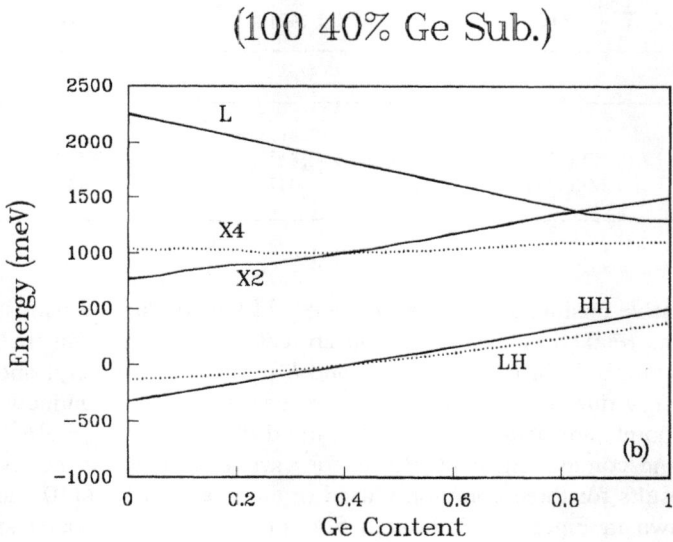

(b)

Figure 5.2 Band alignments of SiGe alloys on the (100) substrate; (a) Si, (b) $Ge_{0.4}Si_{0.6}$, and (c) Ge substrates. All values refer to the valence band edge of the substrate.

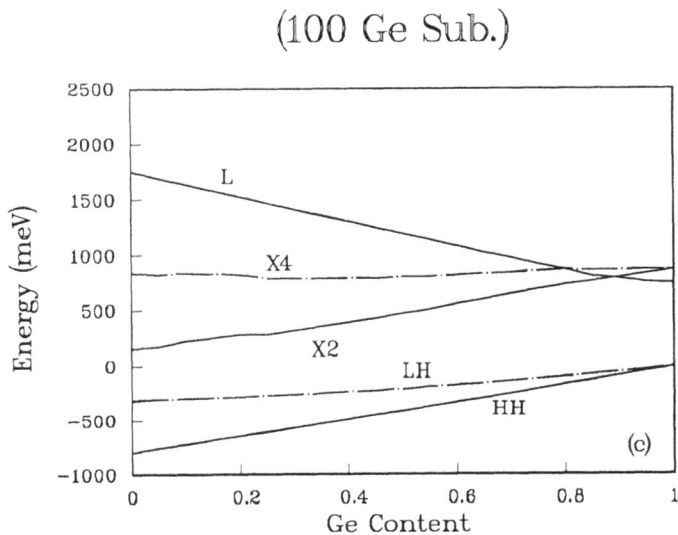

(100 Ge Sub.)

Figure 5.2 continued.

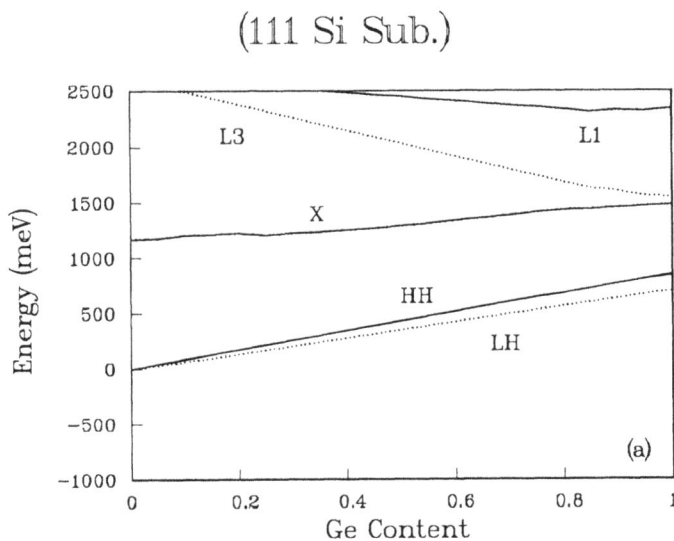

(111 Si Sub.)

Figure 5.3 Band alignments of SiGe alloys on the (111) substrate; (a) Si, (b) $Ge_{0.4}Si_{0.6}$, and (c) Ge substrates. All values refer to the valence band edge of the substrate.

(111 40% Ge Sub.)

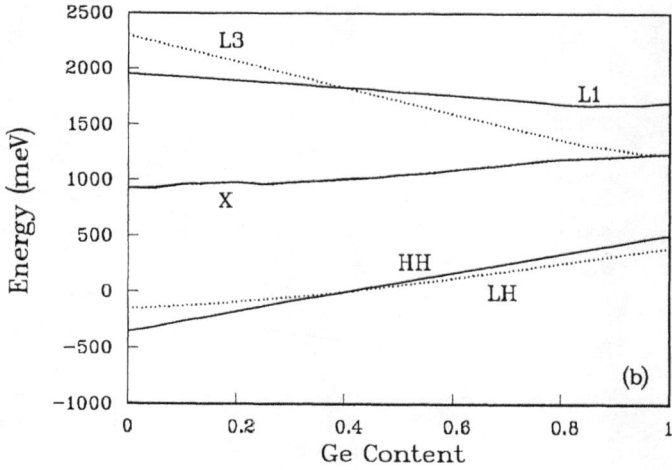

Figure 5.3 continued.

(111 Ge Sub.)

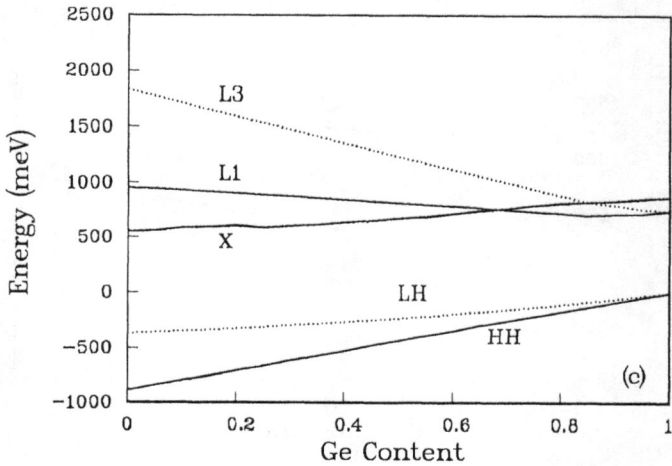

Figure 5.3 continued.

(110 Si Sub.)

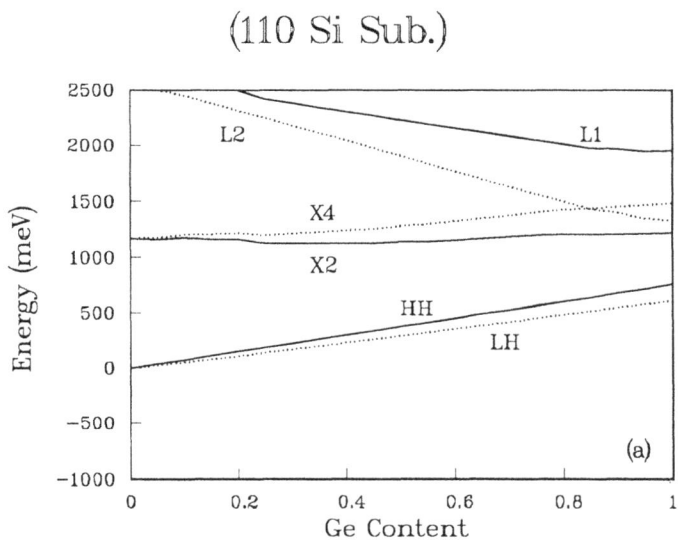

(a)

(110 40% Ge Sub.)

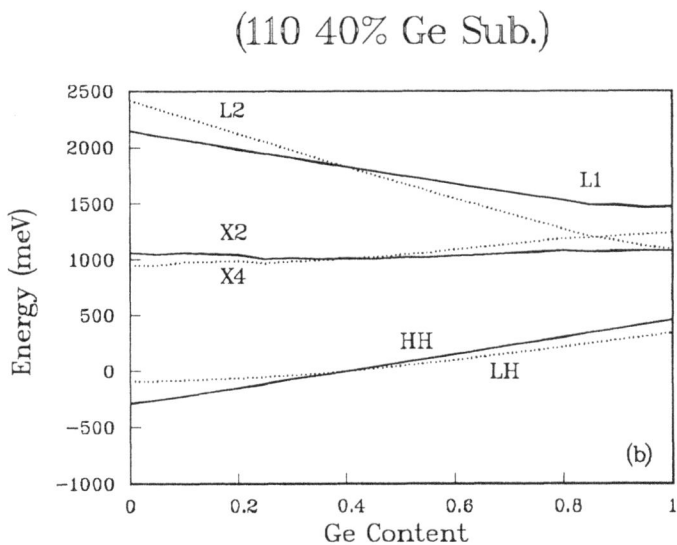

(b)

Figure 5.4 Band alignments of SiGe alloys on the (110) substrate; (a) Si, (b) Ge$_{0.4}$Si$_{0.6}$, and (c) Ge substrates. All values refer to the valence band edge of the substrate.

(110 Ge Sub.)

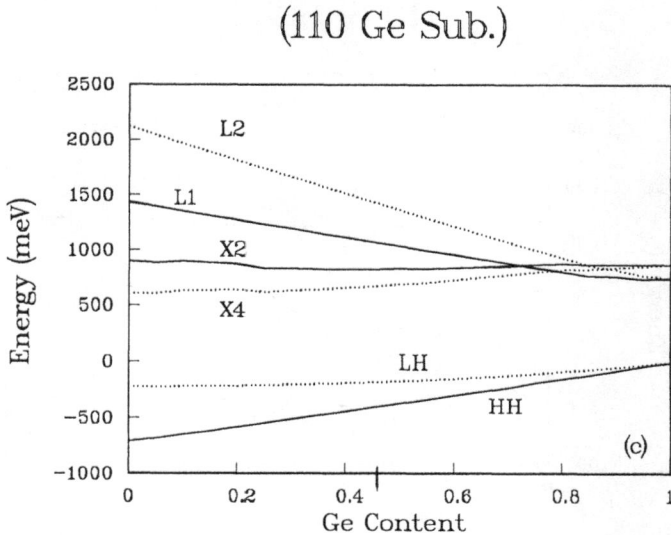

Figure 5.4 continued.

structures shown in Figures 5.2 through 5.4 (or more specifically the band offset data), it is clear that for $Si_{1-x}Ge_x$ grown on Si substrate, the band offset will occur mostly in the valence band. There is negligible conduction band offset. Therefore the $Si_{1-x}Ge_x$ on Si structures favor the valence band engineering or p-type quantum well for intersubband transition. To have a conduction band offset for n-type structures, they should be grown on SiGe substrate so that the Si layers are under tensile strain.

Currently there is a reasonable understanding of the growth of strained SiGe as well as the band alignments. This makes possible the demonstration of a wide variety of quantum devices [44]. In the following sections, we will discuss first the physics and the experimental observations of intersubband transitions in SiGe/Si quantum wells as well as δ-doped quantum well structures. Then the application of such structures for the fabrication of an infrared detector operating near 10 μm will be discussed.

5.4 PHYSICS OF INTERSUBBAND TRANSITIONS

5.4.1 Elements of Intersubband Transition of Γ-Point Quantum Wells

In this section, we will present a very simple treatise of the physics of intersubband transition at the Γ point. The theory is a simple one-band effective mass model and assumes no band coupling. For the spherically symmetric Γ valleys, effective mass reduces to a scalar quantity, and furthermore, for relatively large gap material, the band coupling is small and thus these assumptions are justified. For simplicity and clarity, we further use parabolic quantum wells as the example. The essential features of the polarization selection rules are obtained, and these are independent of the well shape [45]. Later, we will present effective mass tensor formulation to treat the general valleys as in the cases of Si and Ge.

First, we consider the absorption between two states of an isolated parabolic well as shown in Figure 5.5. For a good approximation we can describe the envelope states as those given by that of an infinite parabola. The Hamiltonian (H) of the system in the effective mass approximation (where the effect of the nearest bands are taken into account using an empirical effective mass m^*) can be written as,

$$H = \frac{p^2}{2m^*} + \frac{1}{2m^*}\,\Omega^2 z^2 \tag{5.11}$$

where z is the growth direction and p is the momentum of the electron (or hole) along the z direction; Ω is given in terms of the curvature (K) of the parabola by $\sqrt{K/m^*}$. Energy eigenvalues of the system are given by

$$E_n = \left(n + \frac{1}{2}\right)\hbar\Omega + \frac{\hbar^2 k_\perp^2}{2m^*} \tag{5.12}$$

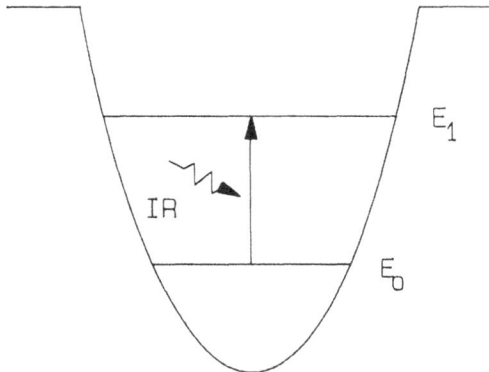

Figure 5.5 Schematic diagram of a parabolic quantum well showing the intersubband transition.

where $k_\perp$ is the wave vector of the electron (or hole) perpendicular to the growth direction and $n = 0, 1, 2, \ldots$.

The wave function $\phi_n(\vec{r})$ of the nth quantized state has the form.

$$\phi_n(\vec{r}) = u_n(\vec{r}) \frac{e^{i\vec{k}_{i\perp} \cdot \vec{\rho}}}{\sqrt{A}} n(z) \qquad (5.13)$$

where A is the area of the well, $\vec{k}_\perp$ and $\vec{\rho}$ are the wave and the position vectors in the xy plane, u is the cell period function near the band extremum, and $n(z)$ is the envelope function due to the quantization along the z-direction.

The transition strength for the intersubband transition can be obtained from the momentum matrix elements. In the dipole approximation, the matrix element for the transition between states $|m\rangle$ and $|n\rangle$ can be expressed as [46]

$$\langle u_n n | \hat{\varepsilon} \cdot \vec{p} | u_m m \rangle \approx \langle u_n | \hat{\varepsilon} \cdot \vec{p} | u_m \rangle_{\text{cell}} \langle n | m \rangle + \langle u_n | u_m \rangle_{\text{cell}} \langle n | \varepsilon_z z | m \rangle \qquad (5.14)$$

where $\hat{\varepsilon}$ is the photon polarization vector, $\vec{p}$ is the momentum of the electron (or hole), and z is the growth direction. The oscillator strength is proportional to the square of the matrix element, eq. (5.14). The selection rules of intersubband transition can be seen from the preceding expansion, which is possible because the spatial dependence of the envelope functions within a unit cell can be neglected. For clarity of subsequent discussions in polarization dependence, we define the xy polarization for which the field is on the quantum well plane, whereas for the z polarization the field is parallel to the quantum well growth direction. In a transition between two quantum states in the same band (i.e., intersubband transition or $u_m = u_n$), the first term in eq. (5.14) vanishes due to the orthogonality of the envelope functions. However, the second term is nonzero as long as $\varepsilon_z \neq 0$, and $|n\rangle$ and $|m\rangle$ are of opposite parity (i.e., the difference of quantum numbers, $\Delta n = \text{odd}$). On the other hand, for transitions involving two different bands (e.g., intervalence band transition), the second term of eq. (5.14) vanishes due to the orthogonality of the Bloch functions. The first term is nonzero only when $|n\rangle$ and $|m\rangle$ have the same quantum number. For example, the transition between the ground states of the valence and conduction bands is allowed. In general, for a transition involving two different bands, the quantum numbers of the envelope states have to be the same (i.e., $\Delta n = 0$). The well-known polarization selection rule $\cos^2\phi$ can be seen from the last term of eq. (5.14). The projection of the photon electrical field to the quantum well direction has a $\cos\phi$ dependence and the square of the momentum matrix yields a $\cos^2\phi$ relation. The key feature of the intersubband transition (at the Γ point) is that only the light incident parallel to the layers can excite the transition.

The oscillator strength (f) corresponding to a transition between states $|m\rangle$ and $|n\rangle$ is given by [47]

$$f_{m \to n} = \frac{2m_0}{\hbar^2} (E_n - E_m) |\langle n|z|m \rangle|^2 \qquad (5.15)$$

The matrix element $\langle n|z|m \rangle$ can easily be obtained by writing z in terms of creation (a^+) and annihilation (a) operators [48] as

$$z = \sqrt{\frac{\hbar}{2m^*\Omega}}\,(a^+ + a) \tag{5.16}$$

where a^+ and a satisfy

$$a^+|n\rangle = \sqrt{n+1}\,|n+1\rangle, \quad \text{and} \quad a|n\rangle = \sqrt{n}\,|n-1\rangle \tag{5.17}$$

The matrix element can be evaluated readily using z from eq. (5.16) as

$$\langle n|z|m\rangle = \sqrt{\frac{\hbar}{2m^*\Omega}}\,[\sqrt{m+1}\,\delta_{n,m+1} + \sqrt{m}\,\delta_{n,m-1}] \tag{5.18}$$

This implies that for a parabolic well, nonzero dipole matrix elements occur only between two adjacent states. For a square well, the dipole matrix element is nonzero between any two states with opposite parity, as stated previously [47]. In the case of absorption, we need to consider only the first term of eq. (5.18). Then the oscillator strength for the transition from n to $n + 1$ is given by

$$f_{n\to n+1} = \frac{m_0}{m^*}(n+1) \tag{5.19}$$

This may be compared with the case of a square well having an infinite barrier, where the oscillator strength for the same transition is given by [47]

$$f_{n\to n+1} = \frac{64m_0}{\pi^2 m^*}\frac{(n+1)^2(n+2)^2}{(2n+3)^3} \tag{5.20}$$

For the transition between the ground state and the first excited state, the oscillator strengths for the parabolic and the square wells are given by m_0/m^* and $0.96m_0/m^*$, respectively. The strength of the absorption can be evaluated using the transition rate (W), which is given by [49]

$$W = \frac{2\pi}{\hbar}\sum_F |\langle \phi_f|V_p|\phi_i\rangle|^2 \delta(E_F - E_I - \hbar\omega) \tag{5.21}$$

where E_I and E_F are the initial and final state energies, ω is the angular frequency of the incident photon, ϕ_i and ϕ_f are the initial and final state wave functions, and the photon interaction potential V_p is given by [49]

$$V_p = \frac{e}{m^*} \frac{\hbar}{2\varepsilon_0 n_r \omega c} \hat{\varepsilon} \cdot \vec{p} \qquad (5.22)$$

where n_r is the refractive index, and $\hat{\varepsilon}$ and $\vec{p}$ are the polarization of the incident photon and the momentum of the electron (or hole), respectively. By evaluating the transition rate using eq. (5.21) and integrating it over the occupied states and the incident photon energy at Brewster's angle, the fraction of energy absorbed per unit length is [50]

$$\frac{e^2 h}{4\varepsilon_0 m^* c} \frac{\rho_s}{l} \frac{f}{n_r(n_r^2 + 1)} \qquad (5.23)$$

where ρ_s is the two-dimensional density of electrons (or holes) in the well and l is the quantum well thickness.

5.4.2 Transition Between Superlattice Minibands

Next, we consider an optical transition of a superlattice made of parabolic quantum wells. Due to the shape of the parabolic potential it is possible to fabricate super-lattices, of which the ground states are isolated and the excited states form a mini-band. This class of structure is particularly important in infrared detector applications because the isolated ground states keep the leakage current low under an external bias (other step quantum wells can be used for the same purpose). In this case, the analysis of the transition process is very similar to that of an impurity-to-band transition where a bound electron (or hole) makes a transition to the conduction (or valence) band. Here we use the tight binding method [48] to find the final state wave function, using the infinite parabolic well wave functions as our basis set. Under these conditions the initial and final state wave functions can be written as

$$\Phi_I = u_0 \frac{e^{i\vec{k}_{i\perp} \cdot \vec{\rho}}}{\sqrt{A}} \phi_{0m}(z) \qquad (5.24)$$

$$\Phi_F = u_0 \frac{e^{i\vec{k}_{f\perp} \cdot \vec{\rho}}}{\sqrt{A}} \sum_n \frac{e^{iqnL}}{\sqrt{N}} \phi_{1n}(z) \qquad (5.25)$$

where A is the area of the well in xy plane, L is the periodicity, N is the number of wells in the superlattice (assumed very large), and ϕ_{0m} and ϕ_{1n} are the ground state and excited state envelope wave functions of an infinite parabolic well at positions rL and nL, respectively.

Taking into account only the nearest-well interactions, the dispersion relation for the first excited state can be written as [51]

$$E_f = E_1 + 2\Delta \cos (qL) + \frac{\hbar^2 k_\perp^2}{2m^*} \qquad (5.26)$$

where E_1 is the first excited state energy of an isolated parabolic quantum well and 2Δ is the bandwidth of the miniband; Δ is given by

$$\Delta = 2m^* \Omega^2 \sqrt{\frac{\beta^3}{\pi}} \int_{-l/2}^{+l/2} z(z - L)\left(z^2 - \frac{l^2}{4}\right) e^{-\beta[(z-L)^2 + z^2]/2} \, dz \qquad (5.27)$$

where $\beta = m^* \Omega / \hbar$ and l is again the width of the parabolic well. For given parameters of the superlattice, Δ can easily be obtained from eq. (5.27) by numerical integration. Using the initial and final state wave functions, we can find the matrix element of the transition to be

$$\langle \Phi_F | \, \hat{\varepsilon} \cdot \vec{p} | \Phi_I \rangle = \sum_n \frac{e^{-iqnL}}{\sqrt{N}} \langle \phi_{1n} | \, \hat{\varepsilon} \cdot \vec{p} | \phi_{0m} \rangle \qquad (5.28)$$

Because ϕ_{0m} is localized at the rth well position, the major contribution to the matrix element occurs when $n = r$. In this case, it can easily be shown using eq. (5.18) that eq. (5.28) reduces to

$$|\langle \Phi_F | \, \hat{\varepsilon} \cdot \vec{p} | \Phi_I \rangle |^2 \approx \frac{\hbar \Omega}{N} (\hat{\varepsilon} \cdot \hat{e}_z)^2 \qquad (5.29)$$

Substituting this into the expression for the transition rate, we find the absorption constant as a function of photon energy ($\hbar \omega$) to be

$$\alpha = \frac{e^2 \hbar}{2m^* c \varepsilon_0 n_r} \frac{\rho_s}{L} \frac{\Omega}{\omega} \frac{(\hat{\varepsilon} \cdot \hat{e}_z)^2}{\sqrt{(E_{max} - \hbar\omega)(\hbar\omega - E_{min})}} \qquad (5.30)$$

where $E_{max} = E_1 - E_0 + 2\Delta$ and $E_{min} = E_1 - E_0 - 2\Delta$.

Figure 5.6 shows the absorption constant as a function of wavelength when the device parameters are chosen so as to obtain a response near 10 μm. The parabolic quantum well is assumed to contain 2×10^{18} carriers/cm^3. The absorption constant

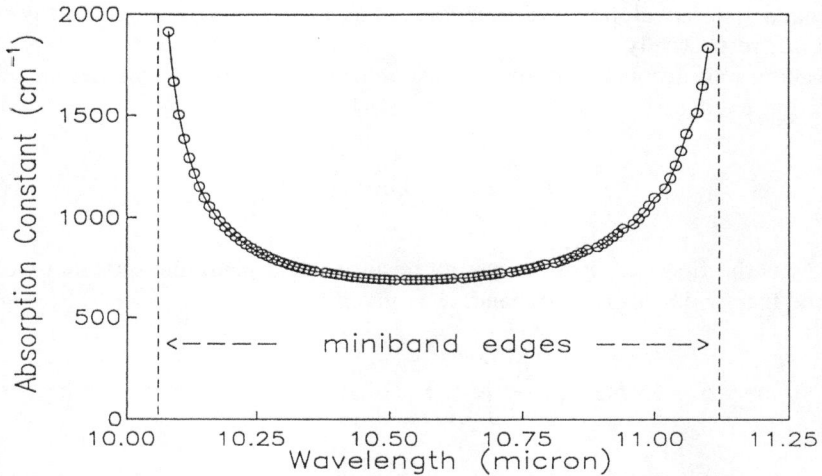

Figure 5.6 Absorption constant as a function of wavelength for a superlattice. Here we assume IR incident at Brewster's angle. Quantum well parameters are chosen to get a response near 10 μm with a 10 meV bandwidth.

shows a singular behavior at the band edges due to one-dimensional van Hove singularities of the joint density of states. This gives a relatively high absorption coefficient near the band edges, as seen in Figure 5.6. The origin of this behavior is due to the exact parallelism between the in-plane dispersion relations of the initial and final subbands. In reality, the nonparabolicity and other effects that alter this parallelism, usually soften the singular behavior, resulting in a finite absorption constant near the band edges. In the following sections, we will discuss many-body effects in intersubband transition, which is particularly important in the case of heavily doped quantum well structures such as δ-doped layers.

5.4.3 Many-Body Effects in Intersubband Transition

The importance of many-body effects has been recognized in the understanding of optical transitions in inversion layers [52–55]. The optical transitions between quantized subbands in an inversion layer show a large discrepancy between the experimentally observed peak positions and the self-consistent calculations using the Hartree approximation [56, 57]. By incorporating many-body effects, such as exchange correlation and depolarization, the calculated values agree with the experimental results [58–60]. Recently, similar behavior has been observed in intersubband absorption in GaAs/AlGaAs quantum wells. The differences between observed and calculated energy positions have been attributed to the lowering of ground state energy as a result of the exchange interactions among electrons in the ground subband [61, 62]. In all of the previous cases, the amount of shift has been relatively small (a few meV) due to the limitations of achieving a large population in the subbands. In the case of δ-doped layers in Si, a large population of carriers occupies the subbands due to the large density of states and high solid solubility of dopants. The many-body effects can play an important role in determining the peak absorption energy. In the following subsections we will describe the estimation of many-body effects in the intersubband transition.

The many-body effects are usually incorporated in the calculation using the local-density functional approximation [60]. In the following paragraphs, we will briefly discuss the steps used in the multiband self-consistent calculation [63]. The Schrödinger equation, including the many-body effects, can be written as

$$\left[-\frac{\hbar^2}{2m^*}\nabla^2 + V_H(z) + V_{xc}(z) \right]\psi_n = E_n\psi_n \tag{5.31}$$

where V_H and V_{xc} are the Hartree and exchange-correlation potentials, respectively. The Hartree potential is given by

$$V_H(z) = \frac{e}{\varepsilon\varepsilon_0} \int (z - z')\rho(z') \, dz' \tag{5.32}$$

where $\rho(z) = e[N_D - n(z)]$, ε is the dielectric constant, N_D is the dopant density,

$$n(z) = n^{2D} \sum_n |\psi_n(z)|^2 \tag{5.33}$$

is the carrier concentration at z, and n^{2D} is the two-dimensional density of carriers. The exchange-correlation potential in the local-density functional approximation using the Hedin and Lundqvist parameterization [64] can be expressed as

$$V_{xc}(z) = -\left[1 + \frac{0.7734r_s}{21} \ln\left(1 + \frac{21}{r_s}\right) \right]\left(\frac{2}{\pi\alpha r_s}\right)R_y^* \tag{5.34}$$

where $r_s = [4/3\pi a^{*3} n(z)]^{-1/3}$, the effective Bohr radius $a^* = 4\pi\varepsilon\hbar^2/m^*e^2$, $\alpha = [4/(9\pi)]^{1/3}$, and the effective Rydberg $R_y^* = e^2/(8\pi\varepsilon a^*)$. In the calculation, the energy levels and wave functions are calculated self-consistently by numerically solving the eq. (5.31). It is also known that the photon-induced many-body contributions, such as depolarization [59] and exciton-like interaction [60] between the ground and excited states, shift the calculated energy position by a substantial amount. These two effects result from the plasma frequency dependence on the charge sheet. For a two-dimensional gas, such effects are estimated by solving the time-dependent Schrödinger equation, self-consistently using the Hartree and exchange-correlation potentials. The peak position, including the shift due to plasma effects, is given by [57]

$$\tilde{E}_{10}^2 = E_{10}^2(1 + \alpha - \beta) \tag{5.35}$$

where $\tilde{E}_{10}$ is the shifted energy and E_{10} is the difference ($E_1 - E_0$) of the subband energy levels calculated by eq. (5.31) self-consistently. The quantities α and β for the transition between the ground state and the first excited state (assuming only the ground state is occupied) are given by [60]

$$\alpha = \frac{2e^2 n_0^{2D}}{\varepsilon\varepsilon_0}\left(\frac{\hbar^2}{2m^*E_{10}}\right)^2 \int_{-\infty}^{+\infty} dz\left[\psi_1(z)\frac{d\psi_0(z)}{dz} - \psi_0(z)\frac{d\psi_1(z)}{dz}\right]^2 \tag{5.36}$$

$$\beta = \frac{2n_0^{2D}}{E_{10}} \int_{-\infty}^{+\infty} dz\psi_1(z)^2\psi_0(z)^2\frac{\partial V_{xc}[n(z)]}{\partial n(z)} \tag{5.37}$$

Next, we will present the experimental studies of infrared transitions in both *n*- and *p*-type quantum well structures in Si.

5.4.4 Electron Intersubband Transition of General Valley *n*-Type Quantum Wells

As mentioned before, electron intersubband transition has been demonstrated for infrared detector application in AlGaAs/GaAs quantum well structures [47, 65]. For GaAs, the isotropic effective mass of the Γ conduction band makes such a selection rule so that intersubband transitions are allowed only for incident light with the electric field polarized along the quantum well growth direction (*z* polarization) as discussed before. For Si, the ellipsoids for orientations other than (001) are tilted, and there are off-diagonal elements in the effective mass tensor [66, 67]. Previously, Yang and Pan [68] have theoretically calculated the absorption coefficient of conduction intersubband transitions of the SiGe/Si quantum wells grown in the (110) and (111) directions for the *xy* and *z* polarizations. The absorption coefficient for the *xy* polarization is shown to be comparable to that of GaAs/AlGaAs in this case. In their calculation, no strain effect was included and all indirect conduction valleys were assumed to be degenerate. Furthermore, there is no discussion of the transition energy (as we will show later, the transition energy estimated from the effective mass is in the range of tens of μm and cannot be used for 8–12 μm applications). Additionally, as we will show, the strain can affect the population of each valley,

and the occupancy of different valleys will in turn determine the polarization behavior of the intersubband transitions. Further, the absorption coefficient changes
according to an appropriate sum of the effective masses of the occupied valleys
projected onto the growth direction. Therefore the strain effect is an important factor
to be taken into account for understanding the physics of the conduction intersubband
transition of SiGe layers.

In the following paragraphs, we will describe the calculation of the oscillator
strength of the conduction intersubband transition for both SiGe strained layers and
δ-doped structures, for various orientations and under different strain conditions. In
the calculation, the one-band effective mass approximation is used and the Hamiltonian used is a tensor form. The polarization behavior is different from that of the
AlGaAs/GaAs intersubband transition, where the conduction minimum occurs at the
Γ point. The results are quite general and can be applied to other material combinations whose conduction minima occur away from the Γ point (as in the case of
AlAs). The results of the oscillator strength can be easily used to predict the actual
absorption coefficient or absorption strength [47].

Oscillator Strength for Intersubband Transitions

In the one-band approximation, the Hamiltonian can be modified in the effective
mass approximation as follows [69]:

$$H = 1/2\, \vec{P} \cdot \overleftrightarrow{W} \cdot \vec{P} + V(z) \tag{5.38}$$

where $\vec{P}$ is a vector momentum operator and $\overleftrightarrow{W}$ is a 3 × 3 inverse mass tensor (instead
of the mass tensor) to account for the anisotropic mass. The coordinate system is
chosen, as shown in Figure 5.7, to have the growth direction along the $\hat{z}$ axis throughout
the chapter. For a given growth direction, the inverse mass tensor for each indirect
conduction valley can be obtained by using a coordinate transformation in which the
Euler's angles are used to describe the relation between the direction of a conduction
valley and the growth direction. The potential energy, $V(z)$, of the SiGe quantum
well may come from an inversion layer, the conduction band offset of the heterostructure [70], or the self-consistent potential due to a δ-doped layer [10]. The preceding expression describes the motion of an electron in the potential well and is
valid if the well width is much larger than the atomic dimension. Then the motion
of an electron can be effectively described by the envelope function. Within this
framework, the envelope function can be written as [69]

$$F(x, y, z) = f(z)\exp[-jz(k_x w_{xy} + k_y w_{yz})/w_{zz}]\exp[j(k_x x + k_y y)] \tag{5.39}$$

where k_i and w_{ij} are the wavevector and elements of the inverse mass tensor, respectively. The envelope function, $f(z)$, is a function of z only and does not vanish

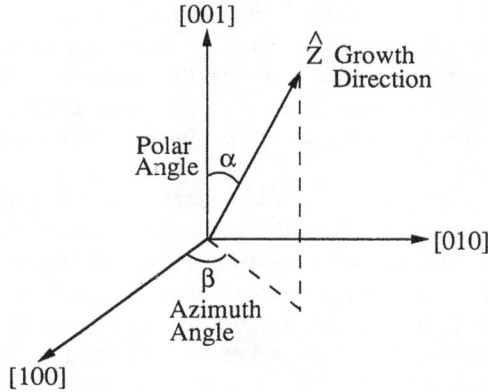

Figure 5.7 Coordinate system chosen to have the growth direction along the $\hat{z}$ axis. The polar angle represents the angle between the (001) direction and the growth direction and the azimuth angle between the (100) axis and the projected direction of the growth direction on the (001) plane.

at the interfaces for a finite potential well. To simplify the calculation, we use the boundary condition that the wave functions vanish at the interfaces; that is, assuming an infinite potential well case. This assumption makes $f(z)$ simply an even or odd function. The approximation of this boundary condition gives the calculated absorption coefficient of the GaAs system within a factor of two of experimental data [68]. This is true for the oscillator strength, as expected. The oscillator strength for the direct conduction valley (conduction band minimum at Γ) has been calculated by West and Eglash [47] and can be modified for the more general indirect conduction valley when considering the optical electric field of the incident light at angle; that is, with a polarization vector $\hat{\varepsilon}$

$$f_{i \to f}^{\gamma} = \frac{2m_0}{\hbar \omega} |\langle \phi_f | \hat{\varepsilon} \cdot \overleftrightarrow{W}^{\gamma} \cdot \vec{P} | \phi_i \rangle|^2 \tag{5.40}$$

where γ is an index of indirect conduction valleys, m_0 is a free-electron mass, ω is the angular frequency of the subband transition, ϕ_i and ϕ_f are the initial and final states, respectively, consisting of the envelope function described earlier and an approximate Bloch wave function for each state. The oscillator strength for any direction of the incident light can be written as

$$f_{i \to f}^{\gamma} = \frac{2m_0}{\hbar \omega} (\varepsilon_x w_{xz}^{\gamma} + \varepsilon_y w_{yz}^{\gamma} + \varepsilon_z w_{zz}^{\gamma})^2 |\langle \phi_f | P_z | \phi_i \rangle|^2 \tag{5.41}$$

Unlike the direct conduction valley case, the xy-polarized optical electric field can cause the electronic motion in the growth direction through the off-diagonal terms of the inverse mass tensor, w_{xz}^γ and w_{yz}^γ, and thus induce the intersubband transitions. Obviously, for the z polarization transition, the electric field acts on the envelope function through w_{zz}^γ and the behavior is similar to the direct gap (Γ band) case.

For the parabolic potential well, the oscillator strength for the transition from n to $n + 1$ can be evaluated similarly to eqs. (5.15) through (5.19) and is given in terms of elements of the inverse mass tensor, w_{ij}^γ, as

$$f_{n \to n+1}^\gamma = \frac{m_0}{w_{zz}^\gamma} (\varepsilon_x w_{xz}^\gamma + \varepsilon_y w_{yz}^\gamma + \varepsilon_z w_{zz}^\gamma)^2 (n + 1) \tag{5.42}$$

For the special isotropic conduction valley at the Γ point, the off-diagonal elements of the inverse mass tensor, w_{xz}^γ and w_{yz}^γ, are zero and the preceding oscillator strength reduces to eq. (5.19).

For an infinite square potential well, the oscillator strength becomes (for comparing with the preceding expression of the parabolic potential well)

$$f_{n \to n+1}^\gamma = \frac{m_0}{w_{zz}^\gamma} (\varepsilon_x w_{xz}^\gamma + \varepsilon_y w_{yz}^\gamma + \varepsilon_x w_{zz}^\gamma)^2 \frac{64}{\pi^2} \frac{(n + 1)^2 (n + 2)^2}{(2n + 3)^3} \tag{5.43}$$

which is the same form as eq. (5.20) for the isotropic conduction band with an infinite square potential well [47], if the off-diagonal terms vanish. As shown in eqs. (5.42) and (5.43), the same mass tensor term is independent of the type of the potential well. That is, the oscillator strengths have the same behavior.

Because the effective mass of the occupied valleys determines the transition rate, we need to analyze the occupancy of the conduction valleys with and without strain. For a relaxed SiGe alloy, the energy gap can be obtained approximately by the virtual crystal approximation; that is, a linear interpolation of the lowest band edge at the Δ and L points of Si and Ge [71], respectively. It has been shown that the relaxed SiGe alloy has Si-like minimum conduction valleys for the Ge mole fraction up to 85%, and beyond 85%, they change to Ge-like. In this unstrained case, the minima of Si- or Ge-like conduction valleys are all degenerate. In quantum wells (even without strain), however, due to the different directional masses, the subband energy levels are split for the degenerate conduction valleys. In the latter case, the conduction valley with the largest directional mass gives the lowest subband energy level and is occupied. For example, for the (110)-oriented Si well under an in-plane tensile strain, the $X4$ valleys have the lowest energy. Upon the determination of the occupied valleys, we can proceed to calculate the oscillator strength. Figure 5.8 shows the ellipsoids for the (001) and (110) growth directions. Here, for the (001) growth direction as shown on the top of Figure 5.8(a), the two ellipsoids in

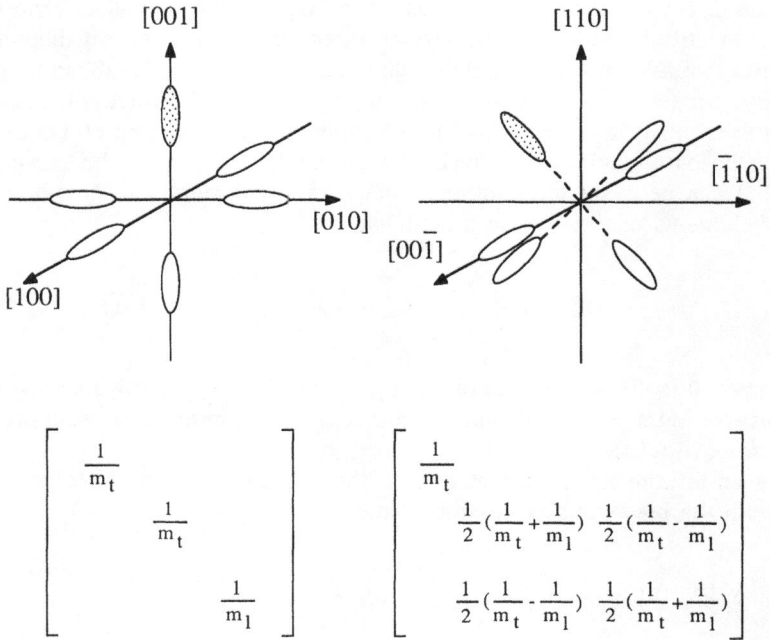

$$
\begin{bmatrix}
\dfrac{1}{m_t} & & \\
& \dfrac{1}{m_t} & \\
& & \dfrac{1}{m_l}
\end{bmatrix}
\qquad
\begin{bmatrix}
\dfrac{1}{m_t} & & \\
& \dfrac{1}{2}\left(\dfrac{1}{m_t}+\dfrac{1}{m_l}\right) & \dfrac{1}{2}\left(\dfrac{1}{m_t}-\dfrac{1}{m_l}\right) \\
& \dfrac{1}{2}\left(\dfrac{1}{m_t}-\dfrac{1}{m_l}\right) & \dfrac{1}{2}\left(\dfrac{1}{m_t}+\dfrac{1}{m_l}\right)
\end{bmatrix}
$$

Figure 5.8 Ellipsoids for (a) [001] and (b) [110] growth directions. The inverse mass tensors illustrated at the bottom are for the corresponding shaded ellipsoids.

the (001) and (00$\bar{1}$) directions are occupied due to the higher mass projected to the (001) growth direction. On the other hand, the four equivalent ellipsoids are occupied for the (110) growth direction as shown in Figure 5.8(b). For the strained case, the occupancy changes depending on the strain condition. If the well is strained, the conduction valleys of the SiGe at the Δ or L point are no longer degenerate, and the determination of occupied valleys cannot be assessed in terms of the directional mass alone. For example, a biaxial compressive strain is present in the SiGe film grown directly on Si. This biaxial compressive strain can be decomposed into a dilation and a uniaxial component; the dilation term simply shifts all the band edges together whereas the uniaxial term lifts the degenerate conduction valleys. The six equivalent Δ valleys split into a twofold and a fourfold degenerate valley [33, 70, 72]. The energy of the conduction valleys under strain can be readily determined from the known deformation potential as discussed in Section 5.3 [73]. The occupancy of these split sets of the valleys will therefore depend on the strain and their effective masses projected to the quantization (or growth) direction. To evaluate the oscillator strength, we need to sum the contribution of each of the occupied valleys according to their occupancies as follows:

$$f_{n \to n+1} = \frac{1}{n_v} \sum_\gamma f^\gamma_{n \to n+1} \eta \qquad (5.44)$$

where n_v is the number of the occupied valleys (six for Si-type and four for Ge-type conduction valley) and η is the fractional occupancy of all equivalent valleys. In general, when the strain and the difference of the projection masses from different equivalent valleys are small, the energy separation of each set of the equivalent valleys is small, and consequently all the valleys may be occupied. The obtained oscillator strength is used to calculate the absorption coefficient, which is simply proportional to the product of the oscillator strength and the two-dimensional density of states [47]. Here the two-dimensional density of states for all the occupied equivalent valleys can be evaluated from the projected effective masses onto the growth plane.

Results and Discussion

Si and Ge δ-Doping Quantum Well. First, we discuss the *unstrained* δ-doped quantum well case for various growth orientation. The oscillator strength of the intersubband transitions between the ground state and the first excited state for a parabolic potential well is shown in Figures 5.9 through 5.12 for relaxed Si, Ge and strained $Si_{0.6}Ge_{0.4}$ on $Si_{0.8}Ge_{0.2}$ and Ge on $Si_{0.2}Ge_{0.8}$, respectively. The two angles shown in the figures indicate the growth direction in relation to (001). For example, the (110) growth direction has a 90° polar angle and an azimuth angle of 45°. Figure 5.9 shows the oscillator strength of Si δ-doped quantum wells (unstrained) for the *xy*- and *z*-polarized electric fields, respectively. For the *xy* polarization shown in Figure 5.9(a), the electric field is polarized on the plane, and the electron motion along the *z* direction is induced through the coupling of the off-diagonal elements, w^γ_{xz} and W^γ_{yz}. This coupling is present for the growth directions other than (001) in Si and therefore can produce the intersubband transitions for the normal incident light. In this normal incidence case, the oscillator strength is averaged over all the angles on the plane as it depends on the optical field direction of the quantum well plane. By varying the Euler's angle (growth direction), the average oscillator strength for other growth directions can be calculated, and the maximum oscillator strength for normal incidence, 0.8243, is found to be near the (023) growth direction. In contrast, for the (001) growth direction, all the off-diagonal terms of the inverse mass tensor are zero and the intersubband transition for the *xy* polarization (i.e., for normal incident light) is forbidden, identical to the case for the direct conduction Γ valley.

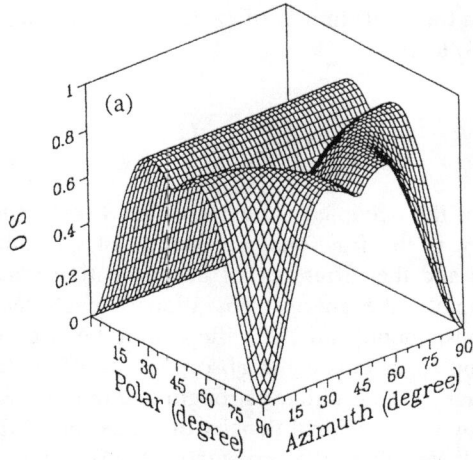

xy polarization (normal incidence)

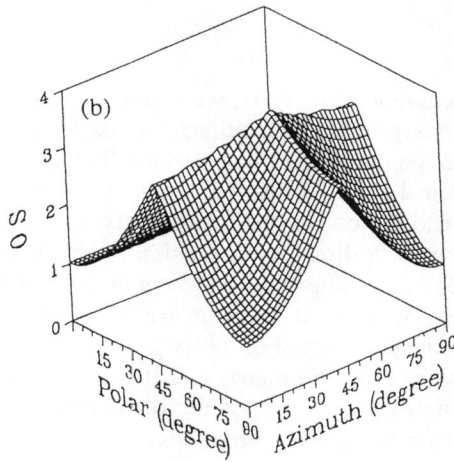

z polarization

Figure 5.9 Oscillator strength for the transition between the ground state and the first excited state in the relaxed Si parabolic well: (a) *xy* polarization and (b) *z* polarization.

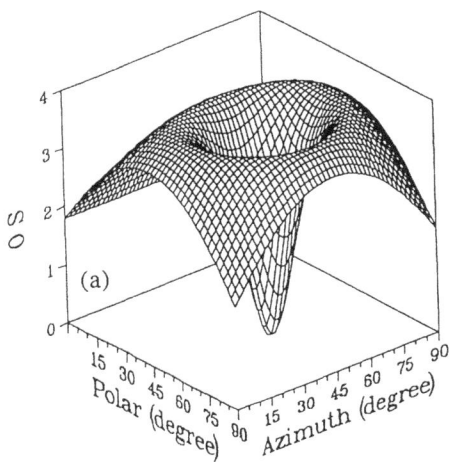

xy polarization (normal incidence)

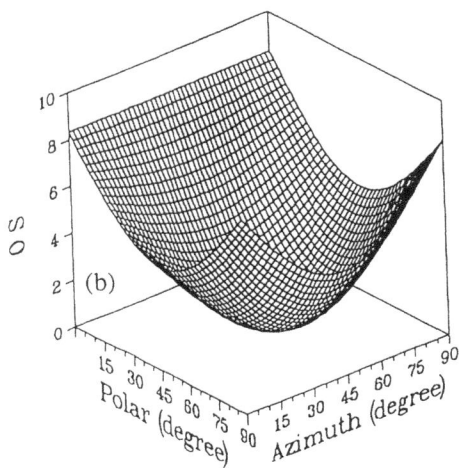

z polarization

Figure 5.10 Oscillator strength for the transition between the ground state and the first excited state in the relaxed Ge parabolic well: (a) *xy* polarization and (b) *z* polarization.

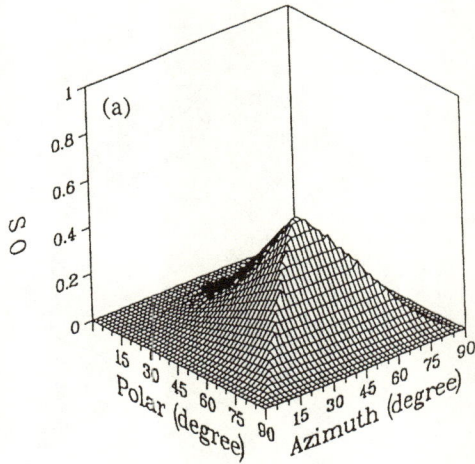

xy polarization (normal incidence)

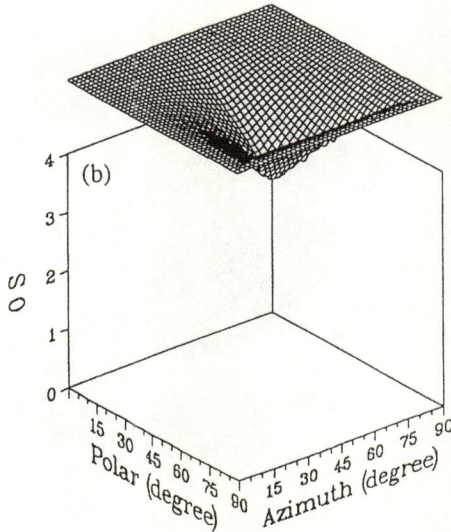

z polarization

Figure 5.11 Oscillator strength for the transition between the ground state and the first excited state in the parabolic quantum well: (a) xy polarization and (b) z polarization. The biaxial compressive $Si_{0.6}Ge_{0.4}$ layer grown on $Si_{0.8}Ge_{0.2}$ is the quantum well, where the Si-like conduction valleys are occupied. The buffer layer is $Si_{0.8}Ge_{0.2}$.

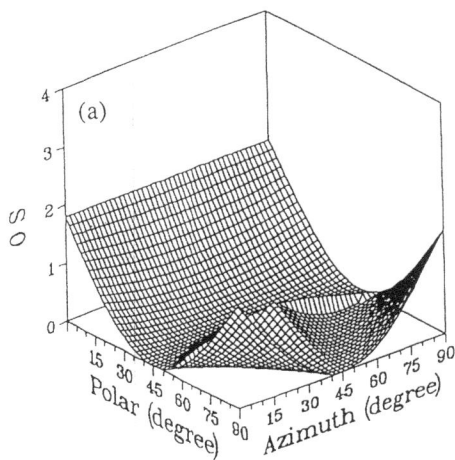

xy polarization (normal incidence)

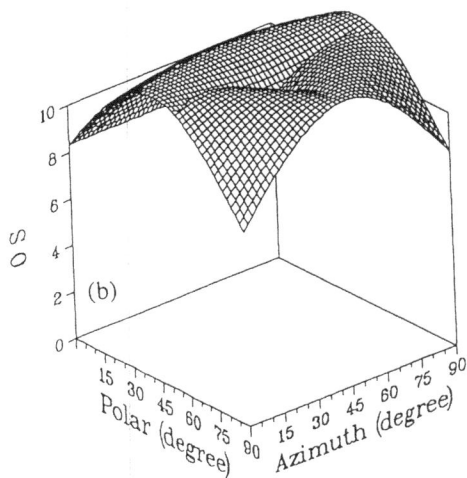

z polarization

Figure 5.12 Oscillator strength for the transition between the ground state and the first excited state in the parabolic quantum well: (a) *xy* polarization and (b) *z* polarization. The biaxial compressive Ge layer grown on $Si_{0.2}Ge_{0.8}$ is the quantum well, where the Ge-like conduction valleys are occupied. The structure consists of $Ge/Si_{0.4}Ge_{0.6}$ on $Si_{0.2}Ge_{0.8}$ buffer layer.

For the z-polarized electric field, the oscillator strength for any growth direction is shown in Figure 5.9(b). It is clear from eqs. (5.42) through (5.43) that the oscillator strength is proportional to w_{zz}^γ for this case. For a given growth direction, different valleys will have different inverse mass parameters w_{zz}^γ. Although the maximum oscillator strength is expected for the largest w_{zz}^γ valleys, these valleys are usually empty due to their high subband energies. Therefore they cannot contribute to the absorption. For the (111) growth direction, in which all conduction valleys at the Δ point are degenerate, all the valleys contribute equally. In this case, the highest oscillator strength peak of 3.8489 in the (111) direction is obtained, which is four times smaller than that of a GaAs quantum well ($f_{\mathrm{GaAs}} \approx 15$). But the higher doping level in Si may make the absorption strength even higher than the GaAs case.

For *unstrained* Ge quantum wells, the oscillator strength, for the xy and z polarizations is shown in Figure 5.10(a) and (b), respectively, for the L, (111) valleys. For the xy polarization, Figure 5.10(a) illustrates that the intersubband transition is forbidden when grown in (111). In this case, similar to the (001) case in Si, the two heavier conduction valleys in the (111) and ($\bar{1}\bar{1}\bar{1}$) directions are populated (and others are empty), and the inverse mass tensor for these valleys has only diagonal terms. For the xy polarization, the maximum oscillator strength of 3.6755 is obtained when the growth direction is near (144). For the z polarization, the dependence of the oscillator strength as a function of the growth direction looks different from that of the Si quantum well (see Figures 5.9(b) and 5.10(b)): when the optical field is polarized in the (001) growth direction, the oscillator strength is maximum for a Ge quantum well, but minimum for Si. This can be understood from the fact that Ge(111) is equivalent to Si(001).

Strained Quantum Wells. Because the strain will affect the energy and thus occupancy, we have calculated the oscillator strength for several strained quantum wells. First, we discuss the strain-symmetrized $Si_{1-x}Ge_x/Si$ quantum well case. In the strain-symmetrized case, a thick relaxed $Si_{1-y}Ge_y$ buffer layer is used, where $y = x/2$ [74], assuming that the layer thicknesses of the Si and $Si_{1-x}Ge_x$ layers are the same. For example, Si and $Si_{0.6}Ge_{0.4}$ layers are grown on a thick, relaxed $Si_{0.8}Ge_{0.2}$ layer. In this case, the Si layers serve as the conduction band wells and experience a biaxial tensile strain on the plane and a compressive strain along the growth direction. For any growth direction, the conduction band valleys closer to the growth direction have energy lower than the others, due to the in-plane tensile strain (or the compressive strain along the growth direction), and these valleys are therefore occupied. Only these occupied valleys are used for the calculation of the oscillator strength. In addition, the inverse mass tensor of the strained Si layers is assumed to be the same to the first order as that of unstrained Si [33].

When the $Si_{0.6}Ge_{0.4}$ layers serve as the quantum wells achieved by δ doping, rather than by the band offset, the wells experience a biaxial compressive strain. The biaxial compressive strain puts the valley energy higher for those conduction Δ valleys closer to the growth direction (via deformation potential). In the SiGe well case, the effective mass value for the SiGe alloys is needed for the calculation of the oscillator strength. However, in view of the lack of the data, we assumed the Δ valley masses for SiGe alloys to be the same as those of Si. This assumption will cause some errors quantitatively but it is expected to be qualitatively correct. For the case of the biaxial compressive $Si_{0.6}Ge_{0.4}$ well, the calculated results of the oscillator strength for the xy and z polarizations are shown in Figure 5.11. When compared with that of relaxed Si shown in Figure 5.9, the oscillator strength is smaller for the xy polarization and larger for the z polarization. This is because the occupied valleys under the biaxial compressive strain change from those of relaxed Si, and therefore different mass tensors for the newly occupied valleys affect the oscillator strength. However, for the (111) growth direction, all six conduction valleys at the Δ point are shifted together at the same energy and therefore remain degenerate. The

Table 5.4
Oscillator Strength for the Parabolic Potential Well for Different
Growth Directions and Strain Conditions

Material	Growth Direction	Number of Occupied Valley	Normal Incidence	Parallel Incidence
Relaxed Si	(100)	2	0.0000(0.0000)	1.0204(0.9803)
	(110)	4	0.7162(0.6880)	3.1418(3.0182)
	(111)	6	0.5197(0.4992)	3.8489(3.6975)*
	(023)	2	0.8243(0.7919)*	2.3259(2.2344)
Relaxed Ge	(100)	4	1.7896(1.7192)	8.3333(8.0056)*
	(110)	2	3.3352(3.2040)	4.4715(4.2957)
	(111)	1	0.0000(0.0000)	0.6098(0.5858)
	(144)	1	3.6755(3.5310)*	2.7162(2.6094)
Si on $Si_{0.8}Ge_{0.2}$		Same as relaxed Si		
$Si_{0.6}Ge_{0.4}$ on $Si_{0.8}Ge_{0.2}$	(100)	4	0.0000(0.0000)	5.2632(5.0562)*
	(110)	2	0.0000(0.0000)	5.2632(5.0562)*
	(111)	6	0.5197(0.4992)*	3.8489(3.6975)
$Si_{0.4}Ge_{0.6}$ on $Si_{0.2}Ge_{0.8}$		Same as relaxed Si		
Ge on $Si_{0.2}Ge_{0.8}$	(100)	4	1.7896(1.7192)*	8.3333(8.0056)
	(110)	2	0.0000(0.0000)	12.1951(11.7155)*
	(111)	3	0.6077(0.5838)	10.9079(10.4789)

The value in the parentheses is the oscillator strength for the square potential well; * indicates the maximum values for that polarized field.

oscillator strength of SiGe wells for the (111) growth direction is exactly that of the relaxed Si well for both the xy and z polarizations discussed previously.

Next, the structure of $Ge/Si_{0.4}Ge_{0.6}$ layers grown on $Si_{0.2}Ge_{0.8}$ (for strain symmetrization) is discussed. For the quantum wells in the strained Ge layers grown on $Si_{0.2}Ge_{0.8}$ buffer layer, Figure 5.12 shows the oscillator strength for the xy and z polarizations. For any arbitrary growth direction, some of the L valleys will still be the minimum under the biaxial compressive strain (compared with the X valley). Generally, the oscillator strength in the biaxial compressive Ge well is smaller for the xy polarization and larger for the z polarization when compared with the relaxed Ge well case shown in Figure 5.10. This is due to the different valley occupancy as discussed before. As shown in Figure 5.12, there are abrupt changes in oscillator strength as a function of the growth direction. The abrupt changes of the oscillator strength are also due to the change of the occupied valleys. For example, at the growth direction of 50° of polar angle and 45° of azimuth angle from (001), the Ge L valleys at the $(\bar{1}\bar{1}1)$ and $(11\bar{1})$ directions are occupied. However, as the growth direction changes to 55° of polar angle and 45° of azimuth angle, the Ge valleys in the $(\bar{1}11)$, $(1\bar{1}\bar{1})$, $(1\bar{1}1)$, and $(\bar{1}1\bar{1})$ directions are occupied instead.

Some of key features of the results for several orientations are summarized in Table 5.4, which gives the occupied valleys and the xy and z polarization results. The values in the parentheses are the oscillator strength for the square potential well, which are very close to those for the parabolic potential well; the dependences of the oscillator strength on the growth direction for the square and parabolic potential wells are the same. From the results, it is clear that the magnitude and angular dependence of the oscillator strength is nearly independent of the shape of the potential well (within a few percent). Our results also show that, for a given growth direction, the oscillator strength for the z polarization is larger than that for the xy polarization. The relative ratio of the two for the relaxed case is in agreement with the calculated absorption coefficient for the inversion layer of MOS [66].

5.4.5 Experimental Observation of Intersubband Absorption of (100)- and (110)-Oriented Si and SiGe

In this section, experimental results are shown for both (100)- and (110)-oriented Si and SiGe multiple quantum wells. Intersubband transitions for n-type (100)-oriented Si have been reported [75–78]. However, all the results show a similar polarization dependence as that of the Γ GaAs/AlGaAs quantum wells, owing to the vanishing off-diagonal elements of the effective mass tensor for (100). For (110), we will demonstrate electron intersubband absorption in $Si_{1-x}Ge_x/Si$ quantum wells grown on (110) Si substrates with the optical field parallel to the layers; that is, detection of incident light normal to the plane of the quantum wells. Previously, in the investigation of the intersubband spectroscopy of the MOS inversion layer, Nee, Claessen and Koch [67] observed an indication of a transition for both perpendicular and parallel incident light directions for (110) MOS inversion layers. They used the effective mass tensor formulated by Stern and Howard [69] and showed that the xy-polarized electric field can induce the intersubband transition through the off-diagonal terms of the inverse mass tensor for (110) samples. One of the problems of using the inversion layer is the low quantum efficiency for IR detector application. Another problem is that the transition energy usually occurs in the far-IR range (tens of meV) because of its large effective mass, thus making it impossible for application in the wavelength range of 8–12 μm ($\sim$100 meV). With the recent advent of molecular beam epitaxy and low-temperature chemical vapor deposition, the successful growth of the strained SiGe layer makes it possible to explore quantum well structures using Si/SiGe heterostructures or δ-doped layers. In what follows, we will discuss experimental verification of the normal incidence transition in the [110] case. The transition energy is shown to be tunable from far IR to a few μm in wavelength. The polarization angle dependences of the absorption spectra are obtained and compared with those observed for samples grown on (001) Si substrates.

In experiments, structures as shown in Figure 5.13(a) are used. The structures are designed to have tensile strain in the Si wells such that the $X4$ valleys·shown in Figure 5.13(b) are mostly occupied and these four valleys give the absorption of normal incidence. Samples used are grown in a Si-MBE system and n-type doping is achieved by Sb thermal evaporation at a low temperature [79, 80]. Details of the growth are described elsewhere [78]. The sample structure consists of an undoped $Si_{1-y}Ge_y$ buffer layer followed by 10 periods of a 50 Å Si well and a 300 Å $Si_{1-x}Ge_x$ barrier, then a 2500 Å $Si_{1-y}Ge_y$ undoped cap layer, as shown in Figure 5.13(a). The Ge composition x of the barrier is varied from 0 to 50% in the experiment, and the Ge content (y) in the buffer layer is chosen to maintain the symmetric strain condition in the multiple quantum wells [74]. More important, this is done to provide a favorable condition for normal incidence detection as discussed in Section 5.4.4 and also will be discussed later. The Si well layers are doped with Sb in the center 40 Å region and are grown below 400°C for the desired sharp and high doping profile. The doping concentration of the Si layers is about 1.3×10^{20} cm^{-3} according to *secondary ion mass spectrometry* (SIMS) profiles.

Three samples are used and they have the same doping concentration in the quantum wells. For sample A, quantum wells are formed only by doping; that is, a doped quantum well structure without the use of SiGe/Si heterostructures. Sample B has $Si_{0.7}Ge_{0.3}$ barrier layers, and sample C has $Si_{0.5}Ge_{0.5}$ barrier layers. Sample A has only 5 quantum wells whereas samples B and C have 10 periods. Details of the measurement procedure are described elsewhere [78]. For the (110) substrate, two waveguides with wedges along ($\bar{1}10$) and (001) directions are prepared for each sample, and these directions are illustrated in Figure 5.13(b).

Figure 5.14 shows transition energy, normalized absorption strength per quantum well, and *full width at half maximum* (FWHM) of three samples with different Ge compositions in barrier layers when the 0° polarization angle is used. The polarization angle ϕ is defined as shown in the inset of Figure 5.15, where the 0°

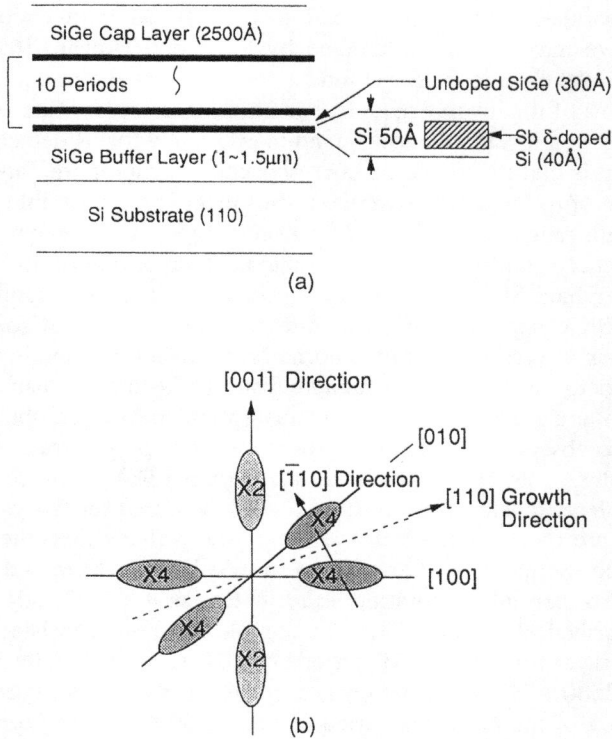

Figure 5.13 (a) Sample structure grown on (110)-oriented substrate. The structure consists of a 1 ~ 1.5 μm undoped $Si_{1-y}Ge_y$/Si buffer layer followed by 10 periods of a 50 Å Si layer doped with Sb in the 40 Å center region, an undoped $Si_{1-x}Ge_x$/Si barrier of 300 Å thick, and finally a 3000 Å undoped $Si_{1-y}Ge_y$/Si cap layer. (b) Growth direction and the relation with all ellipsoids.

polarization angle is shown and the 90° polarization angle represents an optical field of the incident light along the quantum well plane and waveguide wedges. The absorption peak position (transition energy) of sample A is 216 meV (5.8 μm), and those of samples B and C are 226 meV (5.5 μm) and 253 meV (4.9 μm), respectively. The transition energy increases as the Ge composition increases. This is because the conduction band offset of Si/SiGe heterostructures for the $X4$ valleys become large as the Ge composition increases. The potential wells become deeper due to the combined effects of increasing conduction band offset and δ doping, resulting in a larger energy level separation between subbands. With δ doping as in our case, higher transition energies become even larger due to many-body effects depolarization and excitonlike shifts as discussed in Section 5.4.3. This makes the transition wavelength range of 8–12 μm reachable in the $Si_{1-x}Ge_x$/Si quantum well case, in

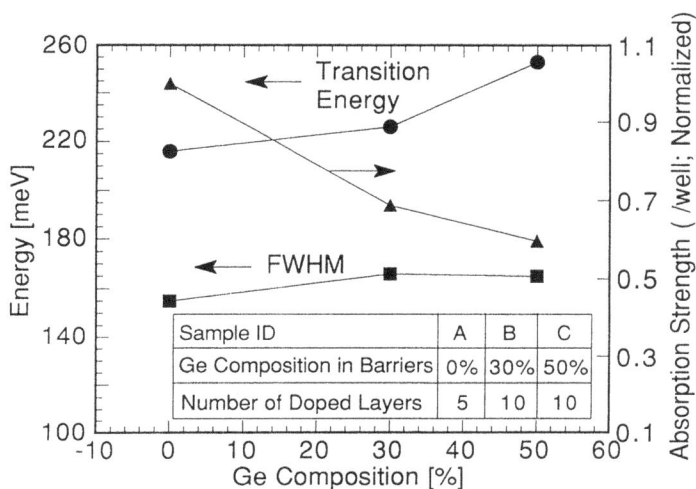

Figure 5.14 Ge composition dependence of transition energy, normalized absorption strength per quantum well, and FWHM for the 0° polarization. The Ge compositions of samples *A*, *B*, and *C* are 0, 30, and 50%, respectively. The peak energy position (transition energy) shifts to a higher energy as the Ge composition is increased. The structure description for the three samples is summarized in the inset.

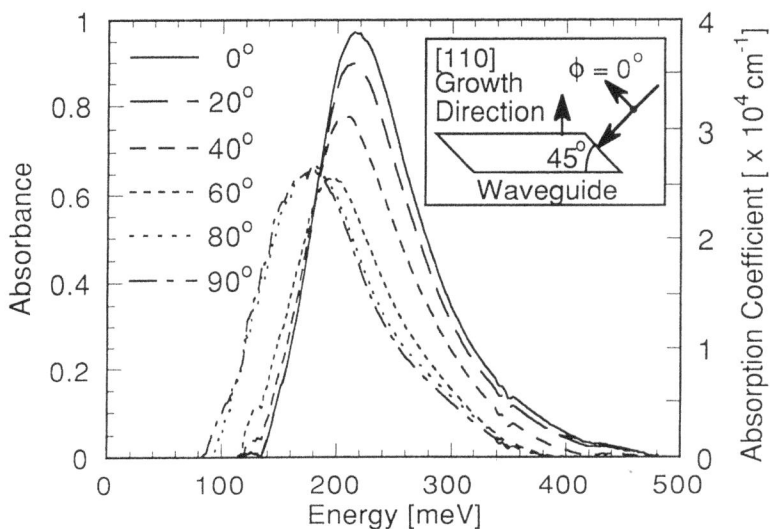

Figure 5.15 Polarization angle dependence of absorption spectra of sample *B*. The polarization angle ϕ is defined in the inset. The absorption spectrum is observed at $\phi = 90°$ due to intersubband transition excited by the parallel field. The peak energy position shifts and an increase of FWHM is also observed as the polarization angle is increased.

contrast with the MOS inversion quantum well case. The transition energies are easily tunable by changing the Ge composition and the doping concentration in the well region. We also observe that the absorption strengths decrease as the Ge composition increases. The full widths at half maximum of the absorption spectra are almost the same with different Ge compositions.

The polarization angle dependence of absorption spectra of sample B is shown in Figure 5.15. Other samples show similar dependences. The absorption strength is at the maximum for the $0°$ polarization angle. The peak value in absorbance corresponds to an absorption of 90% in the sample. As the polarization angle increases, the absorption strength decreases and the peak positions shift to lower energies. The absorption strength is at the minimum for the $50°$ polarization angle. As the polarization angle is increased further, the absorption strength increases again. At large polarization angles, the absorption arises from the intersubband transition of electrons in the $X4$ valleys excited by the parallel field. The *parallel field* is defined as the optical field component along the quantum well planes and the waveguide wedges (i.e., $(\bar{1}10)$ direction in Figure 5.13(b)); and the *perpendicular field* is defined as the optical field along the quantum well growth direction (i.e., (110) direction in Figure 5.13(b)). The absorption peak position at the $0°$ polarization angle is 226 meV (5.5 μm) and 177 meV (7.0 μm) at $90°$, respectively. The substantially large transition energies are due to the combined effect of the band offset and δ doping, as explained before. The reason why the peak position shifts to lower energies at large polarization angles may be in part due to the depolarization effect [67]. It is also observed that the absorption spectra become broader at larger polarization angles. The FWHM of the spectrum at the $0°$ polarization is 166 meV and 189 meV for the $90°$ case, respectively. The polarization angle dependence of absorption strength of sample B for the optical field along both the $(\bar{1}10)$ and (001) wedge directions are summarized with open circles and squares in Figure 5.16. The plotted absorption strength represents the integrated area under the absorption spectrum.

Next, we discuss the conditions to have normal incidence detection. When the polarization angle is $90°$, the intersubband absorption is induced by the parallel field (i.e., normal incident light). The parallel field cannot cause any intersubband transitions for (001) substrates. For (110) substrates, the mass tensor of electrons in the $X4$ valleys (along (100), $(\bar{1}00)$, (010), $(0\bar{1}0)$ directions) has both diagonal and off-diagonal terms due to the four tilted energy ellipsoids as shown in Figure 5.13(b). The diagonal terms represent the electron motion along the growth direction induced by the perpendicular field whereas the off-diagonal terms represent the electron motion along the quantum well growth direction by the parallel field. The projected effective mass of electrons in the four $X4$ ellipsoids to the quantization direction (growth direction) is $0.32m_0$. Similarly the projected effective mass to a quantum well plane along the $(\bar{1}10)$ direction is evaluated to be 0.47 m_0. The mass tensor of electrons in the $X2$ valleys (along (001) and $(00\bar{1})$ directions) has only diagonal terms, and therefore the electrons in the X2 valleys *cannot* give rise to the motion in the

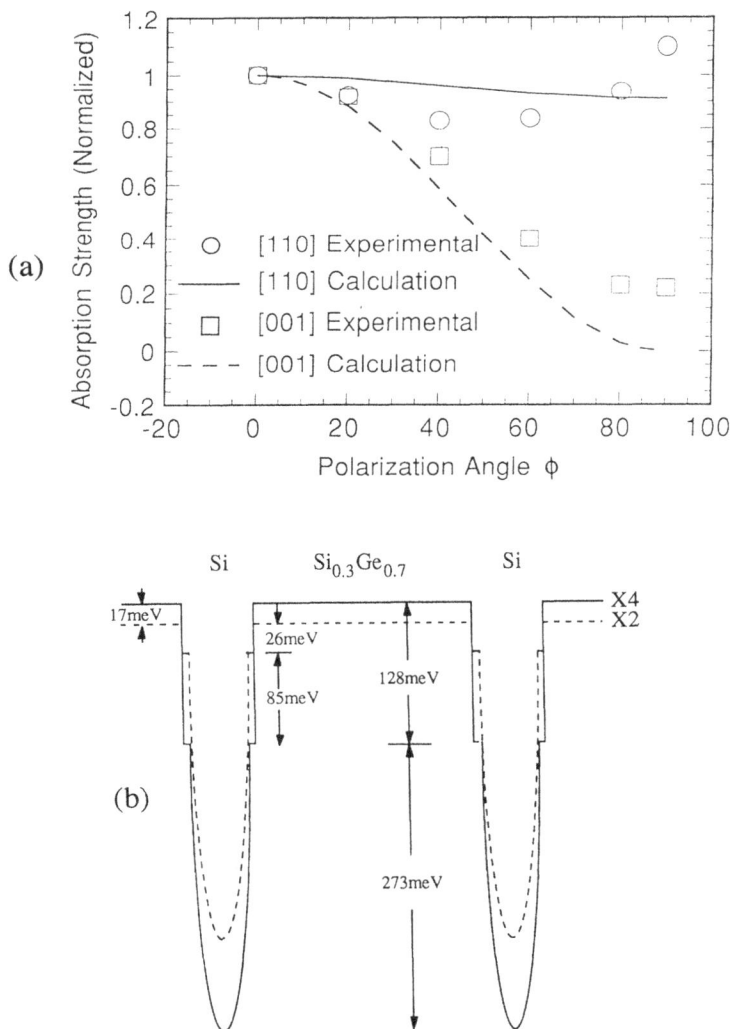

Figure 5.16 (a) Absorption strength as a function of the polarization angle, ϕ, for the δ-doped Si (110) quantum well. The circles and squares are the experimental data Zeindel et al. [75] when the $\hat{x}$ axis is chosen to be along $(00\bar{1})$ and $(\bar{1}10)$, respectively. The solid and dashed lines are the calculated values for the $\hat{x}$ axis along $(00\bar{1})$ and $(\bar{1}10)$, respectively. (b) Band diagram of the symmetrically strained $Si_{0.7}Ge_{0.3}$/Si layers used in the experiment [78].

quantum well growth direction by the parallel field. Consequently, only the perpendicular field can induce the intersubband transition for electrons in the $X2$ valleys as in the case of (001) Si. The effective mass of electrons in the two $X2$ ellipsoids is the transverse mass or $0.19m_0$. Thus, it is important that the $X4$ valleys be occupied to have intersubband transitions excited by the parallel field or, in other words, to have normal incidence detection. The (110) SiGe/Si multiple quantum well structures, shown in Figure 5.13(a), are designed to have a proper strain to satisfy the condition. With the $Si_{1-y}Ge_y$ buffer layer, the Si layers suffer a tensile strain and the $X2$ and $X4$ energy bands in Si layers split, making the $X4$ valleys lower than the $X2$ valleys. The energy level separation of the $X2$ and $X4$ valleys is 85 meV for the $Si_{0.7}Ge_{0.3}$/Si multiple-quantum well structures if the doping effect is not considered. With this energy level separation, most of the electrons will occupy the $X4$ valleys, and this can explain the large absorption strength obtained at $\phi = 90°$.

To confirm the experiment further, we have calculated the absorption for the waveguide structure used in experiments. The calculated absorption results are also shown in Figure 5.16 for the $(\bar{1}10)$ direction (solid line) and the (001) direction (dashed line; i.e., $\cos^2\phi$), respectively. In the calculation of the absorption, the projection of the incident light to the xy plane is assumed to be along the (110) growth direction. It is assumed that carriers occupy only the $X4$ valleys, and the intersubband absorption can be evaluated for all optical field polarizations with the use of different effective masses. When the polarization angle ϕ is changed from 0° to 90° the perpendicular field decreases as $\cos\phi$ and the parallel field increases as $\sin\phi$ as shown in Figure 5.17 for optical field along $(00\bar{1})$ and $(\bar{1}10)$. Figure 5.17 shows the calculated polarization dependence of the oscillator strength of a waveguide structure used for the experiment. Therefore, intersubband absorption due to the perpendicular field is a function of $\cos\phi$ and that due to the parallel field is a function of $\sin\phi$ [81]. The transition energy has also been calculated using the Hartree-Fock potential, including the depolarization and the excitonlike effects [63]. The transition energy obtained is 206 meV for sample B, compared to the value of 85 meV with the Hartree-Fock potential only. This value is in close agreement with the experimentally observed value of 226 meV.

As shown in Figure 5.16, the agreement between experiment and calculation is reasonable, but there is a discrepancy. In the (001) direction case, the deviation at high polarization angles may be due to the difficulty in cutting the wedge at the exact (001) angle in experiments. In that case, the incident light is off the direction and the mass tensor has small off-diagonal terms. In the $(\bar{1}10)$ direction case, the absorption comes from the parallel and perpendicular fields and the attribute of the discrepancy is more difficult to assess. We have calculated the mixed contribution by rotating the sample plane along the (110) direction of the inset of Figure 5.15(a) by 10°. In this case, we introduce a mixed contribution of 95% of $X4$ and 5% of $X2$, and these calculated results are plotted in the solid and dotted lines, giving a reasonably good agreement. On the other hand, it may also be due to different ef-

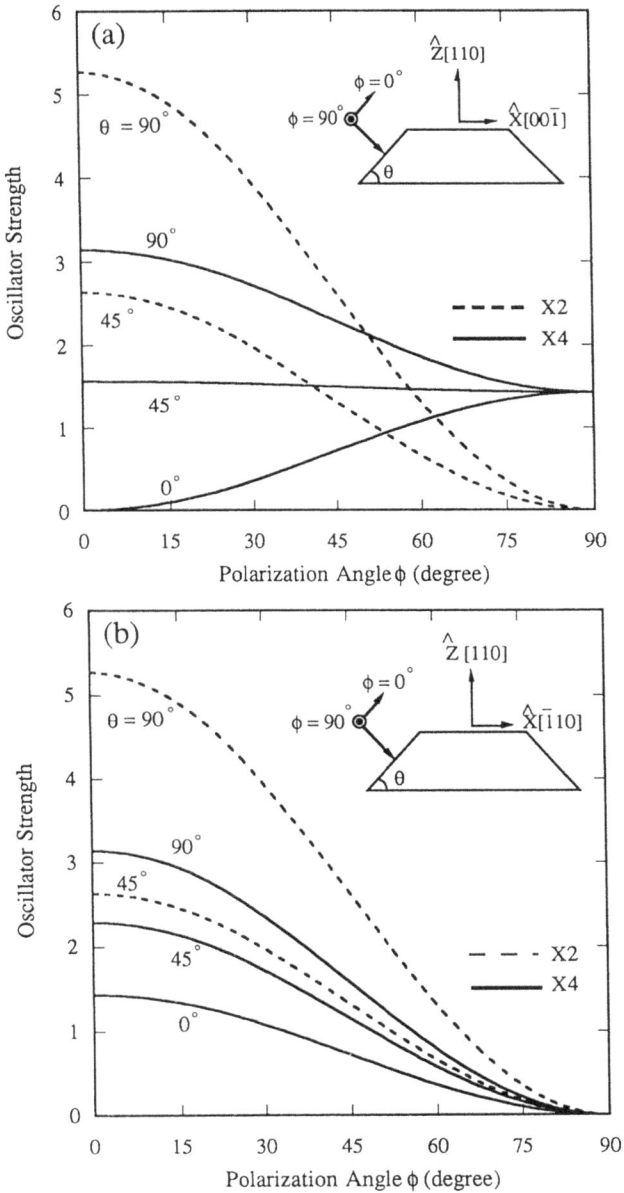

Figure 5.17 Oscillator strength of a waveguide structure for the transition between the ground state and first excited state when the minimum conduction valley is Si-like. The growth direction is along (110) direction, and the $\hat{x}$ axis is chosen along the (a) $(00\bar{1})$ and (b) $(\bar{1}10)$ directions (see inset).

fective masses and the depolarization effect. More work needs to be done in this area.

These results indicate that normal incident detectors can be fabricated by using SiGe/Si strained quantum wells. The normal incidence detection can be realized with the strained quantum well structures grown on substrates other than the (001) orientation for Si-like valleys (X valleys). For Ge-like valleys (L valleys), normal incidence can occur for orientations other than (111). The strain can be engineered to assure that the proper valleys that can provide the normal incidence detection are occupied with electrons.

5.5 *p*-TYPE INTERSUBBAND TRANSITION

The intersubband transitions in SiGe/Si quantum wells were first observed for holes [82] using a multiple-quantum well structure grown pseudomorphically on an Si substrate. For pseudomorphic structures grown on Si, most of the band offset appears at the valence band. More recently, electron intersubband absorption has been observed [83, 78] using a multiple quantum well structure grown on a relaxed SiGe buffer layer to obtain appreciable band offset at the conduction band. In this section, we discuss the experimental observation of hole intersubband infrared absorption in $Si_{1-x}Ge_x$/Si multiple quantum wells grown on Si substrates [82].

All of the samples described here are grown in a molecular beam epitaxy (Si-MBE) system on high resistivity (100 Ω-cm) Si (100) wafers. This reduces the infrared absorption due to free carriers in the substrate. First, an undoped 3000 Å Si buffer is grown on an Si substrate. The multiple-quantum well structure consists of 10 periods of 40 Å $Si_{0.6}Ge_{0.4}$ wells and 300 Å Si barriers. The center 30 Å region of the quantum wells is *p*-doped to about 1×10^{19} cm^{-3} using a thermal boron source [84]. The quantum well structure is capped with a 1000 Å Si layer. The thick barrier layer keeps the average Ge composition of the entire structure sufficiently low to avoid strain relaxation of the $Si_{0.6}Ge_{0.4}$ layers [18].

The measured absorption spectra as a function of photon energy using the waveguide structure is shown in Figure 5.18. The set of curves shows different polarization angles of the infrared beam. Zero degrees correspond to the polarization of the beam along the growth direction. A peak at 153 meV (8.1 μm) with a long tail toward the short-wavelength region is observed. The small features superimposed on the spectra are mainly due to strong infrared absorption near 136 meV by SiO_2 and from the water bands that are in the spectral range of interest. The strength of the intersubband absorption decreases as the polarization of the infrared beam is rotated from the perpendicular direction toward the direction parallel to the plane of layers. Figure 5.19 illustrates the relative absorption strength as a function of the polarizer angle ϕ. The data follows the $\cos^2\phi$ dependence (dashed curve) is in good agreement with the selection rule of the intersubband transition as discussed in Section 5.4. The absorption strength (I_A) due to multiple reflections in the waveguide

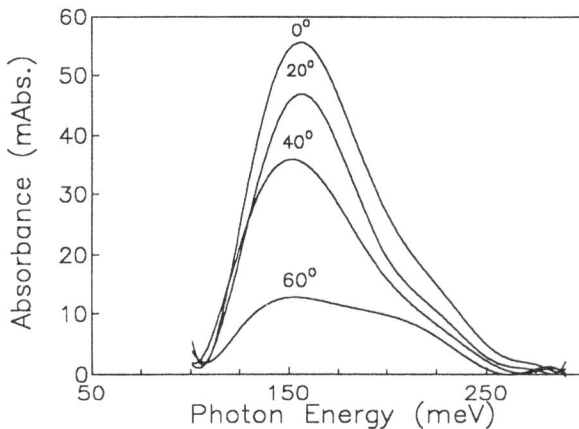

Figure 5.18 Measured absorption spectra as a function of photon energy for different polarizations of the incident infrared beam. The peak at 153 meV is due to intersubband absorption between the heavy-hole ground state and the first excited state. The absorption strength at large polarization angles is shown to decrease in accordance with the selection rules of intersubband transition at the Γ point.

can be found by using eq. (5.23)

$$I_A = \rho_s N_T \frac{e^2 h}{4c\varepsilon_0 n_r m*} f \frac{\cos^2\theta}{\sin\theta} \cos^2\phi \qquad (5.45)$$

where N_T is the total number of quantum wells that the infrared beam passes through due to multiple reflections, ρ_s is the two-dimensional density of holes in the well, I_A is the integrated absorption strength, θ is the angle of incidence extending from the normal to the plane, and the effective mass $m*$ is taken along (100) direction. Here, we ignore the nonparabolic effects on the intersubband transition due to the quantization and the strain present in the quantum well.

The oscillator strength of the transition (f) can be estimated by using the eq. (5.45). For the structure, we used $N_T = 100$, $m* = 0.26 \ m_0$, $n_r = 3.4$, $\theta = 45°$, $\rho_s = 3 \times 10^{12} \ cm^{-2}$, and the integrated absorption strength, $I_A = 5$ meV. These parameters give an oscillator strength of 0.98. This is in good agreement with the calculations as well as the values obtained for intersubband absorption in III–V-based quantum well structures [47, 85], which are in the 0.5 to 1.2 range. It should be noted that this is only an estimate because the nonparabolicity of the hole bands can be severe in the quantum well and the joint density-of-states effective mass can deviate significantly from the bulk value [86].

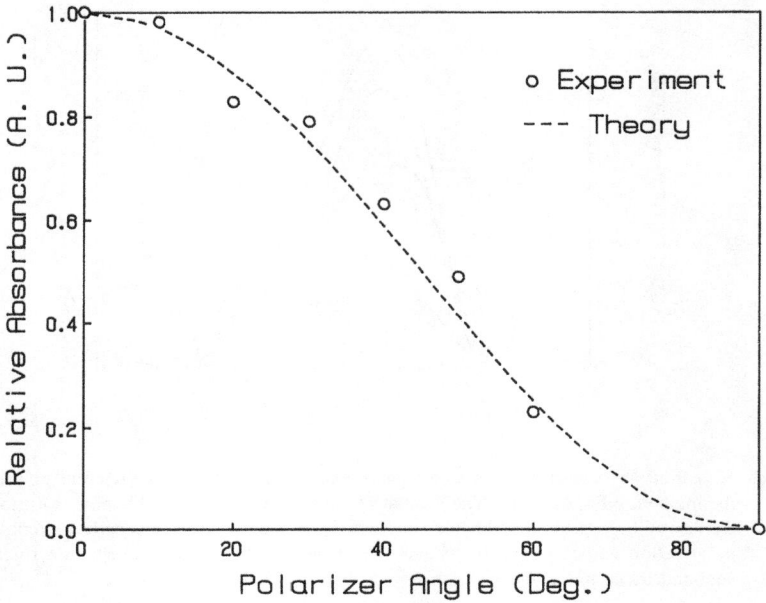

Figure 5.19 Normalized absorption strength as a function of the polarization of the incident infrared beam. Zero degrees corresponds to polarization along the growth direction of the structure. The dashed curve shows the theoretically expected $\cos^2\phi$ dependence.

In assessing the origin of the observed transition, we recognize that the presence of three hole bands (heavy, light, and split-off) with energy separations close to the observed absorption peaks makes the identification of the transitions difficult. However, the strong polarization dependence indicates that the transitions involving quantized states of two different hole bands are relatively small at least for Si. For structures containing large Ge composition as well as high doping in the well, such transitions have been observed [10]. The detailed description of intervalence band transition will be given later in this chapter. Assuming that no transitions are involved for different bands simplifies the deconvolution process. Figure 5.20 shows a detailed band diagram of the quantum wells with the light- and heavy-hole bands and their bound state energies. The bound state energies in the quantum well are calculated by using the procedure discussed in Section 5.4.3. The band offsets under compressive strain obtained from a linear interpolation of those of Van der Walle and Martin [39]. The effective masses of the light and heavy holes along the (100) direction are obtained from Chun and Wang [87], in which the effective masses are calculated by using three band **k · p** and strain Hamiltonians [35, 88]. The band

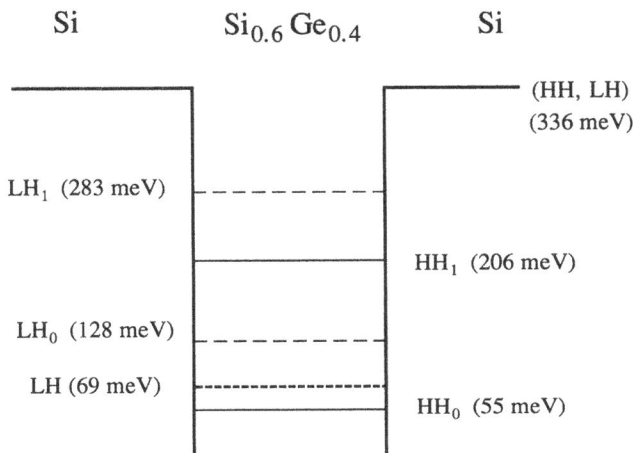

Figure 5.20 Band structure of the Si/Si$_{0.6}$Ge$_{0.4}$/Si quantum well showing the bound states due to both light- and heavy-hole bands. The heavy-hole (HH) and light-hole (LH) band edges are shown in solid and dashed lines, respectively.

offsets for the light and heavy holes are 267 meV and 336 meV, and the effective masses are 0.22m_0 and 0.26m_0, respectively. There are two heavy-hole bound states at 55 meV and 205 meV, and two light-hole states at 128 meV and 283 meV in the quantum well. All of the energies are measured relative to the heavy-hole band edge in the well region.

The light- and heavy-hole bands are degenerate in the Si barrier regions and are split in the well. This is due to different mass quantization and the strain in the Si$_{0.6}$Ge$_{0.4}$ layer [70]. In the quantum well, the light-hole band lies about 70 meV "above" the heavy-hole band, and the split-off hole band is about 290 meV "above" the heavy-hole band (here *above* means higher in terms of hole energy). These energy separations are relatively large compared with kT at room temperature (~26 meV). At thermal equilibrium, most of the holes occupy the heavy-hole ground state. Therefore, the intersubband absorption occurs mainly between quantized heavy-hole states. The absorption peak at 153 meV (8.1 μm) is due to the transition from the heavy-hole ground state to the first excited state. The absorption peak width is about 75 meV, about seven times larger than the intersubband absorption peak widths typically observed in AlGaAs/GaAs quantum well structures [47, 85, 89]. This may be due to the strong nonparabolicity and band mixing effects of the hole bands, particularly in the presence of a strain. The calculated peak position for the transition between the ground state and the first excited heavy hole state is 158 meV. This agrees reasonably well with the experimental value of about 153 meV.

5.5.1 Intersubband Transitions in δ-Doped Quantum Wells

Next, we will discuss the use of p-type δ-doped layers in Si for the study of intersubband transitions. The δ doping in semiconductors can be viewed as an alternative way to achieve quantum well structures without heterojunctions. The potential profile associated with δ doping closely resembles that of a parabolic quantum well due to the finite width of the doped layer [90]. The well thickness and the barrier height can be controlled by the thickness of the doped layer and the doping density. The potential profile and the energy level spectrum in the well are usually obtained by solving Schrödinger's and Poisson's equations self-consistently (Hartree approximation). The existence of quantized states in such a system in Si was probed by using tunneling spectroscopy [75] and optical transitions between quantized states have also been studied [10, 76, 77]. For intersubband detector applications, the population of carriers in such structures are considerably larger than those typically used in heterojunction quantum wells. As a result, the many-body effects play an important role in determining the optical properties of δ-doped quantum wells. In the following section, we describe the intersubband absorption of δ-doped quantum wells, followed by a discussion of the effect of many-body interactions on the optical transitions.

5.5.2 Tuning of Intersubband Transition Energy by Doping

A typical structure used in this study consists of an undoped Si buffer layer followed by 10 periods of 35 Å heavily boron-doped Si layers and 300 Å undoped Si spacers. Four samples (A, B, C, D) with doping concentrations of about 0.7, 2.7, 4.0 and 5.7×10^{20} cm^{-3}, respectively, are prepared to study the dependence of the energy level separation on the population of holes in the subbands. The doping density is calibrated by secondary ion mass spectroscopy. A full width at half maximum of approximately 50 Å is obtained from the depth profile.

The absorption spectra of the samples are taken at room temperature using a *Fourier transform infrared spectrometer* (FTIR). To enhance the absorption strength, waveguide structures are used that are similar to that used for SiGe/Si multiple quantum well structures discussed in Section 5.4.5. The measured absorption spectra of the samples as a function of photon energy are shown in Figure 5.21. Peak positions are found near 210, 259, 348, and 377 meV for samples A, B, C, and D, respectively. It can be clearly seen that the absorption spectra shift toward the high-energy regime with increasing absorption as the doping density is increased. The shift of the absorption peak energy is due mainly to the increase of the potential well depth at high doping densities. The widths of the absorption peaks are more than an order of magnitude larger (as in the n-type case) than those observed in GaAs/AlGaAs quantum well structures, which are typically about 10 meV. The nonparabolicity of

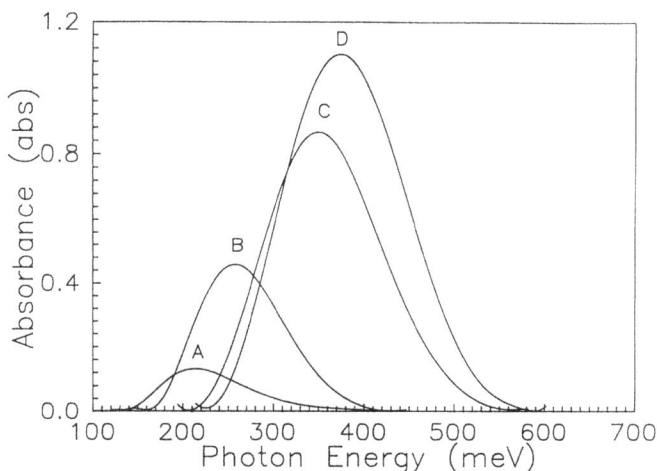

Figure 5.21 Absorption spectra of the four samples as a function of photon energy at 300 K. The set of curves are due to different doping concentrations.

hole bands plays an important role in determining the peak width. In particular, at high doping densities, the hole bands can be filled up to several hundreds of meV.

To further confirm the nature of the transition, Figure 5.22 shows the relative absorption spectra of the sample C as a function of polarization angle ϕ, where $0°$ and $90°$ correspond to the electric field of the photon polarization in the plane of incident and perpendicular, respectively. The reduction of absorption strengths with the polarization angle ϕ as shown in Figure 5.22 is in agreement with the selection rule of the intersubband transitions as in the case of SiGe/Si multiple quantum wells discussed in Section 5.4.5.

To further understand the obtained data, particularly the variation of peak positions as a function of the 2D hole density in the wells, the subband energies are first estimated using a multiband self-consistency calculation (Hartree approximation) [63]. In the calculation, we assume that the doping is uniformly distributed over a 35 Å layer of Si and that the light- and heavy-hole bands are degenerate at the Γ point whereas the split-off band is 44 meV above the light- and heavy-hole bands (here hole energy is taken as positive). The light, heavy, and split-off hole effective masses of $0.2m_0$, $0.30m_0$, and $0.24m_0$, respectively, along the Si (100) direction and density-of-states effective masses of $0.15m_0$, $0.40m_0$, and $0.24m_0$, respectively in the (100) plane are used for the subband energy calculation. The calculated peak positions for the samples A, B, C, and D are 62, 118, 139, and 159 meV, respectively. These values are considerably smaller than the measured peak

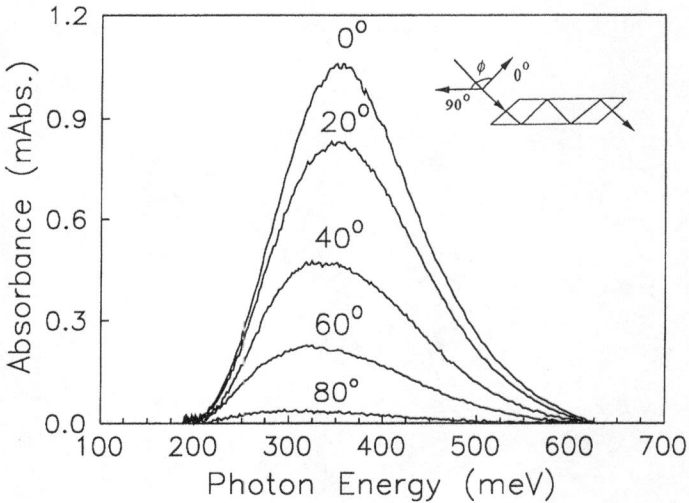

Figure 5.22 Polarization dependent absorption spectra of sample *C* at 300 K. The decrease of the absorbance with increasing polarizer angle is due to the reduction of the component of photon polarization along the growth direction similar to that of SiGe/Si multiple quantum wells.

positions. This discrepancy is due mainly to the failure of the single particle approximation in the calculation. The many-body effects as discussed in Section 5.4.3 play an important role in determining the confining potential as well as the plasma shift (depolarization and exciton-like) of the peak position due to collective transition of the carriers as discussed previously.

To estimate the magnitude of the many-body effects, we first calculated the wave functions and the bound state energies self-consistently, using the Hartree and exchange-correlation potentials as described in Section 5.4.3. The contribution due to depolarization and excitonlike effects are then estimated by using eqs. (5.36) and (5.37). The calculated potential well, eigenstates, and wave functions that correspond to the heavy-hole band for the sample with the 2.7×10^{20} cm^{-3} doping density are shown in Figure 5.23 (here hole energy is taken as positive). Figure 5.24 summarizes the experimental and calculated peak positions, using different levels of approximations. The dashed curve with filled circles shows the measured peak positions and the dotted curve represents the calculated peak positions without the plasma shifts. As mentioned before, these values are considerably smaller than the experimental values. The solid curve in Figure 5.24 shows the incorporation of plasma (i.e., depolarization and excitonlike) shifts to correct the calculated transition energy. This brings the calculated and experimental peak positions reasonably close to agreement. The slightly lower values obtained in the calculation may be due to the omission of the nonparabolicity effect and the occupancy of the excited state in the anal-

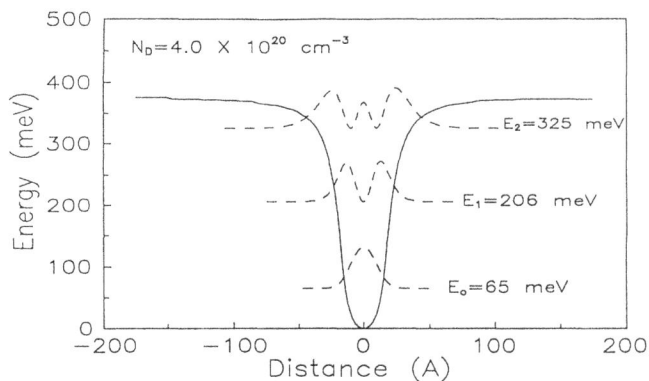

Figure 5.23 Calculated potential well and wave functions for the sample with 2.7×10^{20} cm^{-3}, using a multiband Hartree approximation. Only the heavy-hole subbands are shown.

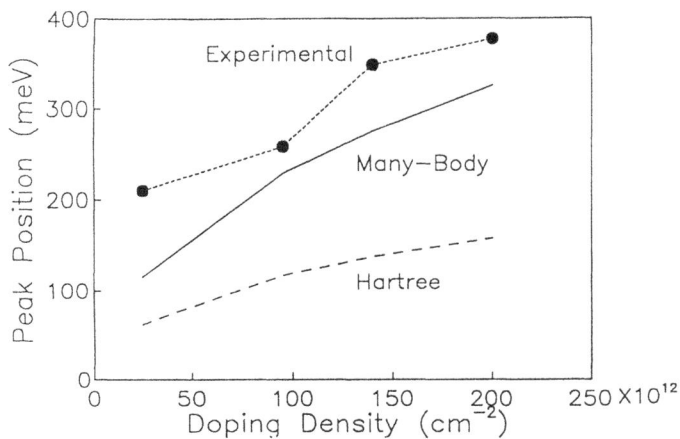

Figure 5.24 Subband separation as a function of doping density: experimental data (dotted curve with filled circles), using the Hartree approximation (dashed curve), and the Hartree approximation with exchange-correlation potential including depolarization and exciton-like shifts (solid curve).

ysis. Next, we will discuss another type of infrared transition that occurs between different hole bands; for example, the light and heavy holes.

5.5.3 Inter-Valence Band Transitions

Many approaches have been proposed to provide normal incident detection other than free carrier absorption, which is limited to the longer wavelength range. We have discussed, for example, the use of off-diagonal elements of the conduction band effective mass tensor (away from the Γ point) for n-type dopants to induce the motion of carriers in the direction normal to the in-plane optical field [68]. Here, we discuss yet another normal incidence transition mechanism based on inter-valence band transition. Earlier on, normal-incident inter-valence band absorption from the heavy-hole and light-hole bands was reported for bulk Ge [91] and using AlGaAs/GaAs multiple quantum wells [92]. In this section, we describe the observation of normal-incidence infrared absorption from inter-valence subband transition in SiGe/Si multiple quantum wells having several different Ge contents. The physics of this transition is also discussed.

Four $Si_{1-x}Ge_x$/Si multiple quantum well structures with different Ge compositions, $x = 0.15$, 0.3, 0.4, and 0.6 are used for our discussion. For all samples, one period of the SiGe/Si multiple quantum well structure consists of a 40 Å $Si_{1-x}Ge_x$ well and a 300 Å Si barrier. The center 30 Å of the $Si_{1-x}Ge_x$ wells is boron doped to about 5×10^{19} cm^{-3} and the Si barriers are undoped. For the $Si_{0.85}Ge_{0.15}$/Si, $Si_{0.7}Ge_{0.3}$/Si, and $Si_{0.6}Ge_{0.4}$/Si samples, there are 10 periods of multiple quantum wells, whereas for the $Si_{0.4}Ge_{0.6}$/Si sample, only 5 periods are used because of the critical thickness limitation of the SiGe strained layers. To enhance the absorption, the quantum well is heavily doped in the center region of 30 Å. To further study the Ge composition dependence, a sample of 10 periods of Si δ-doped quantum wells with 300 Å undoped Si barriers is also used with the same doping profile for comparison. The transmission spectra of the samples are taken at room temperature using a Fourier transform infrared spectrometer. Both waveguide and Brewster's angle geometries are employed in the measurement.

Figure 5.25 shows the measured absorption spectra of the $Si_{0.6}Ge_{0.4}$/Si quantum well sample at two different polarization angles using the waveguide structure. At the 0° polarization angle (electric field having a component in the growth direction as depicted in the inset), an absorption peak occurs at 5.6 μm. At the 90° angle (beam polarized parallel to the plane), this peak vanishes, but another broad peak appears at a shorter wavelength, 3.2 μm. The peak at the 0° polarization is due to intersubband transition between two heavy-hole states [82], as discussed previously in Section 5.5. This is verified from the polarization dependence of the peak ($\cos^2\phi$), a typical behavior of intersubband transition [82]. For the present sample, the heavy-hole intersubband absorption is observed at a somewhat shorter wavelength than that discussed earlier.

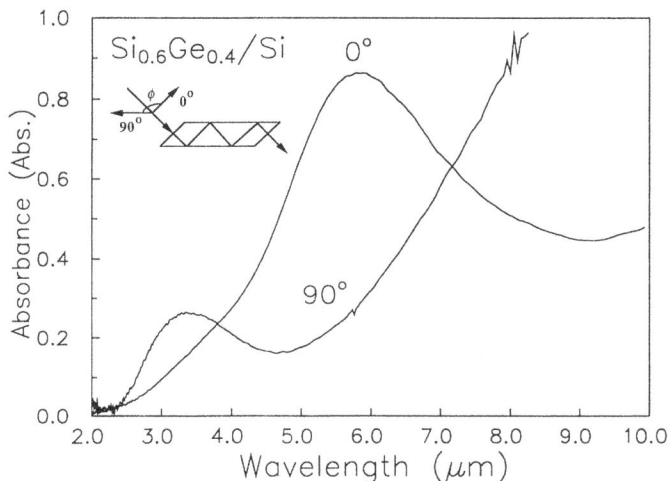

Figure 5.25 Measured room temperature absorption spectra of a $Si_{0.6}Ge_{0.4}/Si$ multiple quantum well sample at 0° and 90° polarization angles. At 0°, a clear peak due to intersubband transition is shown at near 5.6 μm. However, at 90°, this peak vanishes, and another peak appears at 3.2 μm, which is due to intervalence-subband transition. The inset shows the waveguide structure used for the measurement.

The peak at 3.2 μm shows a polarization dependence as illustrated in Figure 5.26 (also using a waveguide structure). At normal incidence (90° polarization), the absorption reaches the maximum, and it drops as the angle decreases. Two other samples, $Si_{0.7}Ge_{0.3}/Si$ and $Si_{0.4}Ge_{0.6}/Si$, show similar absorption spectra as the $Si_{0.6}Ge_{0.4}/Si$ sample discussed previously. For the sample with a lower Ge concentration (15%) and the δ-doped structure in Si [10], absorption spectra reveal clear peaks at 0°, due to the heavy-hole intersubband transition; however, no obvious peaks are observed at 90°. The peaks observed at 90° are also observed at Brewster's angle (without the use of the waveguide structure), but absorption strength is less, as expected due to a single pass of IR through the structure. On the other hand, the heavy-hole inter-subband peaks observed with the waveguide structures are not seen at the Brewster's angle because in this case the absorption is reduced due to both the lack of multiple reflection and the reduction of the polarization angle as the light enters the multiple quantum well structure.

Figure 5.27 shows the absorption (in transmittance) spectra of three different samples at 90° polarization angle. The peak positions are found at 3.4, 3.2, and 2.4 μm for the $Si_{0.7}Ge_{0.3}/Si$, $Si_{0.6}Ge_{0.4}/Si$, and $Si_{0.4}Ge_{0.6}/Si$ samples, respectively. From the figure, it is clear that, as the Ge content in the well increases, the peak position shifts to a shorter wavelength, the absorption strength increases, and the peak broadens. Figure 5.28 summarizes the observed peak position and absorption strength per period as a function of the Ge content in the well.

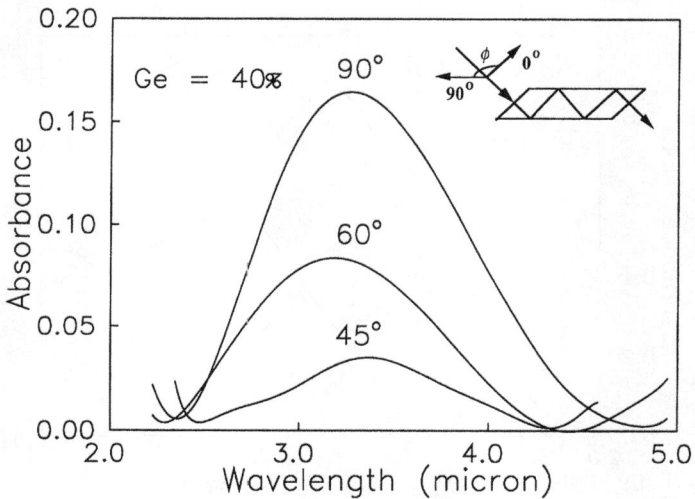

Figure 5.26 Polarization-dependent absorption spectra of the peak shown in Figure 5.25 at 90°. The peak intensity decreases as the angle decreases. The peak is due to inter-valence subband transition caused by the valence-conduction band coupling.

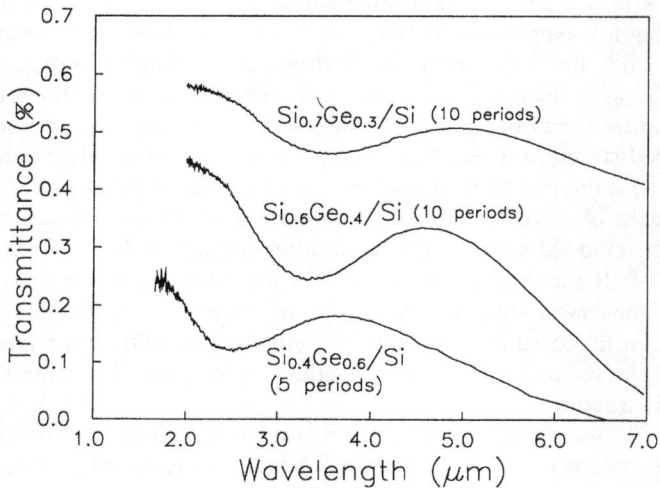

Figure 5.27 Transmission spectra of $Si_{0.7}Ge_{0.3}/Si$, $Si_{0.6}Ge_{0.4}/Si$, and $Si_{0.4}Ge_{0.6}/Si$ samples at normal incidence (90°). It is clear that, as the Ge content increases, the peak shifts to shorter wavelength, absorption strength increases, and the peak broadens. Note that only 5 periods were $Si_{0.4}Ge_{0.6}Si$.

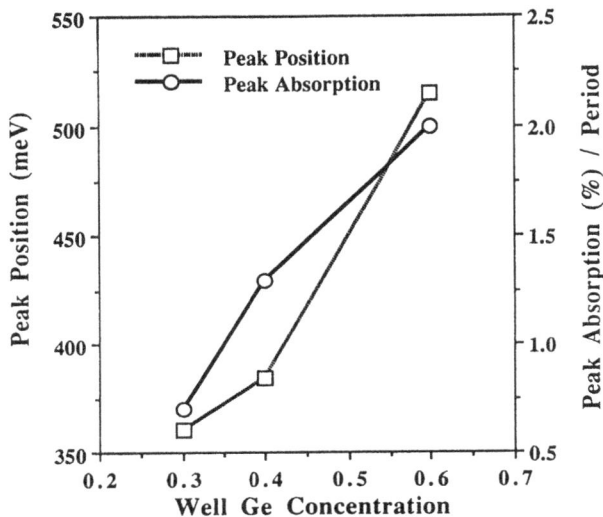

Figure 5.28 Observed peak positions and the peak absorption per period as a function of Ge concentration in the quantum well.

The origin of the absorption process can be understood qualitatively by considering the matrix element of optical transition including band mixing. For intervalence subband transitions (i.e., from the heavy-hole ground state to other valence band subbands), we can see that only the s-like component of the envelope functions contribute to the matrix element. In the following we show the expansion of the matrix element using a four-band approximation.

$$\langle \phi_f | \hat{\varepsilon} \cdot \vec{p} | \phi_i \rangle = \langle U_l F_{ls} + U_h F_{lp} | \hat{\varepsilon} \cdot \vec{p} | U_h F_{hs} + U_l F_{hp} \rangle \qquad (5.46)$$

where U and F represent the Bloch and envelope functions. The indices h and l denote heavy and light hole states and s and p indicate the s and p components of the envelope function, respectively. This equation can be expanded as follows

$$\langle \phi_f | \hat{\varepsilon} \cdot \vec{p} | \phi_i \rangle = \langle U_l F_{ls} | \hat{\varepsilon} \cdot \vec{p} | U_h F_{hs} \rangle + \langle U_h F_{lp} | \hat{\varepsilon} \cdot \vec{p} | U_h F_{hs} \rangle$$
$$+ \langle U_l F_{ls} | \hat{\varepsilon} \cdot \vec{p} | U_l F_{hp} \rangle + \langle U_h F_{lp} | \hat{\varepsilon} \cdot \vec{p} | U_l F_{hp} \rangle \qquad (5.47)$$

For normal incidence, the first term of eq. (5.47) has a non zero contribution whereas the second and third terms vanish. The fourth term is relatively small because it consists of the p-like component of envelope functions originating from the coupling with other bands. The first term of eq. (5.47) can be further simplified into

$$\langle \phi_f | \hat{\varepsilon} \cdot \vec{p} | \phi_i \rangle = \langle U_l | \hat{\varepsilon} \cdot \vec{p} | U_h \rangle \langle F_{ls} | F_{hs} \rangle \tag{5.48}$$

Note that the overlap between s-like components of envelope functions in the previous equation is not zero. This result can be generalized when multiple bands are involved in the transition as

$$M \approx \langle U_f | \hat{\varepsilon} \cdot \vec{p} | U_i \rangle \langle F_{fs} | F_{is} \rangle \tag{5.49}$$

where U_i and U_f are the initial and final Bloch functions, respectively; F_{is} and F_{fs} are the s-like components of the initial and final envelope functions, respectively; $\hat{\varepsilon}$ is the photon polarization vector; $\vec{p}$ is the momentum of the hole; and z indicates the growth direction. The matrix element is zero to the first order if U_f and U_i are both p-like at the Γ point, assuming no band mixing [93]. If, however, the light and spin-orbit split-off Bloch functions have significant band mixing from the s-like conduction band, a substantially large matrix element can result. The band mixing can be significant if the (Γ) bandgap is small enough and if the light hole and spin-orbit split-off bands involved in the transition have a large $k_\parallel$ [94].

For Si, due to the large direct bandgap at the Γ point (4.2 eV), the coupling is weak between the conduction and valence bands. On the other hand, for Ge, because the bandgap at the Γ point is small (0.82 eV), a significantly large coupling is expected. Due to this coupling, the valence band wave functions can have a component of the s-like conduction band state, particularly for large $k_\parallel$. For $Si_{1-x}Ge_x$, the bandgap at the Γ point decreases as the Ge content increases, and therefore, with a sufficient Ge content, the inter-valence subband transition will become stronger. One earlier report in inter-valence band transition in p-type bulk Ge supports this assessment [91].

In the valence bands, all the following transitions are allowed: $HH_0 \rightarrow LH_0$, $HH_0 \rightarrow SO_0$, $LH_0 \rightarrow SO_0$; and $HH_0 \rightarrow$ continuum states. For $Si_{1-x}Ge_x$/Si grown on Si substrate, the $Si_{1-x}Ge_x$ layers are under compressive strain, which causes bands to separate among all three bands (HH, LH, and SO). Therefore, most of holes will occupy the heavy-hole ground state. Energy spectra of the samples are obtained using a three-band self-consistent calculation [63] including many-body effects in a local density functional approximation [95] and strain-induced splittings of the hole bands [96]. The transition energies for $HH_0 \rightarrow LH_0$ are considerably smaller than the experimentally observed peak positions. The observed peak energy positions are close to those of the $HH_0 \rightarrow SO_0$ and $HH_0 \rightarrow$ continuum states close to the barrier height. A theoretical calculation [94] indicates that the extended states with large $k_\parallel$ have a much larger mixing, and these contribute significantly to the normal incidence absorption process. It is possible that a similar mechanism is responsible for the observed normal incidence photocurrent spectrum in a p-type AlGaAs/GaAs multiple quantum well structure [92].

For estimating the absorption strength qualitatively, the couplings of the valence band Bloch states with the conduction band can be expressed by using the first-order perturbation theory [93] as

$$U_i = U_i^0 + \frac{\hbar}{m_0} \frac{\langle U_c^0 | \vec{k} \cdot \vec{p} | U_i^0 \rangle}{(E_g + |E_i|)} U_c^0 \qquad (5.50)$$

where, the index i $(= h, l, s)$ refers to the heavy, light, and spin-orbit split-off holes, respectively, U_c is the Bloch state of the conduction band; E_g is the Γ energy gap; and E_i is the energy of the light and split-off hole band edges measured from the heavy-hole band edge. We can see that as the energy gap reduces, the coupling with the conduction band increases. This explains why the inter-valence subband absorption was observed only for high Ge concentration. From eq. (5.50), we expect that the absorption strength will be proportional to $(E_g + |E_i|)^{-2}$. We have plotted the absorption (measured absorption shown in Figure 5.28) as a function of $(E_g + |E_i|)^{-2}$ as shown in Figure 5.29 for two possible inter-valence subband transitions from

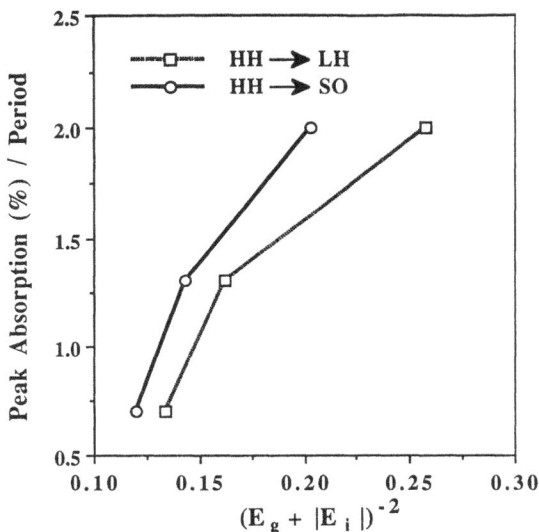

Figure 5.29 Measured absorption strength as a function of $(E_g + |E_i|)^{-2}$, where E_g is the Γ energy gap and E_i is the band edge energy of the light or spin-orbit split-off hole bands measured from the heavy-hole band edge. The observed transition is likely from the heavy-hole state to the quasi-bound state where the wave function is a combination of light-hole and spin-orbit split-off hole states.

heavy-hole to light-or spin-orbit hole subbands. Two curves show a similar approximate $(E_g + |E_i|)^{-2}$ dependence. A detailed calculation including band mixing and strain is necessary to more accurately compare the calculated and experimentally observed transition energies.

The observed polarization dependence shown in Figure 5.26 can be understood by considering the optical matrix element between Bloch functions. The matrix element involved in the present case is proportional to $\langle U_c^0 | \hat{\varepsilon} \cdot \vec{p} | U_{HH} \rangle$, where the Bloch function U_{HH} is only a function of the (x, y) plane [93]. Therefore, the largest absorption is expected when the field is polarized along the (x, y) plane. We will now discuss the use of different types of absorptions in quantum wells for infrared detector applications.

5.6 p-TYPE INTERSUBBAND AND FREE CARRIER DETECTORS

There is a considerable interest in the application of multiple quantum wells for infrared detectors. The use of multiple quantum wells provides flexibility of tuning the response for a desired wavelength as well as a wide selection of material systems with advantageous properties. Most of the work to date is based on intersubband absorption of electrons of III–V-based material systems [97, 98]. The fabrication of GaAs/AlGaAs multiple-quantum well detector arrays with hybrid Si signal processing electronics have also been demonstrated [99, 100]. For p-type intersubband (at Γ point) transition, as discussed earlier, the polarization intersubband selection rule [82] prohibits the detection of normal incident light. However, in p-type structures, strong two-dimensional free carrier absorption was also observed in addition to the intersubband absorption due to heavy doping in the wells. Because the free carrier absorption is not restricted by the photon polarization field, absorption at normal incident is possible. In this section, we describe the photoresponse of the SiGe/Si MQW structure at different polarization angles. The photoresponse is observed at normal incidence (i.e., the polarization of the infrared parallel to the layers) as well as the parallel incidence.

For detector applications, quantum well structures with a single bound state and excited extended states close to the barrier are desirable [46, 101]. This allows the photoexcited carriers to travel without seeing barriers while maintaining a large absorption coefficient [46]. For design purposes, we have calculated the position of the bound and extended states as a function of the well width, using an effective mass approximation. In this calculation, the Ge composition in the well is taken to be 15%. The corresponding band offset is about 130 meV [39], and the heavy-hole effective mass along the growth direction is taken as 0.28 m_0. The energy of the extended states is taken as the maximum of the transmission coefficient. The calculated values as shown in Figure 5.30 indicate that a quantum well 30 Å thick can have absorption near 10 μm with the extended state lying above the Si barrier.

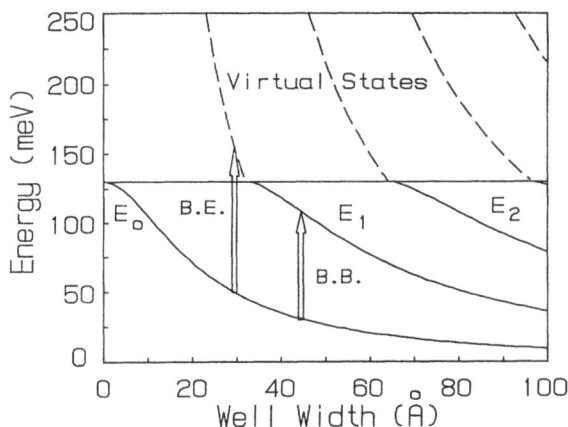

Figure 5.30 Bound and extended states of a $Si_{0.85}Ge_{0.15}$ quantum well as a function of well thickness. The barrier height is about 130 meV. The arrows show the bound-to-bound (B.B.) and bound-to-extended (B.E.) state transitions.

In this structure the temperature of the substrate during growth is kept at about 600°C. This improves the epitaxial quality of the layers, in particular for Si barriers, where most of the photoexcited carrier transport occurs. The quantum well structure consists of 50 periods of 30 Å $Si_{0.85}Ge_{0.15}$ wells (p doped to $\sim 1 \times 10^{19}$ cm^{-3}) separated by 500 Å undoped Si barriers. The entire multiple-quantum well structure is sandwiched between a doped ($p = 1 \times 10^{19}$ cm^{-3}) 1 μm bottom and a 0.5 μm top layer for electrical contacts.

Figure 5.31 shows the room temperature absorption spectrum for different polarization angles as a function of wavelength using a 45° multipass waveguide structure with dimensions 4 mm long and 0.5 mm wide [82]. At the 0° polarization, the incoming photon electric field has components both along the growth direction (z direction) and parallel to the growth plane (xy plane). In this case, the absorption spectrum consists of a relatively broad peak at near 9 μm superimposed on a monotonically increasing background as the wavelength is increased. On the other hand, at the 90° polarization, where the field is polarized only along xy plane, the absorption shows the similar trend to the 0° case, but the absorption strength is stronger and the peak at 9 μm is no longer visible. The latter peak has been shown to emanate from the polarization dependence on the absorption from the heavy-hole ground state to the extended state. The absorption spectrum at the 90° polarization in Figure 5.31 is due mostly to two-dimensional free carrier absorption. In this case, the photon absorption occurs via phonon or impurity scattering to conserve the momentum [102]. As the polarization angle increases, the photon field in the xy plane increases and

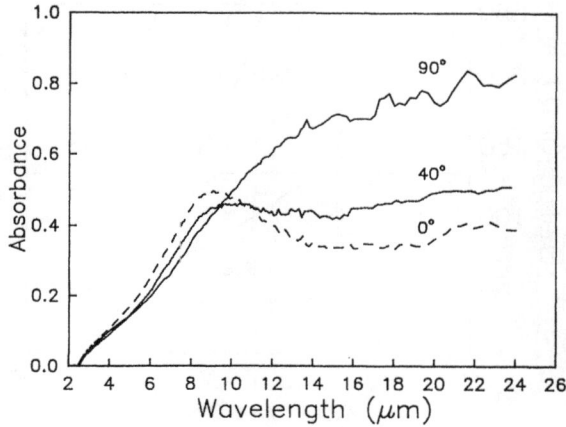

Figure 5.31 Measured FTIR spectrum at 300 K. Absorption spectra at three different polarization angles are shown. At the 0° polarization, a broad peak near 9 μm is seen. This peak is due to a heavy-hole intersubband transition. The 90° case shows a stronger background absorption, most likely due to free carrier absorption; but the peak seen at the 0° polarization is not visible.

the field in the z direction decreases. This causes an increase of free carrier absorption and a decrease of intersubband absorption as the polarization angle increases.

5.6.1 Photoresponse

To study the photoresponse, mesa diodes 200 μm in diameter are fabricated with a 45° facet on the edge of the wafer. The measured leakage current and differential resistance at about 77 K are shown in Figure 5.32 as a function of bias across the device. For the present nonoptimized device, the leakage current at 2 V is an order of magnitude higher than for a typical GaAs/AlGaAs infrared detector [97]. This is due mainly to the higher doping (almost an order of magnitude larger) used in the present device to compensate for the reduction of absorption strength from the larger hole effective mass compared to the electron effective mass in GaAs. The leakage current increases rapidly as the bias across the device is increased and is the result of the enhancement of tunneling and the thermionic components of the current. The spectral dependence of the photocurrent is measured by using a glowbar-source and a grating monochromator. The photon flux is measured using a calibrated GaAs/ AlGaAs quantum well detector. The light is illuminated on the facet at the normal incidence so that the incidence angle on the multiple quantum well structure is 45°, which is shown in the inset of Figure 5.33.

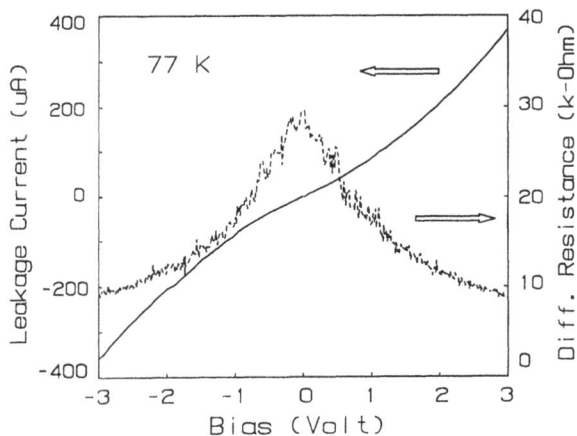

Figure 5.32 Current-voltage and differential resistance measured at 77 K. Positive bias means the top contact positive.

Photoresponse of the SiGe/Si MQW structurel at different polarization angles is discussed next in assessing the normal incidence detection. Figure 5.33 shows the photoresponse at 0° and 90° polarizations at 77 K, with a 2 V bias across the detector. In the 0° polarization case, a peak is found at near 8.6 μm, which is in agreement with the FTIR absorption spectra, except the peak position moves slightly to lower wavelength because of the use of different temperatures in the two measurements. The full width at half maximum is about 80 meV and also is in agreement with the absorption spectra, whereas for the 90° polarization, the peak at near 7.2 μm is observed with a similar FWHM. For the latter, the shift of the peak to a shorter wavelength may be due to the sharing of phonon energy with momentum conserving processes, such as phonon or impurity scattering. The photoresponsivity for both cases is about the same, 0.3 A/W. Figure 5.33 shows the photocurrent along with the result obtained from an unpolarized beam. In the latter case, a peak is found at near 7.5 μm, and the photoresponsivity is about 0.6 A/W, approximately the sum of two polarization cases. The photoresponse is also measured by illuminating the light normally on the backside of the device. The photoresponse with peak responsivity and width similar to the 90° polarization case is observed, as shown by the dashed curve in Figure 5.33.

It is clear that, for the 0° polarization case, the photoresponse is due to intersubband transition between two heavy-hole subbands [103] and also partially due to internal photoemission of holes excited via free carrier absorption. For the 90° polarization case, intersubband transition is forbidden but free carrier absorption is stronger than that of the 0° polarization case, because the entire photon electric field lies in

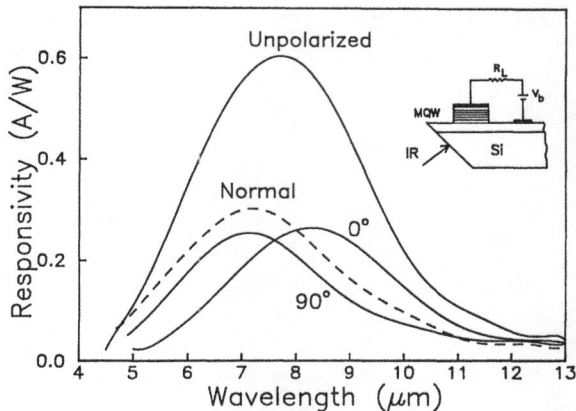

Figure 5.33 Photoresponse at 40 K for two polarization angles with a 2 V bias applied across the detector. Infrared is illuminated on the facet at the normal incidence so that the incident angle on the multiple quantum well structure is 45°, as shown in the inset. Dashed curve shows the normal incidence photoresponse, when the light is illuminated normally on the backside of the detector.

the xy plane. The photoresponse in this case is believed to be due to internal photoemission of free carriers, because the mixing of the hole bands due to the s-like conduction band or nonparabolicity is too small to have significant absorption. We next explain the normal incidence photoconduction process. If the energy of the incoming photon is large enough to create holes having an energy larger than the barrier height, the photoemitted holes can travel above the barriers and be collected under the applied electric field (internal photoemission), and this gives rise to the photocurrent. As the photon energy decreases (wavelength increases), the photocurrent increases due to the large free carrier absorption at longer wavelengths as shown in Figure 5.31. When the photon energy is further decreased so that the energy for generating holes is lower than the barrier height, the photocurrent vanishes as the flow of holes is blocked by the potential barriers. This results in the cutoff of the photocurrent. This type of photoemission is similar to that observed by Lin and Maserjian for an SiGe/Si heterojunction [9]. The difference is that the free carrier absorption is three dimensional in their case. In our case, all multiple layers contribute to the photoemission, therefore a larger quantum efficiency is expected. The internal quantum efficiency (η) of the detector for either polarization is given by $\eta = (1 - e^{-2\alpha l})$ [97]. For the present not yet optimized detector, a quantum efficiency of $\eta \sim 14\%$ is obtained for polarized light. For the unpolarized case, the photocurrent is the sum of both intersubband transition and the photoemission due to free carriers.

Some absorption mechanisms, other than the free carrier absorption may also be responsible for the observed normal incidence detection. The transition between the ground states of the heavy-hole (HH$_0$) and the spin-orbit split-off hole (SO$_0$) bands, as discussed previously, may also be responsible for normal incident detection. Such an intervalence band transition is expected to be relatively weak due to the large Γ bandgap of SiGe when the Ge composition is low.

Another possible mechanism for normal incidence detection may be due to the nonparabolicity of valence subbands away from the Γ point. Intersubband absorption of a normally incident IR beam in InSb inversion layers (metal-insulator semiconductor structure) had been observed [104, 105]. This normal incident absorption was attributed to the nonparabolicity of conduction subbands in InSb. In this case, gate bias was applied to form an inversion layer, and the Fermi-level was well above the subbands inside the inversion layer [104]. For our case with the doping concentration of $p \sim 10^{19}$ cm^{-3}, the Fermi level lies only about 20 meV above the Γ point of heavy-hole ground state (in terms of hole energy) according to our self-consistent calculation. The nonparabolicity may not be significant in the present case, as the Fermi level is relatively small. From the FTIR spectrum (Figure 5.31), no feature of normal incidence intersubband absorption is obvious.

5.6.2 Photoexcited Carrier Transport

To understand the photoexcited carrier transport in the multiple-quantum well structure, we have studied the photoresponse under different bias. The measured responsivity as a function of bias across the device is shown in Figure 5.34. The responsivity rapidly increases as the bias is increased to 2.5 V, where it approaches saturation. The maximum value of R for the present device (which is not optimized) is about 0.3 A/W, comparable to that typically observed in III–V-based quantum well detectors. To understand the dependence of photocurrent on the transport parameters, we calculate the photocurrent ($I_p(E)$) as a function of photogenerated carriers at a given quantum well multiplied by the probability of their reaching the contact [103]:

$$I_p(E) = \sum_{n=1}^{N} \frac{eP_0}{h\nu} (e^{-\alpha nl} + e^{-\alpha(2N-n)l})\alpha l e^{-nL/\nu(E)\tau} \tag{5.51}$$

where P_0 is the incident infrared power, $h\nu$ is the photon energy, α is the absorption coefficient, l is the well width, L is the period of the multiple quantum wells, N is the number of quantum wells, $\nu(E)$ is the drift velocity, and τ is the excited carrier lifetime. Because $\alpha l \ll 1$ the preceding equation can be simplified to

$$I_p(E) \approx \frac{2eP_0}{h\nu}\alpha l \sum_{n=1}^{N} e^{-nL/\nu(E)\tau} \tag{5.52}$$

After evaluating the summation and assuming that for moderate electric fields $L/\nu(E)\tau < 1$, the responsivity $R(E)$ ($= I_p(E)/P_0$) can be found using the drift velocity

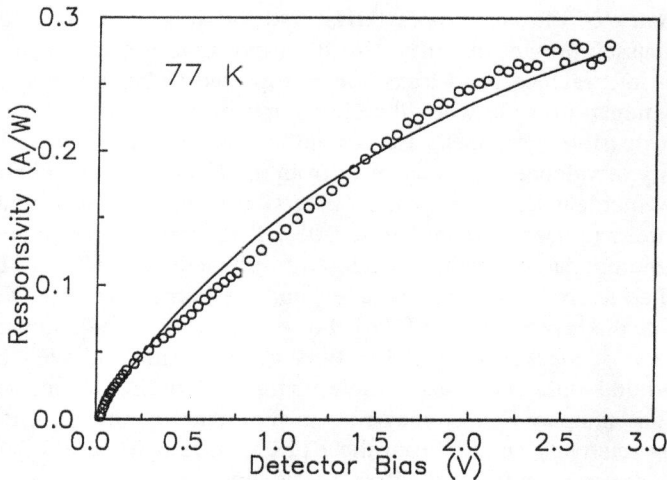

Figure 5.34 Responsivity of the device as a function of applied bias across the device at 77 K. The solid curve is calculated values using eq. (5.53).

for holes $v(E) = \mu E/(1 + \mu E/v_s)$ [106] to be

$$R(E) \approx \frac{2e\alpha\tau}{h\nu} \frac{l}{L} \frac{\mu E}{1 + \mu E/v_s} \qquad (5.53)$$

where μ is the mobility, E is the electric field, and v_s is the saturated drift velocity. To estimate the mobility and the excited carrier lifetime, the experimental data is fitted to eq. (5.53) using the saturated drift velocity [106] of about 1×10^7 cm s^{-1} and the measured absorption coefficient at 9.5 μm (2800 cm^{-1}). The mobility of holes of 1000 cm^2/V s and the excited carrier lifetime of 15 ps seem to fit the experimental data well as shown by the solid line of Figure 5.34. The values of mobility and lifetime are comparable to those obtained for a typical GaAs/AlGaAs detector [97, 99]. The nonpolar nature of Si/Ge should result in a larger intersubband lifetime compared to polar GaAs; however, the alloy and impurity scatterings can have a strong influence on the carrier lifetime. As the responsivity is inversely proportional to the barrier width, it may be possible to further increase the responsivity by reducing the barrier width without degrading the detectivity because the large heavy-hole effective mass keeps the tunneling leakage current low.

5.6.3 Detectivity

The detectivity, D^*, is defined as

$$D^* = \frac{S/N}{P}\sqrt{A\Delta f} \qquad (5.54)$$

where S and N are the signal and the noise in amperes, P is the power received by the detector in watts, A is the area of the detector in cm^2, and Δf is the bandwidth. For the case of the quantum well detector with light incidence at an angle θ, the effective area becomes $A \cos\theta$ and the noise current (shot noise) is given by $\sqrt{4eI_d\Delta f}$, where I_d is the dark current. Using the effective area and the noise current, the detectivity can be written as

$$D^* = \eta(eA \cos \theta/I_d)^{1/2}/2h\nu \qquad (5.55)$$

where $\eta = 1 - e^{-2\alpha l}$ is the quantum efficiency, A is the area of the device and $h\nu$ is the photon energy. The detectivity can be estimated by using eq. (5.55) with the measured leakage current and the quantum efficiency. For the present nonoptimized device, the estimated value is about $D^*(9.5 \ \mu m) = 1 \times 10^9$ cm $\sqrt{Hz}$/W at 77 K. A detectivity of an order of magnitude larger is observed in III–V-based quantum well detectors by optimizing the device parameters. This indicates that, by optimizing device parameters, the sensitivity of the SiGe/Si quantum well detector can be further improved.

5.7 PROSPECTIVE

From the material point of view, the progress will continue for facilitating the processing. For growth of quantum well structures as we discuss here, the future direction in the film preparation may take advantages of molecular flow in MBE and the continuous replenishing feature of CVD as well as the added feature of surface chemistry. The latter will result in improving the doping incorporation and control of the interface abruptness. The present understanding of the fundamental properties, particularly the strain control and its relaxation as well as the film stability during device processing remains limited and further investigations will promote additional application of the strained $Si_{1-x}Ge_x$/Si quantum well beyond the detector discussed here. The fundamental understanding of intersubband transition physics deserves further exploration. Further investigation of the control of the width of the absorption and the transition energy is desired as well.

In the area of detector applications, the effort in using $Si_{1-x}Ge_x$/Si has just begun. There is a need to assess and choose the technology in using either n-type or p-type structures. Much work is needed to optimize the structures for performance. Obviously, the selection will require the knowledge of the specific application. Most important is to fabricate focal plane arrays.

From the technology point of view, we need to make the detector technology and processing compatible to VLSI. Obviously, the integration with shift registers and other electronics will be the next area of focus. When successfully realized, the Si/Ge infrared detectors will then become affordable.

Further quantum well engineering using step quantum wells, coupled quantum wells, and appropriate energy filters may be exploited for improving the detectivity. Many excellent quantum well engineering references are available [45, 107].

5.8 CONCLUSION

In summary, the current status of pseudomorphic growth of $Si_{1-x}Ge_x$ was reviewed, and the present understanding of band splitting and band offsets was assessed. The principle of intersubband transitions was discussed for the general valleys and the Γ valley. A simple parabolic quantum well was used as an example for simplicity. The demonstrations of intersubband infrared absorption in $Si_{1-x}Ge_x$/Si multiple quantum well structures were described. Similarly, the use of δ-doped Si layers for intersubband transitions was illustrated. The nature of the optical transitions, particularly the polarization dependence, was explored in terms of both xy and z polarizations for several mechanisms of normal incidence detection in the cases of the n-type and p-type multiple quantum well structures. In δ-doped quantum wells, the importance of many-body effects in the determination of optical transition energies was illustrated for the Si/Ge case. An exemplified intersubband infrared detector using $Si_{1-x}Ge_x$/Si multiple quantum wells is given. The broad photoresponse was shown to cover a large portion of the 8–14 μm atmospheric window in contrast to a narrowband response of III–V-based detectors. Possible avenue for improvement of the responsivity as well as the detectivity of the device were outlined. The progress in the SiGe intersubband transition physics to date indicate that monolithic integration of SiGe/Si multiple quantum well infrared detectors with Si signal processing circuits should be realizable for focal plane applications.

ACKNOWLEDGMENTS

The authors would like to thank many of their graduate students over the past few years for their dedicated efforts in MBE and other aspects that made writing this chapter possible. This work was supported in part by the Army Research Office (Dr. John Zavada), the Air Force Office of Scientific Research (Dr. Gerald Witt), and SRC (Dr. William Holton).

REFERENCES

[1] Pearsall, T.P., and J.C. Bean, *IEEE Electron Device Lett.*, Vol. EDL-7, 1986, p. 308.
[2] Daembks, H., H.-J. Herzog, H. Jorke, H. Kibbel, and E. Kasper, *IEEE Trans. Electron. Devices*, Vol. ED-33, 1986, p. 633.
[3] Temkin, H., J.C. Bean, A. Antreasyan, and R. Leibenguth, *Appl. Phys. Lett.*, Vol. 52, 1988, p. 1089.
[4] Luryi, S., A. Kastalsky, and J.C. Bean, *IEEE Trans. Electron Devices*, Vol. ED-31, 1984, p. 1135.
[5] Park, J.S., R.P.G. Karunasiri, and K.L. Wang, *J. Vac. Sci. Technol. B*, Vol. 8, 1990, p. 217.
[6] Nayak, D.K., J.C. Woo, J.S. Park, K.L. Wang, and K.P. MacWilliams, *IEEE Electron Device Lett.*, Vol. EDL-12, No. 4, 1991, p. 154.

[7]   Rhee, S.S., J.S. Park, R.P.G. Karunasiri, Q. Ye, and K.L. Wang, *Appl. Phys. Lett.*, Vol. 53, 1988, p. 204.

[8]   Rhee, S.S., G.K. Chang, T.K. Carns, and K.L. Wang, *Appl. Phys. Lett*, Vol. 56, No. 11, 1990, p. 1061.

[9]   Lin, T.L. and J. Maserjian, *Appl. Phys. Letts.*, Vol. 57, 1990, p. 1442.

[10]   Park, J.S., R.P.G. Karunasiri, Y.J. Mii, and K.L. Wang, *Appl. Phys. Lett.*, Vol. 58, No. 10, 1991, p. 1083.

[11]   Kasper, E. and H.J. Herzog, *Thin Solid Films*, Vol. 44, 1977, p. 357.

[12]   Kasper, E. and H.J. Herzog, *Appl. Phys. Lett.*, Vol. 8, 1975, p. 199.

[13]   Manasevit, H.M., I.S. Gergis, and A.B. Jones, *Appl. Phys. Lett.*, Vol. 41, 1980, p. 464.

[14]   Bean, J.C., L.C. Feldman, A.T. Fiory, S. Nakahara, and I.K. Robinson, *J. Vac. Sci. Technol, A*, Vol. 2, 1984, p. 436.

[15]   Meyerson, B.S., K.J. Uram, and F.K. Legoues, *Appl. Phys. Lett.*, Vol. 53, 1988, p. 2555.

[16]   Hoyt, J.S., C.A. King, D.B. Noble, C.M. Gronet, J.F. Gibbons, M.P. Scott, S.S. Laderman, S.J. Rosner, K. Nauka, J. Turner, and T.I. Kamins, *Thin Solid Films*, Vol. 183, 1989, p. 93.

[17]   Van der Merve, J.H., *J. Appl. Phys.*, Vol. 34, 1962, p. 123.

[18]   People, R. and J.C. Bean, *Appl. Phys. Lett*, Vol. 39, 1986, p. 538.

[19]   Hull, R., J.C. Bean, F. Cerdeira, A.T. Fiory, and J.M. Gibson, *Appl. Phys. Lett.*, Vol. 48, 1986, p. 56.

[20]   Miles, R.H., P.P. Chow, D.C. Johnson, R.J. Hauenstein, O.J. Marsh, C.W. Nieh, M.D. Strathman, and T.C. McGill, *J. Vac. Sci. Technol. B*, Vol. 6, 1988, p. 1382.

[21]   Chern, C.H., K.L. Wang, G. Bai, and M.-A. Nicolet, *Mat. Res. Soc. Symp. Proc.*, Vol. 220, 1991, p. 175.

[22]   Kasper, E., *Surf. Sci.*, Vol. 176, 1986, p. 630.

[23]   Parker, E.H.C. ed., *The Technology and Physics of Molecular Beam Epitaxy*, 1990.

[24]   Bean, J.C., S.S. Iyer, K.L. Wang, eds., *Silicon Molecular Beam Epitaxy*, MRS, 1991.

[25]   Ghidini, G., and F.W. Smith, *J. Electrochem. Soc.*, Vol. 131, 1984, p. 2924.

[26]   Meyerson, B.S., F.K. Legoues, T.N. Nguyen, and D.L. Harame, *Appl. Phys. Lett.*, Vol. 50, 1987, p. 113.

[27]   Iyer, S., G. Patton, J. Stork, B. Meyerson, and D. Harame, *IEEE Trans. Elect. Dev.*, 1989, p. 2043.

[28]   Gibbons, J.F., C.M. Gronet, and K.E. Williams, *Appl. Phys. Lett.*, Vol. 47, 1985, p. 721.

[29]   Hirayama, H., T. Tatsumi, A. Ogura, and N. Aizaki, *Appl. Phys. Lett.*, Vol. 51, 1987, p. 2213.

[30]   Hirayama, H., T. Tatsumi, and N. Aizaki, *Appl. Phys. Lett.*, Vol. 52, 1988, p. 2242.

[31]   Hirayama, H. and T. Tatsumi, *Appl. Phys. Lett*, Vol. 55, 1989, p. 131.

[32]   Herring, C. and E. Vogt, *Phys. Rev.*, Vol. 101, 1956, p. 944.

[33]   Balslev, I., *Phys. Rev.*, Vol. 143, 1966, p. 636.

[34]   Hasegawa, H., *Phys. Rev.*, Vol. 129, 1963, p. 1029.

[35]   Hensel, J.C., and G. Feher, *Phys. Rev.*, Vol. 129, No. 3, 1963, p. 1041.

[36]   People, R., J.C. Bean, and D.V. Lang, *in Proc. 1st Int. Symp. Silicon MBE*, Pennington, NJ, 1985, p. 364.

[37]   Ma, Q.M. and K.L. Wang, *Appl. Phys. Lett.*, Vol. 58, No. 11, 1991, p. 1184.

[38]   Van de Walle, C.G., and R.M. Martin, *Phys. Rev. B*, Vol. 34, 1986, p. 5621.

[39]   Van de Walle, C.G., and R. Martin, *J. Vac Sci. Technol. B*, Vol. 3, 1985, p. 1257.

[40]   People, R., *Phys. Rev. B*, Vol. 33, 1985, p. 1451.

[41]   Balslev, I. and P. Lawaetz, *Phys. Lett. A*, Vol. 19, No. 1, 1965, p. 6.

[42]   Hensel, J.C., and K. Suzuki, *Phys. Rev. B*, Vol. 9, No. 10, 1974, p. 4219.

[43]   Rhee, S.S., "Studies of SiGe Tunneling Heterostructures Grown by Si," Ph.D. thesis, University of California at Los Angeles, 1991.

[44]   Karunasiri, R.P.G., and K.L. Wang, *J. Vac. Sci. Technol. B*, Vol. 9, 1991, p. 2064.

[45]   Yuh, P.F., and K.L. Wang, *J. Appl. Phys.*, Vol. 65, 1989, p. 4377.

[46]   Coon, D.D., and R.P.G. Karunasiri, *Appl. Phys. Lett.*, Vol. 45, 1984, p. 649.

[47]   West, L.C., and S.J. Eglash, *Appl. Phys. Lett.*, Vol. 46, 1985, p. 1156.

[48]   Wieder, S., *The Foundation of Quantum Theory*, Academic Press, Inc. New York, 1973.

[49]   Ridley, B.K., *Quantum Processes in Semiconductors*, Oxford University Press, London, 1982.

[50]   Karunasiri, R.P.G., and K.L. Wang, *Superlattices and Microstructures*, Vol. 4, 1988, p. 661.

[51]   Voisin, P., G. Bastard, and M. Voos, *Phys. Rev. B*, Vol. 29, 1984, p. 935.

[52]   Wheeler, R.G., and H.S. Goldberg, *Phys. Rev. Lett.*, Vol. 27, 1971, p. 925.

[53]   Kamgar, A., P. Kneschaurek, and G. Dorda, *Phys. Rev. Lett.*, Vol. 32, 1974, p. 1251.

[54]   Kneschaurek, P., A. Kamgar, and J.F. Koch, *Phys. Rev. B*, Vol. 14, 1976, p. 1610.

[55]   Wheeler, R.G., and H.S. Goldberg, *IEEE Trans. Electron. Devices*, Vol. 22, 1975, p. 1001.

[56]   Stern, F., *Phys. Rev. Lett.*, Vol. 33, No. 16, 1974, p. 960.

[57]   Ando, T., *J. Phys. Soci. Japan*, Vol. 39, No. 2, 1975, p. 411.

[58]   Stern, F., *Phys. Rev. Lett.*, Vol. 30, 1973, p. 278.

[59]   Allen, S.J., D.C. Tsui and B. Vinter, *Solid State Commun.*, Vol. 20, 1976, p. 425.

[60]   Ando, T., *Z. Phys. B*, Vol. 26, 1977, p. 263.

[61]   Bandara, K.M.S.V., and D.D. Coon, *Appl. Phys. Lett.*, Vol. 53, No. 19, 1988, p. 1865.

[62]   Manasreh, M.O., F. Szmulowicz, D.W. Fischer, K.R. Evans, and C.E. Stutz, *Appl. Phys. Lett.*, Vol. 57, No. 17, 1990, p. 1790.

[63]   Karunasiri, R.P.G., J.S. Park, K.L. Wang, and S.S. Rhee, unpublished.

[64]   Hedin, L., and B.I. Lundqvist, *J. Phys. C*, Vol. 4, 1971, p. 2064.

[65]   Levine, B.F., K.K. Choi, C.G. Bethea, J. Walker, and R.J. Malik, *Appl. Phys. Lett.*, Vol. 50, 1987, p. 1092.

[66]   Yi, K.S., and J.J. Quinn, *Phys. Rev. B*, Vol. 27, No. 4, 1983, p. 2396.

[67]   Nee, S.M., U. Claessen, and F. Koch, *Phys. Rev. B*, Vol. 29, 1984, p. 3449.

[68]   Yang, C.L., and D.S. Pan, *J. Appl. Phys.*, Vol. 64, 1988, p. 1573.

[69]   Stern, F., and W.E. Howard, *Phys. Rev.*, Vol. 163, 1967, p. 816.

[70]   People, R., *Phys. Rev. B*, Vol. 32, No. 2, 1985, p. 1405.

[71]   Braunstein, R., A.R. Moore, and F. Herman, *Phys. Rev.*, Vol. 109, No. 3, 1958, p. 695.

[72]   People, R., J.C. Bean, and D.V. Lang, *Proc. Int. Conf. on Physics of Semiconductors*, Vol. 2, 1986, p. 767.

[73]   *Landolt-Börnstein: Numerical Data and Functional Relationships in Science and Technology*, edited by O. Madelung, Springer-Verlag, New York, 1982.

[74]   Kasper, E., H.-J. Herzog, H. Jorke, and G. Abstreiter, *Superlattices and Microstructures*, Vol. 3, 1987, p. 141.

[75]   Zeindl, H.P., T. Wegehaupt, I. Eisele, H. Oppolzer, H. Reisinger, G. Tempel, and F. Koch, *Appl. Phys. Lett.*, Vol. 50, No. 17, 1987, p. 1164.

[76]   Tempel, G., N. Schwarz, F. Müller, and F. Koch, *Thin Solid Films*, Vol. 184, 1990, p. 171.

[77]   Müller, F., G. Tempel, and F. Koch, *Proc. ICPS*, Vol. 2, 1990, p. 1254.

[78]   Lee, Chanho, and K.L. Wang, *J. Vac. Sci Technol. B*, Vol. 11, No. 2, 1992.

[79]   Gossmann, H.J., E.F. Schubert, D.J. Eaglesham, and M. Cerullo, *Appl. Phys. Lett.*, Vol. 57, 1990, p. 2440.

[80]   Iyer, S.S., K.A. Metzger, and F.G. Allen, *J. Appl. Phys.*, Vol. 52, 1981, p. 5608.

[81]   Chun, S.K., and K.L. Wang, Phys. Rev. B, October 1992.

[82]   Karunasiri, R.P.G., J.S. Park, Y.J. Mii, and K.L. Wang, *Appl. Phys. Lett.*, Vol. 57, 1990, p. 2585.

[83]   Hertle, H., G. Schuberth, E. Gornik, G. Abstreiter, and F. Schaffler, *Appl. Phys. Lett.*, Vol. 59, 1991, p. 2977.

[84]   Rhee, S.S., R.P.G. Karunasiri, C.H. Chern, J.S. Park, and K.L. Wang, *J. Vac. Sci. Tech. B*, Vol. 7, 1989, p. 327.

[85]   Levine, B.F., R.J. Malik, J. Walker, K.K. Choi, C.G. Bethea, D.A. Kleinman, and J.M. Vandenberg, *Appl. Phys. Lett.*, Vol. 50, 1987, p. 273.

[86]   Wessel, R., and M. Altarelli, *Phys. Rev. B*, Vol. 40, No. 18, 1989, p. 12457.

[87]   Chun, S.K., and K.L. Wang, *IEEE Trans. Electron Devices*, **39**, 2153 (1992).

[88]   Pikus, G.E., and G.L. Bir, *Soviet Phys. Solid State*, Vol. 1, 1960, p. 1502.

[89]   Mii, Y.J., K.L. Wang, R.P.G. Karunasiri, and P.F. Yuh, *Appl. Phys. Lett.*, Vol. 56, 1990, p. 1046.

[90]   Zenner, A., F. Koch, and K. Ploog, *Surf. Sci.*, Vol. 196, 1988, p. 671.

[91]   Kaiser, W., R.J. Collins, and H.Y. Fan, *Phys. Rev.*, Vol. 91, 1953, p. 1380.

[92]   Levine, B.F., S.D. Gunapala, J.M. Kuo, S.S. Pei, and S. Hui, *Appl. Phys. Lett.*, Vol. 59, No. 15, 1991, p. 1864.

[93]   Kane, E.O., *J. Phys. Chem. Solid.*, Vol. 1, 1956, p. 82.

[94]   Man, P., and D.S. Pan, *Phys. Rev. B.*, October 1992.

[95]   Kohn, W., and L.J. Sham, *Phys. Rev.*, Vol. 140, No. 4a, 1965, p. 1133.

[96]   People, R., *IEEE J. Quantum Electronics*, Vol. QE-22, 1986, p. 1696.

[97]   Levine, B.F., C.G. Bethea, G. Hasnain, V.O. Shen, E. Pelvé, and P.R. Abbott, *Appl. Phys. Lett.*, Vol. 56, No. 9, 1990, p. 851.

[98]   Gunapala, S.D., B.F. Levine, D. Ritter, R. Hamm, and M.B. Panish, *Appl. Phys. Lett.*, Vol. 58, No. 18, 1991, p. 2024.

[99]   Bethea, C.G., B.F. Levine, V.O. Shen, R.R. Abbott, and S.J. Hsieh, *IEEE Trans. Electron Devices*, Vol. ED-38, 1991, p. 1118.

[100] Kozlowski, L.J., G.M. Williams, G.J. Sullivan, C.W. Farley, R.J. Andersson, J.K. Chen, D.T. Cheung, W.E. Tennant, and R.E. DeWames, *IEEE Trans. Electron Devices*, Vol. ED-38, 1991, p. 1124.

[101] Steele, A.G., H.C. Liu, M. Buchanan, and Z.R. Wasilewski, *Appl. Phys. Lett.*, Vol. 59, 1991, p. 3625.

[102] Fan, H.Y., W.G. Spitzer, and R.J. Collins, *Phys Rev.*, Vol. 101, No. 566, 1956.

[103] Karunasiri, R.P.G., J.S. Park, and K.L. Wang, *Appl. Phys. Lett.*, Vol. 59, 1991, p. 2588.

[104] Beinvogl, W., and J.F. Koch, *Solid State Commun.*, Vol. 24, No. 9, 1977, p. 687.

[105] Wiesinger, K., H. Reisinger, and F. Koch, *Surface Sci.*, Vol. 113, 1982, p. 102.

[106] Caugley, D.M., and R.E. Thomas, *Proc. IEEE*, Vol. 55, 1967, p. 2192.

[107] Wang, K.L., and P.F. Yuh, *IEEE J. Quantum Electron.*, Vol. QE-25, 1989, p. 12.

[108] Matthews, J.M., and A.E. Blakeslee, *J. Crystal Growth*, Vol. 27, 1974, p. 118.

[109] Van der Leur, R.H.M., A.J.G. Schellingerhout, F. Tuinstra, and J.E. Mooij, *J. of Appl. Phys.*, Vol. 64, 1988, p. 3043.

[110] Dodson, B.W., and P.A. Taylor, *Appl. Phys. Lett.*, Vol. 49, 1986, p. 642.

[111] Kohama, Y., Y. Fukuda, and M. Seki, *Appl. Phys. Lett.*, Vol. 52, 1988, p. 380.

[112] Arbet, V., S.J. Chang, and K.L. Wang, *Thin Solid Films*, Vol. 183, 1989, p. 57.

[113] Houghton, D.C., C.J. Gibbings, C.G. Tuppen, M.H. Lyons, *J. Appl. Phys. Lett.*, Vol. 56, 1990, p. 460.

Chapter 6

Type III Superlattices for Long-Wavelength Infrared Detectors: The HgTe/CdTe System

T.H. Myers
Department of Physics
West Virginia University, Morgantown

J.R. Meyer and C.A. Hoffman
Naval Research Laboratory, Washington, DC

6.1 INTRODUCTION

Military and commercial requirements continue to drive materials research in the area of long-wavelength infrared and very long-wavelength infrared detectors. HgCdTe is the current infrared material of choice, with excellent performance obtained using either a single element or small arrays of photoconductive detectors. Practical considerations limit the actual utility of HgCdTe for large staring arrays, however. HgCdTe suffers from stringent compositional uniformity requirements due to the observed rapid change in detection wavelength with alloy composition for wavelengths greater than 10 μm. Additionally, excessive tunneling currents due to the intrinsic properties of HgCdTe degrade device performance in the photovoltaic devices required for large arrays, particularly for very long-wavelength infrared. The difficulties associated with HgCdTe have made extrinsically doped silicon the current choice for very long-wavelength infrared detection. Unfortunately, the utility of extrinsically doped silicon is limited, due to the severe weight and reliability penalties for space applications imposed by the required low temperature of operation (<20 K).

The compositional superlattice HgTe/CdTe was first proposed as an alternative material for infrared detection in 1979 by Schulman and McGill [1]. The semimetallic nature of HgTe places this superlattice into a separate class from most of the other II–VI and III–V superlattice systems. Hg-based superlattices are the only heterostructures whose alternating layers combine a direct-gap semiconductor with a symmetry-induced semimetal, the so-called Type III superlattice [2]. Several theoretical studies [1, 3–6] have been performed to elucidate the characteristics of this materials system. In particular, the predictions indicate that detector wavelength will be easier to control than with HgCdTe, tunneling currents will be controllable and reduced with respect to HgCdTe, and very long-wavelength infrared devices fabricated with HgTe/CdTe superlattices will be able to operate at reasonable temperatures.

Unlike some of the other quantum well structures investigated for long-wavelength infrared detection, HgTe/CdTe superlattices use direct illumination for detection and do not rely on complicated schemes to couple the infrared radiation to the quantum well, which permits the construction of simple detector structures and focal plane arrays. In addition, HgTe/CdTe superlattices exhibit large infrared absorption coefficients, allowing a high quantum efficiency to be obtained in a relatively thin device structure. Because detectivity is directly proportional to quantum efficiency, other detector materials that exhibit a low absorption coefficient start out with a distinct disadvantage.

The first HgTe/CdTe superlattice was grown by Faurie, Million, and Piaquet [7], using molecular beam epitaxy in 1982. Since then, numerous experimental studies (as discussed later) have been aimed at understanding and exploiting this materials system. Basic issues such as accurate determination of the valence band offset, demonstration of high-quality materials growth, and development of a successful doping technology had to be resolved before this material could begin to be evaluated for use in infrared detector structures. This chapter briefly describes the basic properties of HgTe/CdTe superlattices, gives a detailed discussion of the materials parameters pertinent to its use as an infrared material, discusses the difficulties associated with the molecular beam epitaxy growth of these superlattices, and presents the results of the first successful demonstration of the use of a HgTe/CdTe superlattice as an infrared detector.

6.2 BAND STRUCTURE CALCULATIONS

The expected advantages of HgTe/CdTe superlattices in infrared detector applications are closely tied to the significant differences between the superlattice band structure and that of the $Hg_{1-x}Cd_xTe$ alloy system. Although a number of theoretical treatments of HgTe/CdTe band structures have been reported [5, 6, 8–12], the results tend to be relatively independent of approach as long as sufficient detail is incorporated. For example, it is essential that a multiband (typically six or eight) formalism be employed, due to the strong interband interactions in a narrow-gap heterostructure. Strain should also be included [13], despite the close lattice match between HgTe and CdTe (0.3%).

Assuming that a valid band structure approach is employed, the accuracy of a given calculation is nonetheless critically limited by the reliability of the input parameters. In particular, it is now known that theoretical predictions for many of the fundamental optical and electronic properties of HgTe/CdTe superlattices depend quite sensitively on whether the valence band offset, ΔE_v, is taken to be small (< 200 meV) or large (> 200 meV). For a number of years, the magnitude of ΔE_v was a subject of considerable controversy. Early magneto-optical results [9, 14] were interpreted as implying an offset of only 40 meV, based on a comparison with theoretical energy gaps and effective masses. The prediction of the common anion rule [1] that the valence band offset at the interface between two compound semiconductors containing the same anion (Te) should be near zero, was consistent with this finding. In subsequent years, other optical [15, 16], magneto-optical [17, 18], and low-temperature resonant tunneling experiments were interpreted as implying ΔE_v < 100 meV (on the other hand, room-temperature tunneling experiments implied offsets of between 110 meV and 390 meV) [19, 20]. The controversy began in 1986, when *x-ray photoemission spectroscopy* (XPS) measurements yielded $\Delta E_v \approx 350$ meV. Similar measurements by a second group gave nearly the same value [21], which was found to be nearly independent of temperature [22]. Since that time, *a priori* theories have yielded $\Delta E_v > 300$ meV (in disagreement with the common anion rule) [23, 24], and it is now recognized that numerous qualitative aspects of the magneto-transport [25], optical [26, 27], and magneto-optical [28–30] data can be understood only in terms of a large valence band offset (examples will be discussed later). It has further been pointed out that due to a double-valued dependence of the energy gap on ΔE_v, the early cyclotron resonance data can be explained equally well by a large offset [31]. Although no comprehensive attempt has been made to refute each of the findings of a small offset in detail, some of the erroneous results are attributable to such factors as the softness of the absorption and photoluminescence edges in small-gap superlattices, misidentification of magneto-optical resonances due to the multitude of distinct transitions contributing in the same spectral region [32], contributions by quasi-two-dimensional carriers occupying only a portion of a given superlattice [33], and perhaps experimental uncertainties in the HgTe and CdTe layer thicknesses. The last is particularly likely as most early determinations relied on x-ray diffraction measurements that yield only the combined period thickness. For more accurate comparisons between experiment and theory, transmission electron microscopy measurements of the individual layer thicknesses must be performed.

Although the exact value of the valence band offset remains somewhat uncertain, there is by now almost universal agreement that it must be large. Following the most common current practice, the band structures presented here will employ the XPS value of 350 meV. Because many of the early guidelines for superlattice infrared detector design were based on the assumption of a small ΔE_v, we will discuss in later sections how those results must be modified to account for the large offset.

Hg-based superlattices owe much of their distinctive character to the fact that they are the only heterostructures whose alternating layers combine a direct-gap semiconductor with a symmetry-induced semimetal, the so-called Type III super-lattice. In CdTe, there is a large positive energy gap (1600 meV) between the electron band (Γ_6) and the heavy- and light-hole bands (Γ_8^h and Γ_8^l), which are degenerate at the valence band maximum. In HgTe, on the other hand, Γ_6 is a valence band as it lies 300 meV *below* the two Γ_8 bands. Γ_8^l then inverts, becoming the conduction band, and the energy gap is identically zero because Γ_8^h and Γ_8^l are still degenerate at the zone center. It is well known that by alloying HgTe and CdTe to form $Hg_{1-x}Cd_xTe$, we can continuously tune the Γ_6–Γ_8 energy separation from -300 meV to $+1600$ meV as the composition x is varied from 0 to 1. Because the effective masses for the two light-carrier states are roughly proportional to $|E(\Gamma_6) - E(\Gamma_8)|$, masses for the electrons and light holes approach zero near the semimetal-to-semi-conductor transition at $x \approx 0.15$ for 77 K.

Although the physical mechanisms and symmetry considerations are much different in the two cases, quantum confinement effects on the superlattice energy bands are in many ways quite analogous to composition-variation effects on the Γ_6, Γ_8^h, and Γ_8^l bands in the alloy [6]. When the HgTe wells are thin, quantum confinement induces a large positive energy gap (E_g) between the conduction band E1 and the two lowest-order valence bands HH1 and LH1 (higher order valence bands such as HH2 and HH3 may also be present). By temporarily setting the valence band offset and strain to zero, the two hole bands are degenerate at the zone center similar to the alloy. Next consider weakening the quantum confinement by increasing the well thickness, d_W. The energy gap then shrinks, until at some point all three bands meet at the Γ point. For still larger d_W the gap becomes negative and the two light-carrier bands invert, with one becoming a conduction band and E1 a valence band as in semimetallic $Hg_{1-x}Cd_xTe$. However, in this region E1 continues to respond to reductions in the quantum confinement by moving to lower energies in the manner of an electron band rather than to higher energies like a hole band. Similarly, even though LH1 is now a conduction band it is unaffected by quantum confinement as long as there is no valence band offset. As in $Hg_{1-x}Cd_xTe$, the in-plane effective masses for both E1 and the lighter of the two hole bands are approximately proportional to $|E_g|$. In the thick-well limit, the gap between E1 and the two hole bands approaches the Γ_6–Γ_8 separation in bulk HgTe; that is, about -300 meV.

The effects of strain and a valence band offset are now introduced and super-imposed on the variation with d_W discussed previously. Strain in the superlattice has the effect of moving the LH1 band to higher energies with respect to HH1 [13]. On the other hand, the ordering of the two bands can be reversed by assuming a large valence band offset. Because quantum confinement has a more pronounced effect on light-carrier bands, LH1 moves below HH1 whenever ΔE_v exceeds ≈ 200 meV. For this reason, we expect the experimental electronic properties to be extremely sensitive to whether the valence band offset is large or small. Many of the observed phenomena can be understood only in terms of the superlattice band structure when HH1 is taken to be the dominant valence band [25].

Figure 6.1 illustrates theoretical band edge dispersion relations [34] for a (100) HgTe/Hg$_{0.15}$Cd$_{0.85}$Te superlattice in which the wells are thin enough (58 Å) to introduce a positive energy gap of 69 meV between E1 and HH1 (quantum confinement moves LH1 considerably below the scale of the figure, to -92 meV at the zone center). The barriers are taken to have a small but significant mercury content (about 15% for (100) orientation) [35], because the MBE depositions are carried out with the Hg-source shutter left open during the entire growth run. Note that, near the zone center, the electron and hole dispersion relations in the plane ($k_\rho \equiv (k_x^2 + k_y^2)^{1/2}$) are nearly mirror images of one another (LH1 fully complements HH1, in that its mass is heavy in the plane but light in the growth direction [6]). However, m_{pp} is exceptionally nonparabolic [25], becoming infinite and then negative (electronlike) at energies only 20 meV below the valence band maximum. One consequence of

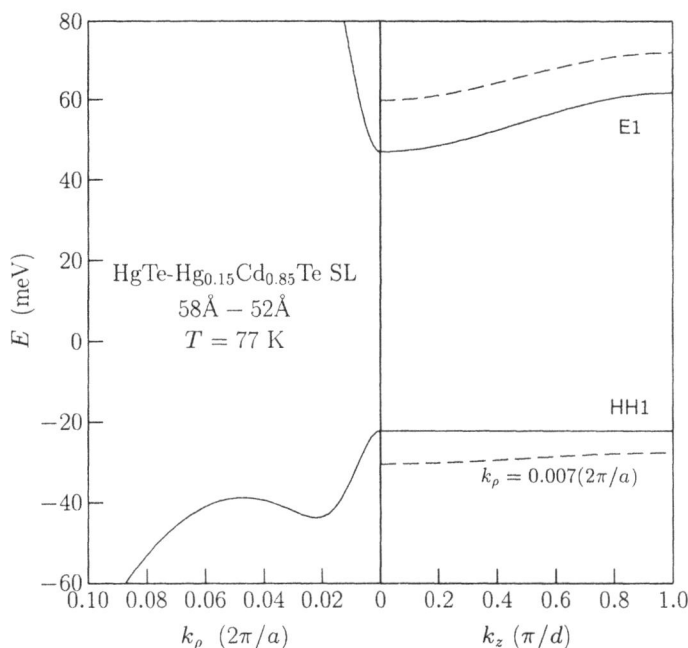

Figure 6.1 Calculated band structure for a (100) HgTe/Hg$_{0.15}$Cd$_{0.85}$Te superlattice with well and barrier thicknesses $d_W = 58$ Å and $d_B = 52$ Å ($T = 77$ K). The dashed curves in the right panel are the k_z dispersion at $k_\rho = 0.007(2\pi/a)$ rather than $k_\rho = 0$, illustrating that whereas m_{pz} is nearly infinite for the holes at the valence band maximum, those at an energy $\approx k_BT$ farther down display finite dispersion along k_z.

this is that $\langle m_{p\rho} \rangle$ (the average over the thermal distribution) varies rapidly with temperature, becoming over an order of magnitude heavier when T is increased from 20 to 90 K.

As in any superlattice, the dispersion relations in the growth direction, k_z, vary strongly with barrier thickness (the functional dependences will be discussed in a later section). Although $E1$ is clearly anisotropic for the present value of d_B (52 Å), m_{nz} is nonetheless relatively light (about 0.1 m_0). At the same time, HH1 is nearly dispersionless in k_z at the valence band maximum. However, the dashed curves in the right panel of Figure 6.1, which give $E(k_\rho = 0.007(2\pi/a), k_z)$ rather than $E(k_\rho = 0, k_z)$, illustrate that there are holes within $k_B T$ of the valence band maximum that have significant dispersion in k_z ($m_{pz} \approx 0.7 m_0$ in this example). The electron mass in the growth direction has little dependence on k_ρ (compare the solid and dashed curves for $E1$).

With increasing well thickness and hence reduced quantum confinement, the $E1$ band moves to lower energies. At $d_W \approx 80$ Å, $E1$ and HH1 begin to anti-cross and the superlattice becomes semimetallic [25]. A typical semimetallic band structure is illustrated in Figure 6.2, which shows the in-plane and growth-direction dispersion relations for a (100) HgTe/Hg$_{0.15}$Cd$_{0.85}$Te superlattice with $d_W = 86$ Å and $d_B = 29$ Å. The growth-direction wave vector at which the two bands anti-cross is denoted as k_{zc}. Note that when $k_z > k_{zc}$, $E1$ is the conduction band and HH1 is the valence band as in Figure 6.1. However, for $k_z < k_{zc}$ the roles are reversed. It is also apparent from a comparison of the two dotted curves in the left panel of the figure that the in-plane dispersion for HH1 at the valence band maximum has a very strong dependence on the growth-direction wave vector because $m_{p\rho}(k_z) \propto E_g(k_z)$, and $E_g(k_z)$ is a rapidly varying function. Holes with $m_{p\rho}$ varying by over an order of magnitude coexist in the superlattice at the valence band maximum, a phenomenon, which has been designated *mass broadening* [11, 25]. In fact, it is mass broadening, which leads to the k_z dispersion of HH1 at energies below the valence band maximum (dashed curve in the right panel). Comparison of the dotted curves in the left panel shows that, although $E(k_\rho, k_z = k_{zc}) = E(k_\rho, k_z = \pi/d)$ when $k_\rho = 0$ (no dispersion), the energy at $k_z = k_{zc}$ is much lower than that at $k_z = \pi/d$ when $k_\rho = 0.006(2\pi/a)$ (significant dispersion). Mass broadening is also responsible for the relatively light m_{pz} in Figure 6.1 at $E \approx k_B T$, although in that case the effect is somewhat smaller due to the weaker variation of $E_g(k_z)$. Finally, note that the miniband width of $E1$ is considerably larger than in Figure 6.1 due to the thinner barriers in Figure 6.2 (29 Å versus 52 Å).

Considering the large number of unusual features in the band structures of Figures 6.1 and 6.2, it should not be surprising that the observable free carrier properties of HgTe/CdTe superlattices are strongly influenced, and in some cases dominated, by the complexities in the band structure. Explicit properties that can be calculated from the band structure, such as E_g as a function of layer thickness, are discussed in later sections.

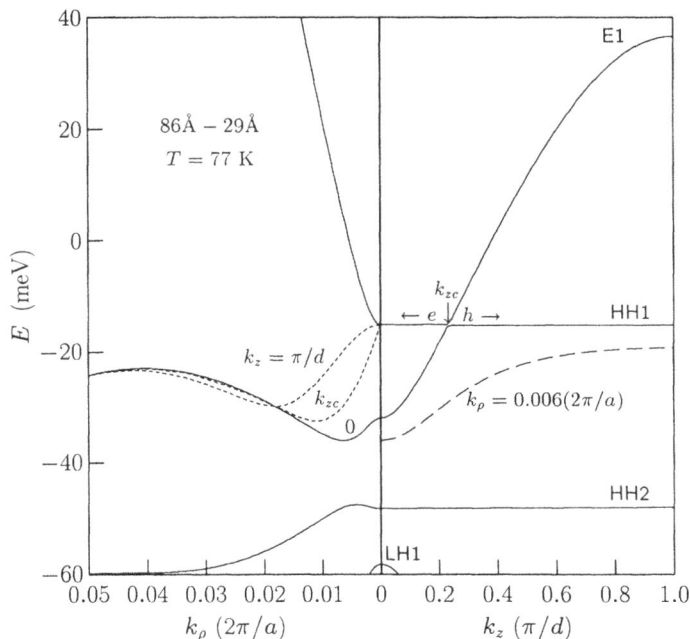

Figure 6.2 Calculated band structure for (100) HgTe/Hg$_{0.15}$Cd$_{0.85}$Te superlattice with well and barrier thicknesses $d_W = 86$ Å $d_B = 29$ Å ($T = 77$ K). The dotted curves in the left panel are $E(k_\rho)$ for the uppermost valence band at $k_z = k_{zc}$ and $k_z = \pi/d$, whereas the dashed curve in the right panel is the k_z dispersion for the same band at $k_\rho = 0.006(2\pi/a)$.

6.3 BASIC MATERIALS PROPERTIES

This section discusses properties of HgTe/CdTe superlattices that, although of importance for ultimate detector use, are more general in nature. An understanding of the basic materials properties is a prerequisite for utilizing this material in specific applications, such as for an infrared detector material.

6.3.1 Peculiarities of HgTe/CdTe Superlattice Growth

Although HgTe/CdTe superlattice growth has been reported using *metallo-organic vapor deposition* (MOCVD), [36–38], the relatively high temperatures and long MOCVD growth times required preclude the use of this technique for superlattice growth. The low temperatures used in molecular beam epitaxy and monolayer-by-monolayer control qualify it as the premier technique for HgTe/CdTe superlattice growth. The first HgTe/CdTe superlattice was grown using MBE by Faurie et al.

in 1982 [7], quickly followed by growth of superlattices in other laboratories [39, 40].

Many of the general principles of III–V semiconductor growth by MBE discussed in Chapter Two also apply to II–VI materials. However, significant differences specific to the growth of Hg-based materials must be addressed to achieve high quality. This section briefly reviews the aspects of MBE growth pertinent to HgTe/CdTe superlattices.

In general, separate elemental sources of Hg, Cd, and Te can be used for MBE growth. Unlike the III–VI compounds, some of the II–VI compounds (such as CdTe) dissociate congruently. The use of CdTe as the source of Cd and Te during MBE results in a source that is significantly more stable than the use of either Cd or Te alone [41]. The congruent dissociation of CdTe also leads to the appropriate flux at the surface of the growing layer to give stoichiometric CdTe. The ability to provide a stoichiometric flux from a single source is particularly important, because the low temperatures involved with HgTe/CdTe superlattice growth are insufficient for self-regulated growth to occur, as is typical for the III–V compounds. In particular, any slight excess Te overpressure at temperatures less than 200°C will lead to the formation of Te precipitates [42].

Evaporation of HgTe, however, is highly congruent [43]. Hg evaporates preferentially from the surface at relatively low temperatures, leaving behind the Te. Therefore, separate sources of Hg and Te are required for HgTe growth. A relatively high overpressure of Hg is also required to prevent HgTe from dissociating at the growth temperatures employed. In addition, the condensation coefficient of Hg has been measured to be on the order of 10^{-3} for growth temperatures in the range of 160–190°C, commonly used for MBE of Hg-based materials [44]. These effects require a high flux of Hg to grow high-quality HgTe, as well as to keep it from dissociating. A practical consequence is that large amounts of Hg are introduced into the MBE growth chamber, leading to the need for Hg containment in an UHV environment, and the need for novel Hg sources capable of providing a flux with long-term stability [45, 46].

Further difficulties are associated with the Hg overpressure required for HgTe growth. In general, high-quality growth occurs only for a relatively narrow range of Hg overpressure [47, 48], leading to the need for precise control over both the Hg flux and the substrate temperature (as the condensation coefficient is a strong function of temperature). Furthermore, the Hg condensation coefficient is orientationally dependent, leading to different optimal fluxes for growth along the different crystal axes. For example, it has been demonstrated that (111)*B* [49] and (211)*B* [50] growth require about a factor of four less Hg flux than is needed for (100)-oriented growth, and the optimal Hg flux for growth on the (110) [51] orientation is somewhere in between. A final potential difficulty is that the optimal Hg flux required for high-quality growth may change as a function of layer thickness. Although preliminary data suggest that this may be the case for the growth of the HgCdTe alloy [52], it has not yet been demonstrated whether the same holds true for HgTe/CdTe super-lattice growth.

During the growth of an HgTe/CdTe superlattice, it is difficult to completely interrupt the Hg flux during the brief time that the CdTe layer is grown. Furthermore, degradation due to Hg outdiffusion would occur without an Hg overpressure. There-fore, the Hg flux is typically uninterrupted during CdTe growth. The constant Hg flux has the additional benefit of ensuring a group II overpressure during growth, decreasing the probability of Te precipitate formation. This approach does, however, lead to the incorporation of Hg into the CdTe layers. All HgTe/CdTe superlattices grown by MBE are in actuality $HgTe/Hg_{1-x}Cd_xTe$ superlattices, where x ranges from 3 to 9% on the (111)B orientation, is about 10% on the (211)B orientation, and is about 15% on the (100) orientation [35, 53, 54].

Finally, photon-assisted MBE should be discussed. Photon-assisted MBE is an energy-assisted growth technique in which the substrate is illuminated with above bandgap photons during the entire growth process. Such illumination has been shown to alter the MBE growth kinetics, as detailed in several articles [55–57]. The primary use of photon-assisted MBE has been to enhance the doping, as discussed in a later section. Two other effects of photon-assisted MBE growth are discussed here. In 1987, Myers et al. [54] reported a significant improvement in the structural quality of MBE-grown Hg-based materials under illumination. Although the illumination intensity, about 80 mWcm^{-2}, is not large enough to significantly alter the surface temperature of the growing layer [58], higher quality growth was observed at tem-peratures 10–15°C lower than commonly used. Without the illumination, the struc-ture quickly degraded at lower temperatures. Although the origin of this effect is still not clearly understood, it has also been observed in other laboratories [57, 59, 60].

The most important long-term effect of photon-assisted MBE may be to lower temperatures during growth. Because the Hg condensation coefficient is a strong function of temperature, lower temperatures imply lower Hg flux, ameliorating the practical difficulties encountered with the large amounts of Hg typically used. Also, growth at 170–200°C leads to conditions that are Hg-deficient, that is, leaving a small density of Hg vacancies in the material. This effect complicates the substitu-tional p-type doping on the Te site. Lowering the growth temperature shifts the growth conditions into the Hg-rich regime and also minimizes interdiffusion effects.

Arias et al. [60] reported the growth of high-quality HgTe/CdTe superlattices at 155°C with material characteristics suitable for device applications. Lansari et al. [61] have further developed the use of lower substrate temperatures for Hg-based layers, reporting the growth of high-quality HgTe/CdTe superlattices on the (100) orientation at temperatures as low as 125°C. Additionally, this latter study suggests that low-temperature growth may lead to a reduction in the twin-related defect struc-ture occurring for the (100) orientation. Certainly photon-assisted MBE is one area that deserves continued development for the growth of Hg-based materials by MBE.

6.3.2 Structural Properties and Defects

The ultimate performance of a detector is often limited by the presence of defects in its base material. This section discusses the structural properties of HgTe/CdTe superlattices in terms of three types of defects: extended defects, or dislocations; twinning-related defects; and point defects. Growth on alternative substrates will not be discussed as there are problems precluding their use in the growth of superlattices. For example, the use of GaAs always leads to a dislocation density of greater than 10^7 cm^{-2} (see [62]), and growth on InSb, which is very closely lattice matched, typically results in a high In-doping level in the layer [63].

Dislocations can be formed due to conditions used for the layer growth, be nucleated at the substrate-layer interface, or propagate from dislocations in the substrate itself. The lowest dislocation density layers in the (Hg, Cd)Te system are grown using CdTe or CdZnTe substrates. Indeed, the dislocation density in MBE-grown HgCdTe layers appears to be limited by dislocations propagating from the substrate [59]. In HgTe/CdTe, the repeated interfaces and slight residual strain appear to have the additional beneficial effect of causing the dislocations to bend and form dislocation half-loops, preventing them from propagating into the growing layer [64, 65]. The transmission electron microscopy micrograph in Figure 6.3 shows numerous multiple repeats of a basic 30 Å HgTe/50 Å CdTe superlattice unit [66]. In general, the close lattice match between the substrate and the superlattice results in few dislocations originating at the interface. However, this micrograph was chosen because it shows a dislocation propagating from the substrate into the superlattice and then bending over to connect with another dislocation. A half-loop is formed which effectively annihilates both dislocations. This technique has been used to obtain very dislocation-free material [48], leading some groups to claim that their HgTe/CdTe superlattice layers are typically of higher quality than alloy layers grown by MBE [64]. The high structural quality of superlattices grown by MBE is further illustrated by the narrow full width at half maximum of 28 arcseconds observed in x-ray diffraction rocking curve measurements for the best superlattices [54], as shown in Figure 6.4. Such a full width at half maximum is comparable to that typically measured for bulk CdTe or CdZnTe, and is typical of the full width at half maximum observed for superlattices grown in the GaAs-AlGaAs system.

Twinning often occurs in the II–VI compounds, especially in CdTe [67] with the principal twin direction around a (111) direction. Hg-based epilayers grown on the (111)B orientation are always quite specular. However, transmission electron microscopy and x-ray analysis indicate that the layers typically contain a large density of twin lamelle parallel to the surface [48, 50, 64, 68] corresponding to twinning around the growth direction. Although this effect appears to be somewhat suppressed in the growth of superlattices [64], there is still a significant probability that twinning will take place [48]. The presence of a high density of twin lamelle has been shown to lead to poor detector performance in the alloy [69]. The inability to control twin-

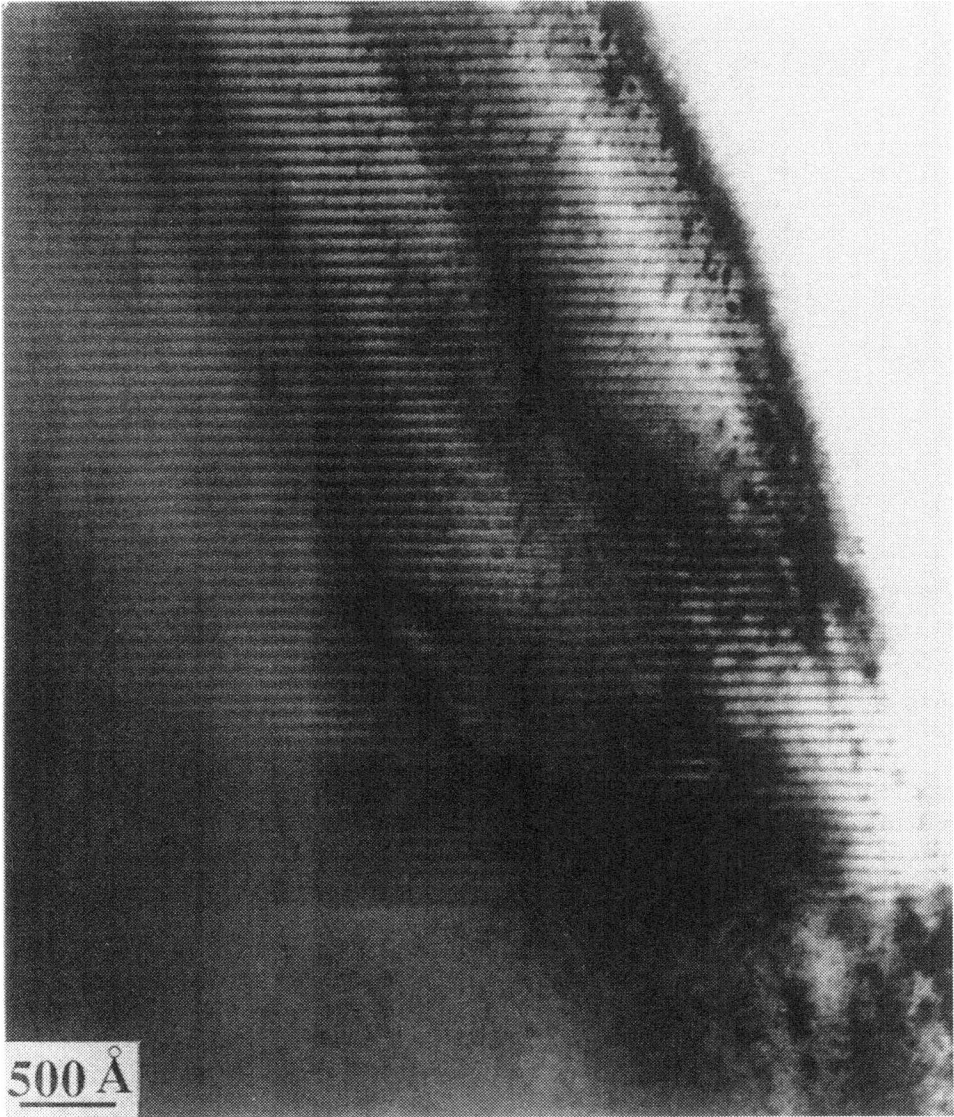

Figure 6.3 A transmission electron microscopy micrograph of an HgTe/CdTe superlattice. The dark and light bands represent HgTe and CdTe, respectively. The measured thicknesses were d_W = 30 Å and d_B = 50 Å. Note the dislocation threading from the substrate, which is quickly bent over and annihilated by forming a half-loop with a second dislocation. Dislocations were not observed away from this interface.

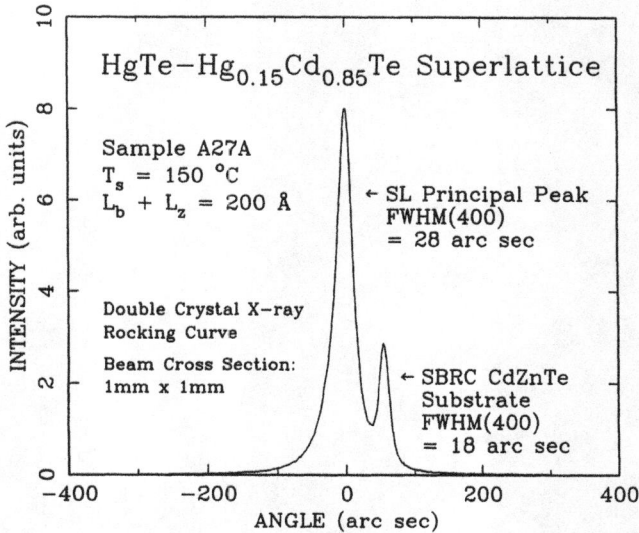

Figure 6.4 Double-crystal x-ray diffraction rocking curve obtained for an HgTe/CdTe superlattice. The relatively narrow full width at half maximum of the superlattice diffraction peak represents structural quality comparable to that found in the best III–V superlattice structures.

ning on the (111)*B* orientation precludes its use in the growth of HgTe/CdTe superlattices for detectors.

On (100)-oriented substrates, twinning is also present and manifests itself in the form of pyramidal hillock structures across the epilayer surface. Transmission electron microscopy studies indicate that the pyramidal defects are again caused by a twin about the (111) direction, rotating a (111) plane into the growth direction [50, 64]. Because this (111) plane requires a different optimal Hg overpressure for high-quality growth, growth in the pyramidal hillock tends to be highly defective and usually results in polycrystalline material. The presence of these defects has been shown to be deleterious to infrared detector performance [70]. The polycrystalline material in the hillock serves as a source of significant noise current in a detector structure and can cause a short circuit across a photovoltaic junction.

A large fraction of the studies performed on HgTe/CdTe superlattices were performed on the (100) orientation, and many of the resulting measurements reflected high-material quality outside the pyramidal hillocks. Typical hillock densities are in the high 10^4 cm^{-2} range, with the lowest reported for MBE growth being about 1 $\times$ 10^4 cm^{-2} [48, 59, 61]. When their density is low, the hillocks tend to be uniform in size. The lateral dimensions suggest that they originate at the substrate-epilayer interface, possibly due to particulates, substrate defects, or substrate inclusions. Hillocks also appear to nucleate during growth, possibly due to fluctations in the Hg overpressure [48] or pressure bursts from the Hg source leading to Hg droplets reaching the substrate [59]. Hillock densities currently obtained for MBE growth of HgTe/ CdTe superlattices on the (100) orientation preclude a reasonable yield for detectors. This orientation will therefore not be used for superlattice-based detectors unless new approaches, such as low temperature growth by photon-assisted MBE [61], eliminate the formation of pyramidal hillock defects.

A major advance in the growth of HgTe/CdTe superlattices occurred when Koestner and Schaake [50] proposed the use of the (211)B orientation as a solution to the problem of defect formation due to twinning. To form during growth, the twin nucleus must become stable due to dissolution in the untwinned matrix within a few monolayers. On the (111) orientation, which is most prone to twinning, the nucleus can reach this critical dimension within one monolayer. The twin nucleus also passes the critical dimension very quickly on the (100) orientation. However, on the (211)B orientation, which is 19.5° off the (111)B toward the (001), the critical dimension is not achieved for many monolayers of growth so the nucleus has a large probability of dissolving. The use of the (211)B orientation has resulted in much improved structural properties in Hg-based layers grown by MBE [50, 59, 71, 72], leading to the growth of outstanding HgTe/CdTe superlattices exhibiting long minority carrier lifetimes and photoluminescence from multiple intersubband transitions [73]. Growth on the (211)B orientation has also lead directly to the fabrication of the first high quantum efficiency HgTe/CdTe superlattice infrared detector [66].

Finally, a limited amount of data indicates that the point defects play a role in HgTe/CdTe superlattices. From studies of metal-insulator semiconductor structures fabricated on (100) superlattices, Koestner, Goodwin, and Schaake [72] report the presence of high tunneling currents in MBE layers that do not contain high dislocations or twin-related defects. Because extended defects do not appear to be responsible, we must consider the possibility of point defects. One possibility is that a Te antisite defect may be kinetically trapped in MBE (100) HgCdTe layers due to the low growth temperatures [74]. Significantly, (100)-oriented layers are typically n-type as grown, distinct from (111)B- and (211)B-oriented layers, which are p-type due to Hg vacancies. The simplest explanation is that, for (100) growth, Te antisites form when Te fills the Hg vacancies present during growth. This antisite leads to n-type doping and also increased midgap states that enhance trap-assisted tunneling currents. Midgap states would decrease carrier lifetimes in undoped (100) material, explaining the very low minority carrier lifetimes [75–77] observed in (100) superlattices, as discussed later. Although evidence for Te antisite formation is not prevalent in growth on the (211)B orientation, it should be considered a potential problem in all Hg-based MBE growth.

6.3.3 Structural Stability and Interdiffusion

Stability with respect to interdiffusion has been of concern since the growth of the first of HgTe/CdTe superlattices [78]. For early growths, it was suspected that interdiffusion was the underlying mechanism causing the measured E_g to be much larger than predicted [3]. To understand interdiffusion at the low temperatures used for superlattice growth, it is necessary to measure extremely small interdiffusion coefficients, which are on the order of 10^{-20} cm^{2}s^{-1}. There have been many interdiffusion studies of the HgTe/CdTe system at temperatures above 400°C [79]. Extrapolation of these results to lower temperatures indicates that significant interdiffusion should take place at the growth temperatures employed. However, it is now known that this extrapolation leads to erroneous values of interdiffusion, as different mechanisms apply at lower temperatures.

Initial studies of interdiffusion in HgTe/CdTe superlattices were performed using x-ray diffraction [80, 81]. The superlattices were heated in-situ in a diffractometer, and the superlattice diffraction satellite peaks monitored. By monitoring satellite peak intensities as a function of time, the change in the interface profile could be inferred from Fourier analysis [80]. Although these studies did not take into account the composition dependence of the interdiffusion coefficient [82], it was felt that the results should be representative. Figure 6.5 shows the values obtained

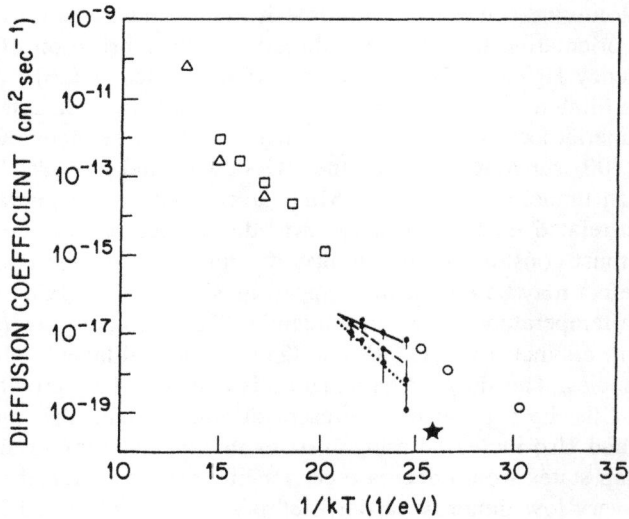

Figure 6.5 Arrhenius plot of interdiffusion coefficients deduced from linear diffusion analysis. The high temperature data was taken from selected high temperature studies [79]. The open circles represent low-temperature data determined by x-ray diffraction [80, 81]. The star represents a value determined by transmission electron microscopy [83], which is in disagreement with the x-ray values. The closed circles represent depth-dependent values as determined by high-resolution transmission electron microscopy [84] using chemical vector mapping. The solid line represents values obtained near the free surface of the superlattice, and the dotted curve represents values obtained at a depth of 7000 Å.

in the x-ray study, which appear to be in good agreement with the extrapolation from high temperature data.

A study performed by Otsuka et al. [83] did not agree with these results, however. In their study, the first 30 and the last 30 periods of two separate superlattices grown at 175°C were examined by transmission electron microscopy. One superlattice was 2 μm thick, whereas the other was 4 μm thick. In both cases, the estimated interface roughness was about one monolayer, and no difference was observed in the sharpness of the interface between the initial and final superlattice layers. Assuming that the interfacial roughness was due to interdiffusion rather than the growth procedure, an estimated upper bound for the interdiffusion coefficient of 5×10^{-20} cm^{2}s^{-1} is obtained. This value is also shown in Figure 6.5, and it is significantly smaller than the x-ray results. Million et al. [64] reported identical results for HgTe/CdTe superlattices examined using high-resolution transmission electron microscopy and lattice imaging. Again, for a 5 μm thick superlattice grown at 190°C, no measurable difference was observed between the layers at the free surface and that at the substrate interface.

A resolution of this discrepancy is presented by the elegant measure of low-temperature interdiffusion performed by Kim, Ourmazd, and Feldman [84]. High-resolution transmission electron microscopy was combined with chemical vector pattern recognition [85] to image individual HgTe/CdTe interfaces and directly map the chemical interdiffusion. This method yielded composition profiles at each interface with high spatial resolution and chemical sensitivity. Various superlattices were annealed under Ar overpressure at temperatures that ranged from 200–265°C. The most interesting observation was that the magnitude of interdiffusion was a function of interface depth beneath the surface. The interdiffusion coefficient, as determined by the simple composition-independent model, decreased by two orders of magnitude as one went from the surface to 7000 Å below the surface. These data are also included in Figure 6.5. Note that the surface interdiffusion coefficient is consistent with the x-ray results, whereas the "bulk" interdiffusion coefficient is closer to the upper bound measured by Otsuka et al. [83] and Million et al. [64].

By using the more correct composition-dependent interdiffusion coefficient, nonlinear analysis of the composition profiles [84] revealed that the depth dependence was apparently due to different magnitudes of the preexponential factor and not a depth-dependent activation energy. The depth dependence of the preexponential factor, and hence of the depth-dependent interdiffusion, can be interpreted as arising from an inhomogeneous, depth-dependent concentration of point defects. The origin of these defects is most likely due to the nonequilibrium nature in which these superlattices were annealed. An Hg overpressure was not used in either Kim et al. [84] or the x-ray study [80], leading to surface degradation due to Hg outdiffusion and thereby enhancing the interdiffusion coefficient. During growth, the superlattices are in equilibrium and the point defects are not generated, leading to stable interfaces, as observed by Otsuka et al. [83] and Million et al. [64].

It is encouraging that measurable interdiffusion has *not* been detected by transmission electron microscopy in as-grown, high-quality superlattices [64, 83], indicating that little interdiffusion takes place during growth at reasonable temperatures. In addition, the superlattices are reasonably stable after growth as well. Using a nonlinear interdiffusion analysis, Kim et al. [84] examined projected superlattice stability by arbitrarily defining the stability time as the time required for the Hg concentration at the interface to change by 25%. A projected stability time greater than 14 days was obtained for "bulk" superlattice interdiffusion at a temperature of 120°C, which is 40–50°C higher than normally used for device processing. Therefore, as long as surface degradation is minimized and equilibrium maintained, significant interdiffusion will not take place in any postgrowth device processing.

6.3.4 Absorption Coefficient

One of the most critical optical properties for the design and operation of detectors containing HgTe/CdTe superlattices is the absorption coefficient as a function of photon energy, $\alpha(h\nu)$. When energy gaps appropriate for infrared detection were first predicted by Schulman and McGill [1], there was concern that the α would be weak. Subsequent numerical calculations close to the band edge [86, 87] indicated that large α should be the case for HgTe/CdTe superlattices and predicted absorption coefficients of 10^4 cm^{-1} or greater for a cutoff wavelength longer than 10 μm. Such infrared absorption was first demonstrated experimentally by Jones et al. [88] who showed a relatively high value of α of about 10^4 cm^{-1} at room temperature for a HgTe/CdTe superlattice. By comparison, the absorption coefficient of HgCdTe at comparable cutoff wavelengths is about $1-2 \times 10^3$ cm^{-1}. The large α implies that superlattice absorbing layers can be significantly thinner than in the equivalent alloy, leading to layer thicknesses on the order of a few micrometers. The large α is not a peculiarity of the superlattice itself [86], but reflects the high absorption coefficient associated with HgTe due to its intrinsically high density of states. A calculation [5] of optical properties in the superlattice indicates that, not only is absorption stronger than in the alloy, but it is also predicted to increase with decreasing wavelength. The latter is based on the calculation of the variation in oscillator strength for the first intersubband transition, which is predicted to increase by more than a factor of two as the HgTe well thickness increases from 20 to 40 Å.

Most reported measurements on HgTe/CdTe superlattices are at room temperature despite the importance of $\alpha(h\nu)$ in the 40–77 K temperature range typically used for infrared detectors. These studies were focused on extracting the bandgap energy as a function of temperature and superlattice parameters [15], and α was at best determined as an intermediate step. Three studies report measurements of α at 77K: two $\alpha(h\nu)$ measurements [73, 89], and an average α obtained from quantum efficiency measurements on a metal-insulator semiconductor device structure [77].

The latter superlattice had a cutoff wavelength (λ_c) of 4.5 μm as determined by the photoresponse of the device, in good agreement with calculated values based on measured d_W and d_B, and gave an average α of about 9 $\times$ 10^3 cm^{-1}, which is consistent with the previous predictions. The former measurements give a near bandgap α ranging between 3 and 9 $\times$ 10^3 cm^{-1} for various long-wavelength infrared superlattices.

Figure 6.6 shows $\alpha(h\nu)$ near the absorption edge for one such HgTe/CdTe superlattice with λ_c of about 10.4 μm at 77 K, as reported by Yang et al. [89]. The onset of absorption is due to the first heavy hole-to-conduction band transition. These transitions will be designated H1. Also shown is the absorption coefficient for HgCdTe of equivalent λ_c. Absorption in the superlattice is about twice as strong as in the equivalent alloy over most of the spectral region shown, in reasonable agreement with predictions. These near band-edge values indicate that thicknesses on the order of 5 μm or greater will be required for the absorbing layer in a long-wavelength infrared device based on such a superlattice.

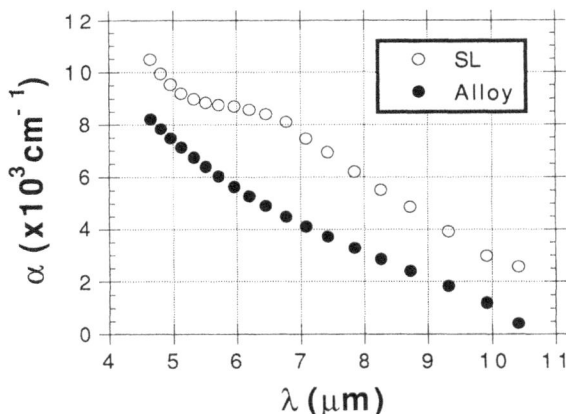

Figure 6.6 Comparison of absorption coefficient at 77 K versus wavelength for an HgTe/CdTe superlattice [89] and the equivalent HgCdTe alloy with cutoff wavelengths of 10.4 μm.

The superlattice absorption coefficient in Figure 6.6 shows a steady increase from the band edge at 10.4 μm to about 7 μm followed by a plateau until a wavelength of about 5 μm is reached, after which α again exhibits a steady increase. This structure is due to transitions involving the first light hole-to-conduction band transition (designated L1). The spatial periodicity of typical infrared superlattices are such that features due to numerous different intersubband transitions are observed over a relatively narrow wavelength range. Transitions other than H1 and L1 are clearly evident in the data shown in Figure 6.7 (e.g., H2) [90]. Also shown are

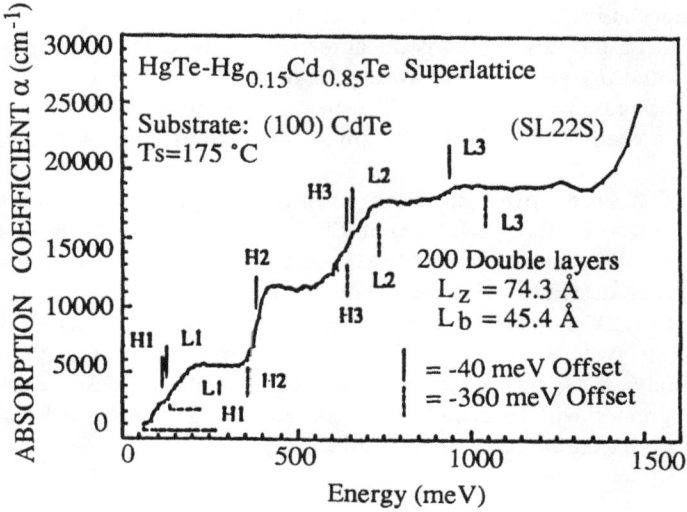

Figure 6.7 Room-temperature absorption coefficient versus photon energy for a HgTe/CdTe superlattice [90]. Structure due to intersubband transitions is clearly seen. Also shown are calculated values for the onset of absorption due to each intersubband transition for a valence band offset of −40 meV (solid) and −360 meV (dashed).

calculated energies for the onset of absorption in the various subbands. Most recent measurements of $\alpha(h\nu)$ have been aimed at the corroboration of theoretical predictions, to determine the valence band offset and understand the degree of sophistication required in modeling.

The first evidence of higher energy absorption features was reported by Leopold, Wroge, and Broerman [91] in transmission spectra, which were not reduced to α. The substructure was in general agreement with the theoretical calculations in the same study. However, the softness of the measured features and the lack of sophistication in the model used did not allow extraction of band structure parameters from the data. The first detailed study attempting to correlate theoretical predictions and measured structure in α was reported by Patten et al. [92]. The rises and plateaus observed correlated well with the predictions of the more sophisticated model used. Again, the measured features in α were soft, and decisive information about the valence band offset could not be determined. The superlattices were (111)B oriented on (100)GaAs, and as discussed previously, most likely contained numerous twins resulting in a degradation in the superlattice properties. Indeed, one strong conclusion from this study was that the sharpness of the transitions was a direct indicator of the quality of the superlattices [92].

Growth on the (100) [26, 29, 89, 93] and (211) [73] orientations have since resulted in superlattices exhibiting sharp structure due to intersubband structure, such as shown in Figure 6.7. When models of sufficient sophistication (i.e., 6 to 8 bands) are used to account for the mixing of higher level bands, then the measured features in $\alpha(h\nu)$ can be explained. However, Figure 6.7 illustrates that most of the features cannot be used to distinguish between the cases of large and small valence band offset. The features that best distinguish the two cases are the separation between the H1 and L1 transitions (proportional to ΔE_v) and, to a lesser extent, the smaller bandgap (H1) obtained for the larger valence band offset.

Figure 6.8 shows $\alpha(h\nu)$ reported by Harris et al. [73] for a superlattice with a low temperature λ_c of 9.5 μm. The steep absorption edge near 130 meV is due to the H1 transition, which defines the bandgap for the superlattice. Note that the figure also shows photoluminescence measured for the superlattice. The photoluminescence peak at 130 meV is quite strong, with a narrow full width at half maximum of 10 meV. The peak position corresponds to the tail of the absorption edge and clearly indicates the bandgap. There is a small rise in the absorption near 260 meV due to the L1 transition, which is much less pronounced in α as the H1 transition. However, this superlattice was of sufficient quality that photoluminescence was also observed for the L1 transition, as indicated in the figure. This was the first report of a photoluminescence signature peak for a second-order transition, and clearly indicates a separation of 131 meV between the H1 and L1 transitions. Theoretical predictions are consistent with this measurement only when a $\Delta E_v > 350$ meV is used. The H2 transition is also strongly observed in α near 330 meV.

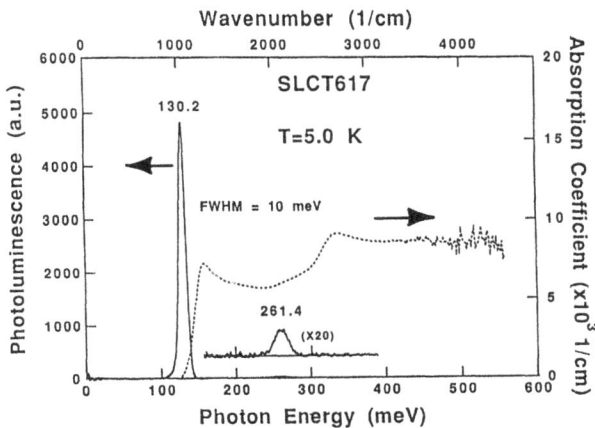

Figure 6.8 Photoluminescence and absorption coefficient versus photon energy for an HgTe/CdTe superlattice at 5 K.

The H1 photoluminescence was observed up to 140 K [73]. The presence of this peak allowed an accurate determination of H1 as a function of temperature, as is shown in Figure 6.9. An estimate of ΔE_v was obtained from this data by comparison with theoretical calculations. The experimental values for H1 fall within the curves calculated using temperature-independent values of ΔE_v of 390 and 440 meV. The relatively weak temperature dependence of ΔE_v indicated by these results is in agreement with other optical [29, 89] and x-ray photoemission studies [22, 94]. It should be pointed out that this is in contradiction with some optical measurements [15, 17] and theoretical calculations [95].

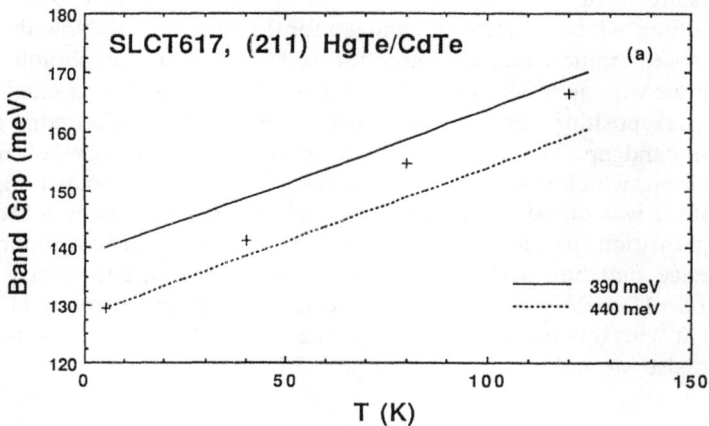

Figure 6.9 Temperature variation of E_g for an HgTe/CdTe superlattice. This variation was determined by monitoring the photoluminescence signal and is consistent with a large and fairly temperature-independent value for the valence band offset.

Figure 6.10 shows $\alpha(h\nu)$ measured at 80 K for a different superlattice from the same study. The sharp rise in absorption near 87 meV corresponds to the H1 transition, yielding a $\lambda_c \approx 14$ μm. The near-edge absorption coefficient quickly rises to about 4×10^3 cm^{-1} when $h\nu \approx 90$ meV (13.7 μm), indicating that detectors based on such a superlattice need to be only about 5 μm thick to give adequate light absorption. Also seen in the figure are steps characteristic of the L1, H2, and H3 transitions.

Current theoretical models can accurately predict the absorption coefficient in an HgTe/CdTe superlattice, particularly near the bandgap. Both theoretical predictions and experimental measurements indicate that, because $\alpha(h\nu)$ is larger in the superlattice than in the equivalent alloy, devices can be fabricated with relatively thin layers ($\approx$ 5 μm thickness).

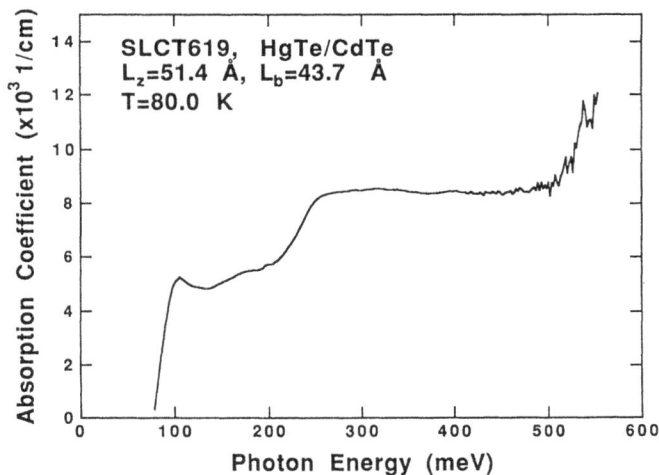

Figure 6.10 Absorption coefficient versus photon energy at 80 K for an HgTe/CdTe superlattice with $\lambda_c \approx 14 \ \mu m$.

6.3.5 Magneto-Optics

Magneto-optical experiments [14, 28, 96–98] have played a leading role in both the initial confusion and the eventual resolution of the valence band offset controversy. The confusion was due in part to the large number of distinct classes of transitions that can contribute in the same far infrared spectral region [32], some of which have resonance energies with multivalued dependences on such parameters as ΔE_v [31]. On the other hand, magneto-optical measurements provide a wealth of information about the heterostructure electronic properties if properly interpreted; such as for confirming the reliability of the band structure models and accuracy of the calculated energy gaps. Because many of the most interesting and unique magneto-optical effects are not closely related to infrared detector considerations, we will restrict the present discussion to a brief survey of cyclotron resonance results for effective masses along the growth axis, which do have direct bearing on detector performance.

When the magneto-optical experiment is carried out in Faraday geometry (optical propagation and magnetic field both parallel to the superlattice growth direction), the electron and hole cyclotron orbits lie in the plane and the resonance field yields a characterization of m_ρ. In Voigt geometry (magnetic field in the plane), on the other hand, the cyclotron orbits must tunnel through the superlattice barriers. The measured mass is then $(m_\rho m_z)^{1/2}$, and by comparing the Faraday and Voigt results we can in principle determine both masses. For a semimetallic p-type superlattice at $T = 4.2$ K, Perez et al. [28] found the hole mass to be extremely anisotropic (m_{pz}/

$m_{pp} = 280$), consistent with the band structure in Figure 6.2. Such a result can be understood only if the valence band offset is large and HH1 is the dominant valence band, because the mass ratio for holes in LH1 would be less than unity for the relevant barrier thicknesses. Berroir et al. [29] have similarly obtained $m_{nz}/m_{np} \approx 40$ in an n-type semimetallic sample, which is also consistent with Figure 6.2.

Dobrowolska et al. [96] and Manasses et al. [98] have investigated the variation of m_{nz} with temperature in superlattices that are semimetallic at low T but semiconducting at higher T. Data for temperatures in the range 150–190 K, at which $E_g \approx$ 30 meV, are shown as the open circles in Figure 6.11 (the theoretical curves will be discussed in a later section). As expected, m_{nz} is over a factor of two smaller in the superlattice with thinner barriers.

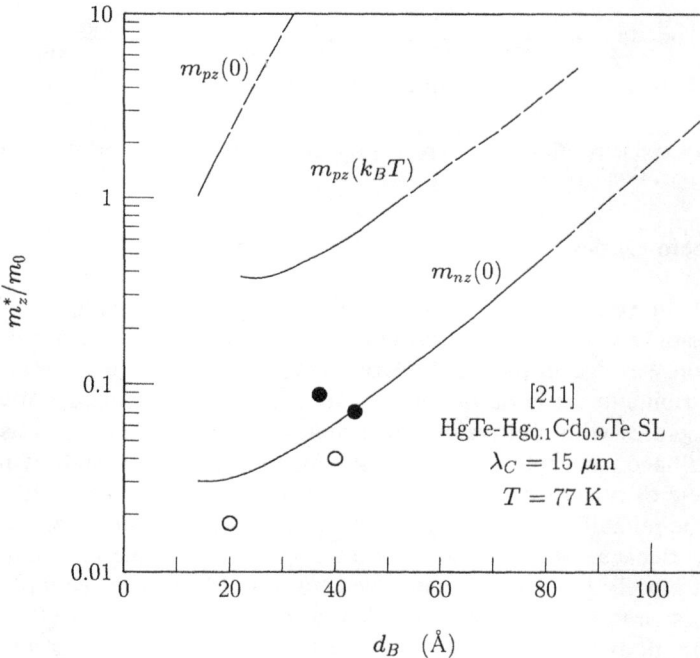

Figure 6.11 Theoretical (curves) and experimental (circles) data for growth direction electron and hole effective masses versus barrier thickness for (211) superlattices at $T = 77$ K. In the theory, the well thickness is varied with d_b such that the energy gap remains constant at 83 meV. The hole mass is seen to be extremely dependent on whether it is evaluated at the top of the valence band ($E = 0$) or at an energy k_BT away. The analogous shift of the electron mass is less than 14%. The electron cyclotron resonance data are from two (211) super-lattices with $E_g \approx$ 90 meV (filled circles) [73, 99] and two (100) superlattices with $E_g \approx$ 30 meV [96, 98].

The experiments most directly relevant to detector design are those reported by Hoffman et al. [99] and Harris et al. [73], who studied electron cyclotron resonance in (211) superlattices with energy gaps on the order of 90 meV at the temperatures of measurement (4.2 K). Data from those experiments are represented by the filled circles in Figure 6.11. Taken together, the available results from four different groups form a consistent (though quite limited) picture. Agreement with the theoretical curve is seen to be within a factor of two.

6.3.6 Intrinsic Carrier Concentration and In-Plane Transport

It is well known that free carrier transport experiments provide one of the most useful probes of band-edge properties in a semiconductor. This is particularly true of HgTe/ CdTe superlattices, because field-dependent Hall and resistivity measurements in conjunction with a mixed-conduction analysis allow us to simultaneously obtain concentrations and mobilities for multiple-carrier species [11]. It is relatively rare for the transport in this system to be dominated by a single type of carrier with a discrete mobility, due to such effects as "mobility broadening" (a consequence of mass broadening), the thermal generation of minority carriers, and the frequent presence of quasi-two-dimensional carriers populating only a few layers of the superlattice [99] (induced by band bending at the superlattice-substrate interface).

It has become apparent in recent years that, because the magneto-transport data for a narrow-gap semiconductor are sensitive to thermally generated minority carriers at quite low temperatures, analysis of the temperature dependence of the intrinsic concentration $[n_i \equiv (np)^{1/2}]$ represents one of the most reliable methods for determining the energy gap. From the law of mass action, if $n_i(T)$ is normalized to $T^{3/2}$ and plotted against T^{-1} on a semilogarithmic scale, to a first approximation the slope of the line obtained is proportional to the zero-temperature extrapolation of the gap, E_g^0. Figure 6.12 illustrates this for five HgTe/CdTe superlattices with well thicknesses varying between 51 Å and 104 Å [11, 100]. With increasing d_W the slope decreases rapidly, until at $d_W = 78$ Å the superlattices are clearly in the semimetallic regime where $n_i/T^{3/2}$ is nearly constant down to $T = 15$ K. There is then a wide range of well thicknesses (104 Å is shown) for which the energy gap remains near zero as the $E1$ and $HH1$ bands anti-cross, as discussed in the section on band structure. Not shown on the figure is that for even larger d_W a second semiconducting region is reached [31, 100]. This latter phenomenon represents another observation that can be understood only in terms of a large valence band offset. It is important to note the the intrinsic carrier concentrations for HgTe/CdTe superlattices are in the appropriate range for infrared detector fabrication.

In-plane electron and hole mobilities in HgTe/CdTe superlattices have been reported by a number of investigators [11, 53, 78, 99, 101, 102]. The principal trends followed by these data are now generally understood in terms of the dominant

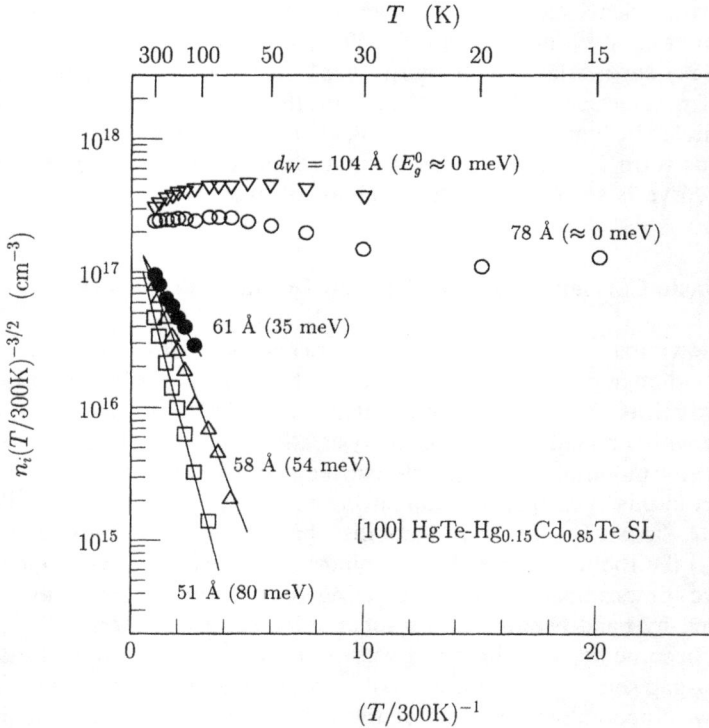

Figure 6.12 Experimental [11, 100] intrinsic carrier density (normalized by $T^{3/2}$) versus inverse temperature for five HgTe/Hg$_{0.15}$Cd$_{0.85}$Te superlattices. Well thicknesses and energy gaps derived from the slopes of the straight-line fits are indicated.

scattering mechanisms, statistical considerations, and the distinctive nature of the superlattice band structure. Recent calculations [103] have accounted for *ionized impurity* (II) scattering, *electron-hole* (EH) scattering, *acoustic* (AC), *polar-optical* (PO), and *nonpolar-optical* (NPO) phonon scattering, and interface roughness scattering were included in obtaining the transition rates. Because the last mechanism is somewhat less familiar than the others, the evidence for its importance in limiting the electron mobility in HgTe/CdTe superlattices will be considered separately.

Interface roughness scattering in a superlattice should be treated by using a formulation that allows for roughness-induced energy fluctuations in all three spatial dimensions [103]. The scattering rate then varies approximately as $\tau^{-1}(k_\rho, k_z) \propto (\Delta \Lambda \partial E / \partial d_W)^2$, where Δ is the magnitude of the fluctuations (in monolayers), Λ is the lateral correlation length for the roughness, and $E(d_W, k_\rho, k_z)$ is the superlattice energy level. In the simplified approximation of a particle in a box with infinite barriers, $E \propto$

d_w^{-2}, giving $\partial E/\partial d_w \propto d_w^{-3}$ and a mobility proportional to d_w^6 (this relation is only approximately correct when the more exact dispersion relations are accounted for in narrow-gap structures with strong nonparabolicity [104]). For a number of HgTe/CdTe superlattices studied experimentally [11, 76, 96, 99], Figure 6.13 plots electron mobilities as a function of well thickness, where a given symbol represents samples grown as a series under similar conditions. Although we do not expect a universal dependence of μ_{IR} (where IR indicates interface roughness) on d_w because Δ and Λ may be quite sensitive to the particular growth parameters, it is significant that two different series of thin-well superlattices follow the signature d_w^6 dependence (curves). If Δ is taken to be one monolayer, these data imply $\Lambda \approx 80$ Å for the filled triangles [99] and $\Lambda \approx 300$ Å for the filled circles [11]. The former superlattices were oriented along the (211) axis whereas the latter were (100), although at this time it is unclear whether orientation plays a crucial role in determining the value of Λ. The curves in the figure are terminated at $d_w \approx 65$ Å, as the scattering rate is expected to change qualitatively when the semiconductor-to-semimetal transition

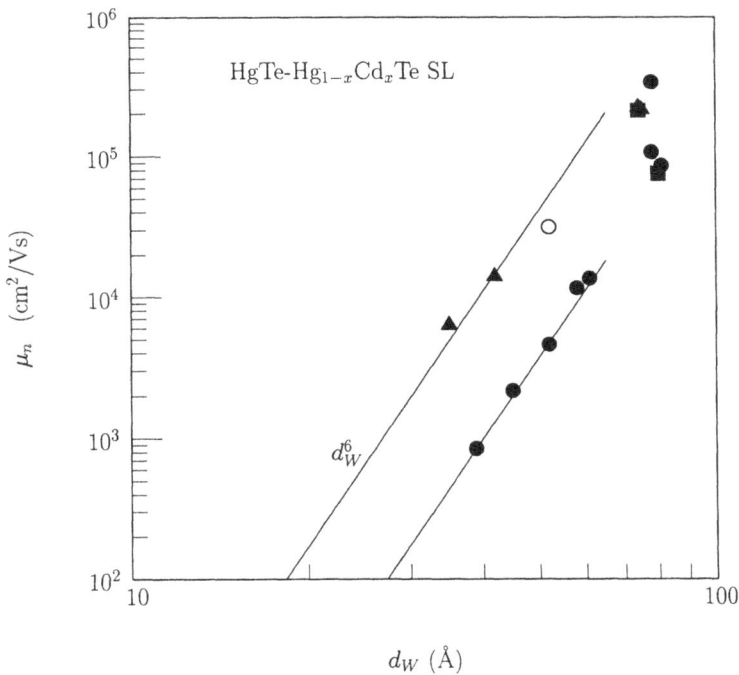

Figure 6.13 Experimental low-temperature electron mobilities versus well thickness for a series of HgTe/CdTe superlattices, compared to curves representing $\mu_n \propto d_w^6$.

is crossed (the electrons then occupy the HH1 band, whose energy levels are less sensitive to fluctuations in d_W). Note that the experimental variation of mobility with d_W is much less systematic in the superlattices with larger d_W.

For one of the (211) superlattices from Figure 6.13 (with $d_W = 45$ Å and $E_g^0 = 88$ meV), Figure 6.14 compares the experimental [99] mobility as a function of temperature (points) with the calculated [103] dependence (curves). The sample is taken to be uncompensated, as its low-temperature electron density of 9.6×10^{15} cm^{-3} is already at the upper end of the range of unintentional doping levels usually observed in HgTe/CdTe structures. The theoretical curves indicate that interface-roughness scattering dominates the electron mobility over the entire temperature range between 4.2 and 260 K. The low-temperature mobility due to ionized impurity scattering exceeds the experimental value by nearly an order of magnitude, and only at temperatures approaching ambience does polar optical phonon scattering finally be-

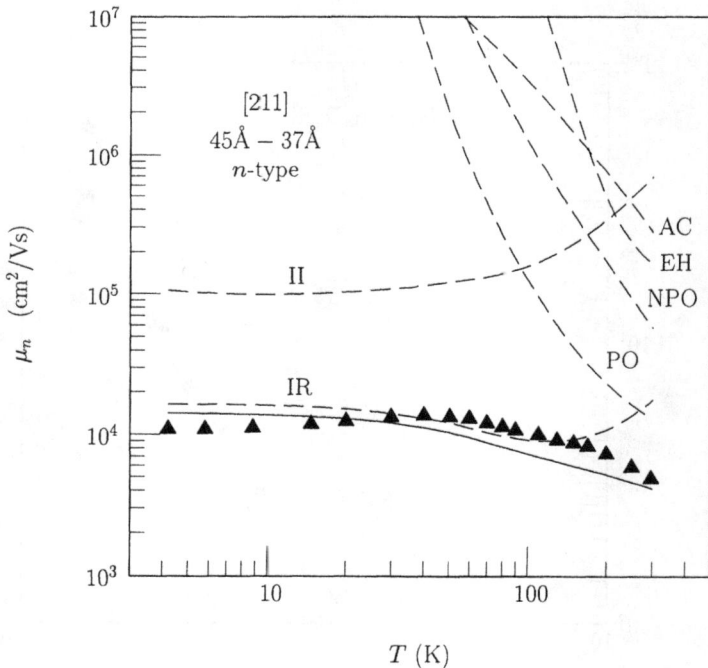

Figure 6.14 Theoretical (curves) and experimental [99] (points) electron mobilities versus temperature for a (211) HgTe/CdTe superlattice with $N_D - N_A \approx 9.6 \times 10^{15}$ cm^{-3} and the indicated well and barrier thicknesses [103]. The solid curve is the net mobility, and the dashed curves represent the various individual scattering mechanisms.

come stronger than the interface roughness contribution. The solid curve, which represents the net mobility for all mechanisms, is seen to accurately reproduce the experimental dependence on temperature.

The identity of the dominant electron scattering mechanism depends sensitively on d_W, because interface-roughness scattering becomes negligible in semimetallic superlattices (see Figure 6.13). Experimental [11] and theoretical [103] electron and hole mobilities for a p-type sample with $d_W = 78$ Å are plotted in Figure 6.15. The net acceptor concentration is 4.2×10^{15} cm^{-3}, and again the sample is assumed to be uncompensated. Because of the vanishing energy gap, thermally generated minority electrons were observable at temperatures as low as 10 K. The calculation yields that electron-hole scattering dominates μ_n at all temperatures up to 280 K. For holes, carrier-carrier scattering is much less important, and the low-temperature mobility is dominated by ionized impurity scattering. However, phonon scattering becomes stronger by the time T reaches 40 K. It was pointed out in the section on

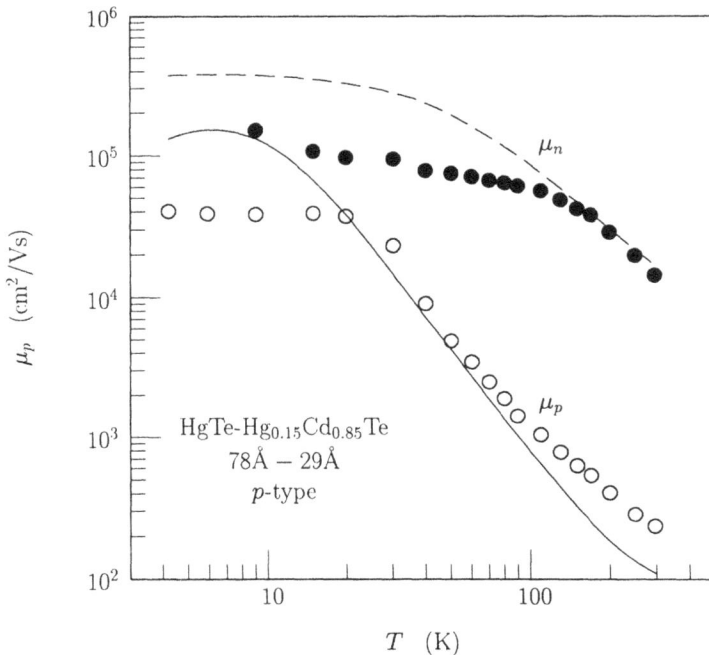

Figure 6.15 Experimental [11] (points) and theoretical (curves) electron and hole mobilities versus temperature, for a p-type ($N_A - N_D \approx 4.2 \times 10^{15}$ cm^{-3}) semimetallic superlattice with the indicated layer thicknesses. (After Meyer et al.) [103].

band structure that the average hole effective mass increases rapidly with temperature, which accounts for the strong decrease of μ_p with temperature. In both theory and experiment, μ_n and μ_p are comparable at low temperatures, but μ_p becomes two orders of magnitude smaller by the time T reaches 300 K. No adjustable parameters were used in the calculation for this superlattice indicating the high degree of accuracy in the predictive powers of current HgTe/CdTe superlattice band-structure models.

If high intentional doping levels are introduced, ionized impurity scattering naturally becomes the dominant mechanism for both electrons and holes at low temperatures. As discussed later, modulation doping of the CdTe barriers with indium has yielded high-mobility n-type superlattices with donor densities ranging up to 3 $\times 10^{17}$ cm^{-3}. Figure 6.16 illustrates low-temperature mobility results for eight mod-

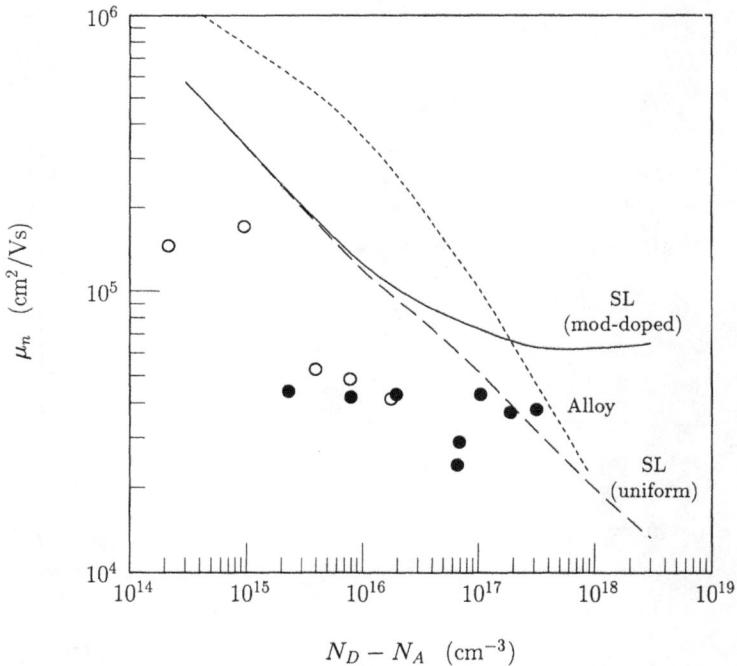

Figure 6.16 Experimental (points) and theoretical (curves) electron mobilities (4.2 K) versus net donor concentration for n-type HgTe/CdTe superlattices. The filled circles represent modulation-doped superlattices [102], and the open circles are for unintentionally doped samples [99, 100, 133]. Theoretical curves for both modulation doping (solid) and uniform doping (dashed) are shown. The dotted curve is the corresponding dependence for an Hg$_{0.8}$Cd$_{0.2}$Te alloy [105].

ulation-doped superlattices (filled circles) [102], along with five unintentionally doped samples (open circles) [99, 100]. It is apparent that, whereas μ_n decreases significantly with increasing impurity concentration in the lightly doped regime, there is little variation with $N_D - N_A$ at higher doping levels. This sharply contrasts the case of the $Hg_{1-x}Cd_xTe$ alloy (dotted curve) [105], for which the low-temperature mobility strongly decreases with N_D over the entire range up to 10^{18} cm^{-3}. The dependence in the alloy is due primarily to the increased density of scattering centers that accompanies the larger electron concentrations, and theory shows that a similar variation is expected in the superlattice when the wells and barriers are both doped uniformly (dashed curve). However, when the calculation is repeated for the case of modulation doping (solid curve), the screening of the impurity potentials becomes more effective at high n. The screening length is then shorter than the superlattice period, and the theoretical mobility becomes relatively independent of density at high N_D. In fact, μ_n is actually predicted to *increase* with N_D at the highest modulation doping levels. Although the theoretical μ_n overestimates the experimental values by about a factor of two, the functional dependence on $N_D - N_A$ is reliably reproduced without any adjustable parameters.

6.4 PROPERTIES DIRECTLY AFFECTING DETECTOR PERFORMANCE

Various materials parameters directly affect the viability of the use of the HgTe/ CdTe superlattice system as an infrared detector material. Some, such as extrinsic doping, must be controlled to allow the formation of detector structures. Others, such as minority carrier diffusion lengths, determine the ultimate performance of the detector. This section discusses HgTe/CdTe superlattice properties critical to device fabrication and performance.

6.4.1 Extrinsic Doping and Carrier Concentration Control

Extrinsic doping is a critical issue for developing a photovoltaic infrared detector technology. Indeed, the absence of a reliable procedure for p-type doping was one of the primary factors delaying the implementation of HgTe/CdTe superlattices in detector structures. The only viable approach to extrinsic doping of HgTe/CdTe superlattices is in-situ doping. Postgrowth doping is precluded as this requires process temperatures significantly above the growth temperature, which would lead to an unacceptable amount of interdiffusion. Similarly, doping by ion implantation may cause interdiffusion due to the implant itself, and the typical thermal anneal to remove implant damage could not be performed for the same reason. The development of in-situ doping techniques was therefore critical, not only because they allow the tailoring of sophisticated device structures, but also because conventional alternative methods are all inapplicable to HgTe/CdTe superlattices.

The doping of Hg-based materials during MBE is complicated because the growth typically takes place under Hg-deficient conditions, resulting in the formation of Hg vacancies. These vacancies aid incorporation of dopant atoms on metal (Hg, Cd) sublattice sites, but cause the incorporation of dopants on the Te sublattice to be more difficult. For example, initial doping studies of MBE-grown HgCdTe indicated that arsenic and antimony, which we expect to be p-type dopants when substituted for Te [106], were found to produce n-type doping [107]. This can be understood by antisite doping; that is, the incorporation of the As and Sb on to the metal sublattice rather than the Te sublattice. The ease of incorporation of both types of dopants on to the metal sublattice motivated research into extrinsic doping, using dopants that substituted, on the metal sublattice, only Group I elements as acceptors and Group III elements as donors.

In-situ n-type doping was first reported in MBE grown HgCdTe using indium [108]. Doping efficiencies were high, around 50–70%. However, a drastic memory effect was initially reported for In doping [107]. After growing an In-doped layer, residual In was reported to significantly contaminate subsequently grown layers. This effect led to the investigation of an alternate n-type dopant, silicon [107]. Unfortunately, Si was found to produce highly compensated layers with transport properties insufficient for device fabrication [107]. Subsequent investigators then reexamined the use of In and did not observe the memory effect [48, 109, 110]. The lack of evidence for a memory effect coupled with a high doping efficiency has lead to indium becoming the n-type dopant of choice for HgTe/CdTe superlattices.

Figure 6.17 illustrates that the indium dopant concentration is easily controlled through varying the In source temperature [109, 111]. The n-type concentration closely tracks the In vapor pressure expected for the In oven temperature. In general, the n-type carrier concentration can be controlled in long-wavelength infrared superlattices from the typical background level of about 1×10^{15} cm^{-3} up to $\approx 1 \times 10^{17}$ cm^{-3} by using In doping, with mobilities in the 10^4 cm^{2}V^{-1}s^{-1} range. This electron concentration is appropriate for use in the fabrication of infrared detectors. Refer back to Figure 6.16 for the dependence of mobility on carrier concentration for n-type HgTe/CdTe superlattices.

Interestingly, the control of indium doping in HgTe/CdTe superlattices is sufficiently high to allow the growth of modulation-doped superlattices [112]; that is, superlattices with the dopant present only in the barrier layers. Modulation doping both with and without a setback was discussed previously [113] to lead to enhanced in-plane mobilities and observation of the quantum Hall effect. However, infrared detectors based on superlattices rely on transport through the barrier layers, and calculations have not been performed to evaluate the effect of modulation doping on such transport. Because impurity scattering is still the dominant mechanism and the minority holes spend more time in the wells, it is conceivable that modulation doping will increase the mobility by lowering the effect of impurity scattering. The potential

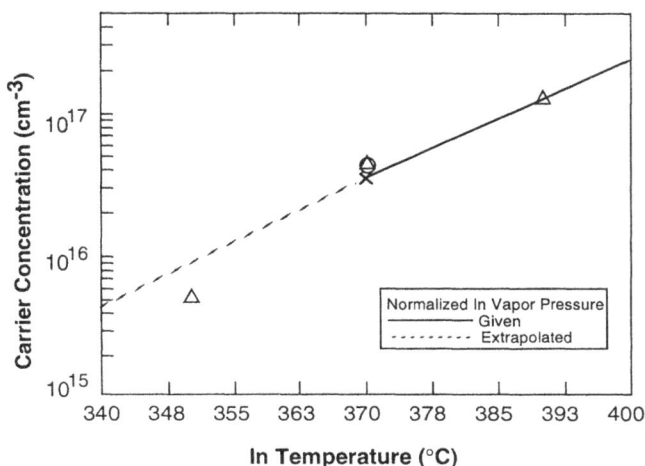

Figure 6.17 Carrier concentration versus In oven temperature. The carrier concentration of *n*-type HgTe/ CdTe superlattices can be controlled through substitutional doping with indium. Note that the carrier concentration tracks well with the expected In vapor pressure, which was normalized at 10^{17} cm^{-3}.

factor of two increase in minority carrier hole mobility may prove important for photogenerated charge collection, as discussed in a later section. At present, the *n*-type regions HgTe/CdTe superlattices grown for fabrication into infrared detectors do not employ modulation doping.

P-type doping of Hg-based materials grown by MBE has proven to be a difficult problem. The initial problems with the Group V elements (As, Sb) [108] led to the investigation of Group I elements, which readily incorporate on the metal sublattice. Lithium was shown to be a highly effective *p*-type dopant in HgCdTe grown by MBE, with doping levels increasing monotonically with Li flux. Doping levels of up to 10^{18} cm^{-3} were obtained. However, Li has an appreciable diffusion coefficient even at room temperature [108], precluding its use in a device structure. Although not investigated, it is anticipated that similar results would be obtained for sodium and potassium. The Group IB elements (copper, silver, and gold) are also *p*-type dopants. However, Cu is not a feasible dopant, as its diffusion coefficient rivals that of Li [114]. Au also exhibits a large enough diffusion coefficient to present difficulties [114]. From initial studies, it appeared that the diffusion coefficient of Ag was low enough that diffusion effects could be controlled [109, 110, 115]. Unfortunately, the use of Ag has never clearly produced the doping profiles required for device formation, possibly due to "short-circuit" diffusion effects.

To date, the only solution has been to develop MBE techniques for the use of p-type dopants with low diffusion coefficients, such as As. In 1989, Harper et al. first reported p-type doping by As incorporation in CdTe, using photon-assisted MBE [116]. Han et al. [117] soon demonstrated that thin CdTe layers grown by photon-assisted MBE and doped with As could be used to produce modulation-doped HgCdTe layers. In this study thin As-doped CdTe was grown between thick HgCdTe layers, producing p-type material that closely resembled homogeneous p-type HgCdTe. Such modulation doping procedures were shown to produce p-type layers that were stable with respect to As diffusion.

Arias et al. [60] were the first to report p-type doping of HgTe/CdTe super-lattices using As. P-type doping was achieved both by photon-assisted MBE and by conventional MBE, with better properties obtained for the layers grown by photon-assisted MBE [60]. The As incorporation was found to be relatively independent of whether both the HgTe and the CdTe layers were grown under As flux, or only the CdTe layers were exposed. This result implies that the As atoms are incorporated mainly into the CdTe layers due to the stronger Cd-As bond as compared to the Hg-As bond. Other results have since confirmed that the As appears to be incorporated only into the CdTe layers [66]. Therefore all As-doped HgTe/CdTe superlattices are modulation doped, independent of whether the As flux is present during the growth of the HgTe layers.

Arias et al. [60] also reported enhanced p-type Hall mobilities which ranged from 1000 to 16,000 $cm^2V^{-1}s^{-1}$ for hole concentrations of 9×10^{15} to 4×10^{16} cm^{-3}. Hole freeze-out was not observed, probably due to the fact that the As atoms are not present in the HgTe wells. SIMS analysis of these same samples indicated As concentrations in the 2×10^{17} to 1×10^{18} cm^{-3} range for As source temperatures between 150 and 170°C, indicating that a low percentage of As is actually electrically active as acceptors. Subsequent studies have indicated higher activation efficiencies for As-doped HgTe/CdTe superlattices [66, 73]. Hole concentrations as large as 10^{17} cm^{-3} have been reported [118].

Harris et al. [66] also reported that As doping levels were highly dependent on the substrate temperature during growth. For example, a 30°C reduction of the substrate temperature was found to result in an order of magnitude increase in As incorporation for a constant As flux. This temperature dependence implies that the As sticking coefficient is strongly dependent on the substrate growth temperature.

In summary, In and As have been demonstrated to be effective n- and p-type dopants, respectively, of HgTe/CdTe superlattices. Doping profiles can be con-trolled to produce sharp junctions, as is illustrated in Figure 6.18. This SIMS profile shows a 1 μm thick As doped cap layer on a 3 μm base layer doped with In, rep-resenting a dopant profile suitable for infrared detector fabrication. Indeed, this SIMS profile was taken from a small piece of an HgTe/CdTe layer that was subsequently processed to produce high quantum efficiency HgTe/CdTe photovoltaic detectors [66].

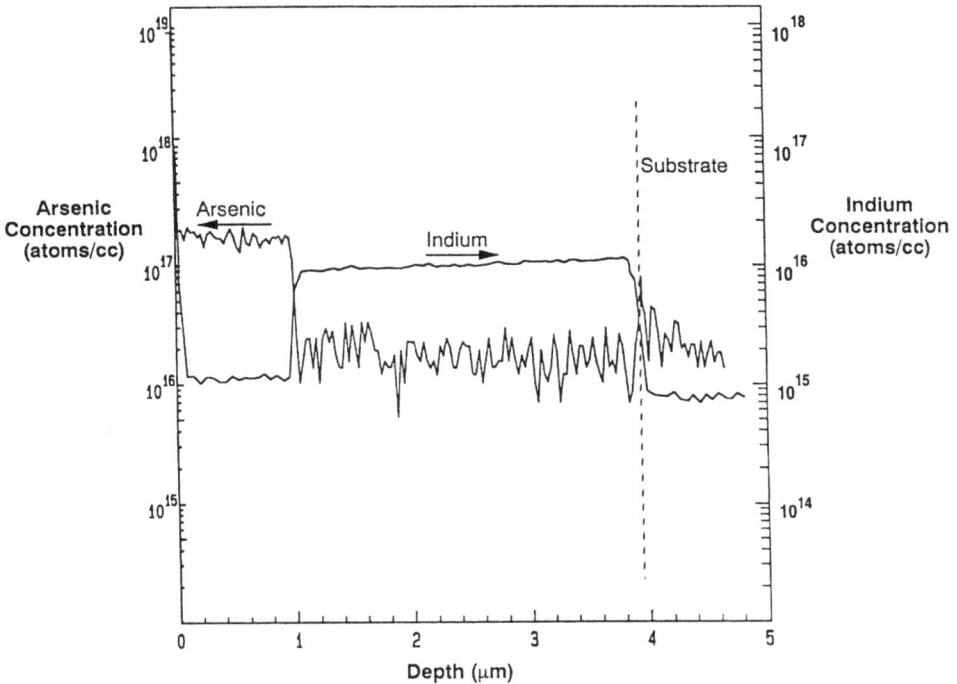

Figure 6.18 Indium and arsenic SIMS depth profiles obtained from the same HgTe/CdTe superlattice. This profile indicates that in-situ doping of HgTe/CdTe superlattices can be controlled to produce structures appropriate for fabrication into photovoltaic infrared detectors.

6.4.2 Control of Bandgap in HgTe/CdTe Superlattices

The superlattice bandgap, E_g, is controlled by the thickness of the well and barrier layers [4]. By contrast, E_g in the alloy is controlled by the composition x, given by the relative Hg-to-Cd ratio. At some finite value of x, the bandgap of the alloy goes to zero, leading to a singularity in the cutoff wavelength, λ_c. Therefore, to adequately control λ_c for longer wavelengths, the composition of the alloy must be controlled to progressively higher precision. Assuming a small valence band offset, ΔE_v, λ_c was originally predicted to be quite insensitive to layer thickness for HgTe/CdTe superlattices. However, the presence of a large ΔE_v causes E_g to go to zero in this superlattice system much faster than originally believed, decreasing the range of d_W available for spanning the infrared spectral range. The superlattice semiconductor-to-semimetal transition then again leads to a singularity in λ_c *versus* d_W for the superlattice, a result that has been confirmed experimentally [99]. Although still easier to control than in the equivalent alloy, λ_c is not as insensitive to d_W for longer wavelengths, as predicted using $\Delta E_v \approx 0$ meV.

For the typical detector operating temperature of 77 K, Figure 6.19 shows calculated superlattice energy gaps and cutoff wavelengths (taken to be the wavelength at which the photon energy is equal to the energy gap; that is, $\lambda_c = 1.240/E_g$) as a function of well thickness, with the barrier thickness held constant at 50 Å. The results were obtained from an eight-band $\mathbf{k} \cdot \mathbf{p}$ formalism [10] using $\Delta E_v = 350$ meV, and results for both the (100) (dashed) and (211) (solid) orientations are given. Also shown is the early result of Smith, McGill, and Schulman [119], which was based on a valence band offset of zero. It is evident that this and other [86, 120] early estimates of the appropriate well thicknesses for infrared detectors must be revised considerably due to the larger ΔE_v now known to be present.

Using the calculated dependencies of λ_c as a function of d_W from Figure 6.19, Figure 6.20 shows plots of $d\lambda_c/dd_W$ for the (211) and (100) orientation, along with

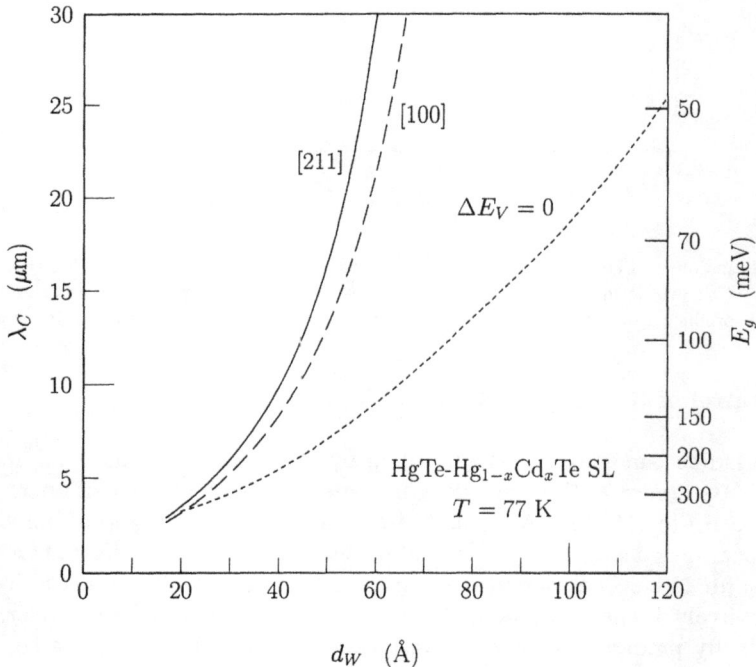

Figure 6.19 Cutoff wavelength and energy gap ($\lambda_c \equiv hc/E_g$) versus well thickness for (100) (dashed curve) and (211) (solid curve) superlattices with $d_B = 50$ Å, at 77 K. The dotted curve represents an early theoretical result of Smith et al. [119] ($d_B = d_W$), which was based on a valence band offset of zero.

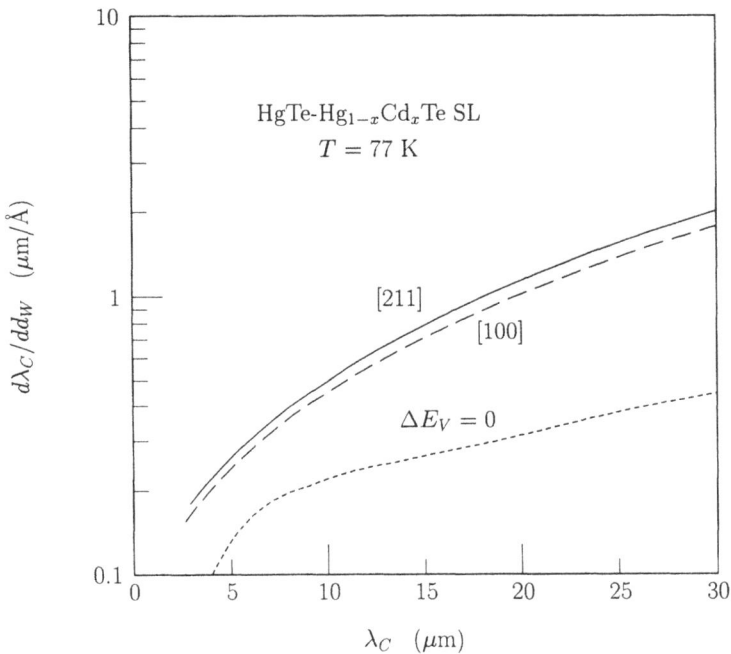

Figure 6.20 Derivatives of $\lambda_c(d_w)$ versus cutoff wavelength, from the curves in Figure 6.19.

the prior result for a vanishing valence band offset. This figure shows that the sensitivity of λ_c to d_w is approximately two to four times greater than had been previously believed.

The procedure outlined by Reno and Faurie [41] will be used to examine the consequences of the dispersion of λ_c with layer thickness. For MBE, lateral variation of layer thickness or alloy composition is fixed by the geometry of the system. From reflection high-energy electron diffraction oscillations [121, 122] average HgTe and CdTe growth rates can, in principle, be determined to a higher degree of accuracy than fluctuations that occur during growth. The main uncertainty in the control of cutoff wavelength, $\Delta\lambda$, then results from the variations in flux throughout the growth of a layer due to fluctuations in the MBE source temperatures. Calculations are presented for both the alloy and the HgTe/CdTe superlattice system.

Typical temperature control of an MBE furnace leads to variations in temperature, ΔT, on the order of 0.1°C. For this value of ΔT, the variation in flux from an effusion cell can be calculated theoretically [123] by using published vapor pressure data. However, because experimental results tend to exceed the calculated values by a factor of two [41] the theoretical flux variations have been doubled in the following analysis, giving fluctuations in flux of $\Delta F_{Te}/F_{Te} \approx 0.006$ and $\Delta F_{CdTe}/F_{CdTe} \approx 0.003$. Assuming that ΔT for each effusion cell is independent of the other cells, then the change in composition, Δx, for the alloy due to temperature fluctuations is given by [41]

$$\Delta x = x(1 - x)[(\Delta F_{Te}/F_{Te})^2 + (\Delta F_{CdTe}/F_{CdTe})^2]^{1/2} \qquad (6.1)$$

Using eq. (6.1) along with the empirical expression for E_g in the alloy determined by Hansen, Schmit, and Casselman [124] and $\lambda_c = 1.240/E_g$, $\Delta\lambda$ for the alloy can be calculated and is shown in Figure 6.21. For the HgTe/CdTe superlattice system, λ_c is not a strong function of the CdTe barrier thickness, d_B, for thicknesses of 50 Å or greater. Therefore, the variation in λ_c is primarily due to changes in the HgTe well thickness, d_W, and Δd_W is given simply by $d_W(\Delta F_{Te}/F_{Te})$. Using the values of $d\lambda_c/dd_W$ shown in Figure 6.20, Figure 6.21 indicates $\Delta\lambda_c$ calculated for a zero and a 350 meV valence band offset. Only the (211) value is shown, as it represents the worst case for a large ΔE_v. As can been seen, variations in λ_c are still more easily controlled in the superlattice for long wavelengths than the alloy. However, the improvement in control is smaller than originally predicted using $\Delta E_v \approx 0$.

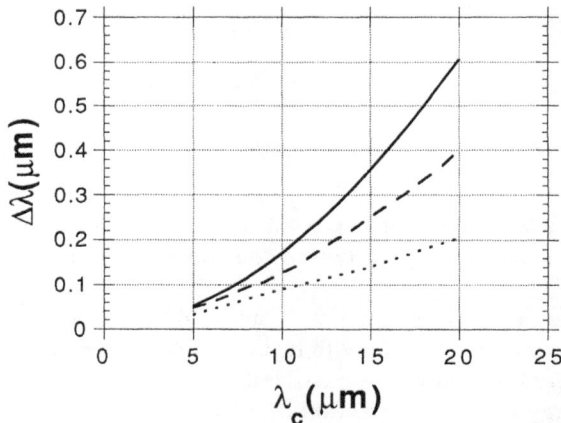

Figure 6.21 Predicted variation in λ_c as a function of λ_c due to fluctuations in MBE effusion cells for HgCdTe (solid curve) and HgTe/CdTe superlattices with zero (dotted curve), and 350 meV (dashed curve) valence band offsets.

6.4.3 Carrier Transport, Tunneling, and Charge Collection Efficiency in the Growth Direction

In addition to reducing the fluctuations in cutoff wavelength, another significant advantage of HgTe/CdTe superlattices for infrared detectors is that we can tailor the electron and hole transport and tunneling in the growth direction through varying d_B. Although in-plane transport in this superlattice system has been the subject of numerous investigations (see Section 6.3.6), growth direction transport has received less attention. Furthermore, those calculations [125, 126] and measurements [19, 20, 125, 127] dealt primarily with resonant tunneling through single and double heterostructures, which have only marginal relevance to infrared detector considerations.

Only Reed et al. [20] have discussed measurements of electron transport along the growth axis through a superlattice miniband, in a structure with a thick CdTe barrier separating two 10-period superlattice regions. Their results showed a negative differential resistance feature at voltages large enough that the minibands on opposite sides of the barrier are no longer aligned in energy. Although ohmic conduction along the z-axis was demonstrated at low voltages, the results presented were not sufficiently detailed to allow conclusions concerning growth direction mobilities. At present, the best predictors of μ_{nz} and μ_{pz} are the calculated effective masses, which have been confirmed by experimental cyclotron resonance data to be reliable in certain limited regions, as discussed in Section 6.3.5.

The magnitude of the valence band offset affects both m_{nz} and m_{pz}, although the hole mass is far more sensitive as ΔE_v determines the ordering of the LH1 and HH1 bands [6]. For a small offset, strain causes LH1 to be highest in energy, and we obtain $m_{pz} \approx m_{nz}$. However, when ΔE_v is large, quantum confinement moves LH1 to an energy significantly below HH1 and $m_{pz} \gg m_{nz}$. This is illustrated by the theoretical calculations shown previously in Figure 6.11. Good agreement with the experimentally determined electron cyclotron masses indicated by the points in the figure confirms the reliability of the theory.

The figure shows that both m_{nz} and m_{pz} can be tuned over a considerable range by varying d_B. As had been predicted by the early zero-offset calculations, m_{nz} considerably exceeds the electron mass in the alloy with the same bandgap ($0.0065m_0$ for $E_g = 83$ meV). As pointed out by Kinch and Goodwin [86] (see also Smith et al. [119]), this will lower tunneling currents by up to several orders of magnitude in infrared devices fabricated with HgTe/CdTe superlattices because tunneling currents depend exponentially on the inverse of the effective mass.

In photovoltaic superlattice detectors, the growth-direction electron and hole masses also determine to a large extent the photogenerated charge collection efficiency, which depends on the minority carrier diffusion length: $\Lambda \propto (\mu_z \tau_r)^{1/2}$. Here μ_z is the mobility along the growth axis, and τ_r is the recombination lifetime. Note that for any $d_B \geq 20$ Å, holes at the valence band maximum [$m_{pz}(0)$] have an extremely large mass, implying that they are essentially localized in the low temperature limit. This would indicate a mobility far too small to allow efficient collection of the photogenerated charge. However, as discussed in Section 6.2.2, the valence band in HgTe/CdTe superlattices displays a "mass-broadening" effect [25], which causes the in-plane mass to depend strongly on growth direction wavevector and m_z to depend strongly on energy. Figure 6.11 indicates that $m_{pz}(k_BT)$ is orders of magnitude smaller than $m_{pz}(0)$, having a value comparable to the heavy hole mass in the HgCdTe alloy of similar λ_c. Therefore, mass broadening should give a mobility μ_{pz} large enough in the temperature range of most infrared detectors to allow high collection efficiency as long as d_B is not too thick.

However, a further complication is that we expect the miniband conduction model to break down above some critical barrier thickness. The miniband approach is valid only when the tunneling time $\tau_t \approx h/(2\pi W)$($W$ is the miniband energy width) is shorter than the scattering time, τ_s. When $\tau_t > \tau_s$, the basis states are confined to a single quantum well, with only occasional tunneling to an adjacent well. "Growth-direction mass" then becomes an invalid concept, and the transport is primarily by hopping from one well to the next. Because hopping mobilities tend to be far lower than miniband mobilities [128], we expect poor collection efficiencies at any d_B for which the minority carrier transport must depend on that mechanism. In Figure 6.11, the miniband and hopping regimes are indicated by the solid and dashed portions of the curves, respectively, where we have estimated τ_s from the experimental magneto-transport and magneto-optical data discussed previously [25, 99]. Inhibited hole transport in the n-doped collection region may explain the reduction with decreasing temperature of the high quantum efficiency observed for the only published HgTe/CdTe superlattice photovoltaic detector (midwave infrared with $d_B \approx 50$ Å) [66].

The hole transport problem is complicated, and further experiments will be required to clarify the range of d_B for which m_{pz} is large enough to assure efficient collection of the photogenerated carriers. It may ultimately be found that detectors with a p-type collection region are preferable, because the prediction that m_{nz} and μ_{nz} at intermediate d_B is in a convenient range for high quantum efficiency can be made with a reasonable level of confidence. Detector fabrication must also take into account the high in-plane mobilities measured for small bandgap HgTe/CdTe superlattices [25, 99], which can lead to appreciable in-plane diffusion of photogenerated carriers.

6.4.4 Minority Carrier Lifetime

Minority carrier lifetime (or excess carrier recombination time), τ_r, is one of the most important parameters in determining infrared detector performance. As discussed earlier, τ_r has a large influence on photogenerated charge collection. In addition, τ_r is inversely related to the magnitude of dark current generated in an infrared detector. Interestingly, in spite of the considerable amount of fundamental research devoted to the HgTe/CdTe superlattice system in recent years, there are currently only five reports of minority carrier lifetimes [73, 75, 76, 77, 129]. Lifetimes in the neighborhood of 10 ns were reported for undoped (100)-oriented superlattice as measured by photoconductive decay [75, 76] or calculated from observed dark currents in metal-insulator semiconductor structures [77]. Such short lifetimes are inappropriate for low-doped n-type material in high-performance infrared detectors and were indicative of midgap trapping states, possibly due to the presence of the Te antisite defect discussed previously. Only recently [73, 129] have there been reports of excess carrier lifetimes of several hundreds of nanoseconds, which approaches numbers typically observed in good quality HgCdTe of comparable E_g and doping levels.

Figure 6.22 displays experimental lifetime measurements reported by Reisinger et al. [129] for a superlattice that was not intentionally doped. Hall measurements indicated that the conductivity was n type, with a carrier concentration of about 3 $\times$ 10^{15} cm^{-3} at 77 K. Lifetimes near 400 ns are recorded at temperatures below about 100 K for this superlattice with $\lambda_c \approx 8$ μm. Figure 6.22 also shows calculated lifetimes for bulk HgCdTe of equivalent E_g and similar doping levels for a number of recombination processes, including radiative, Auger AM1, Auger AM7, and

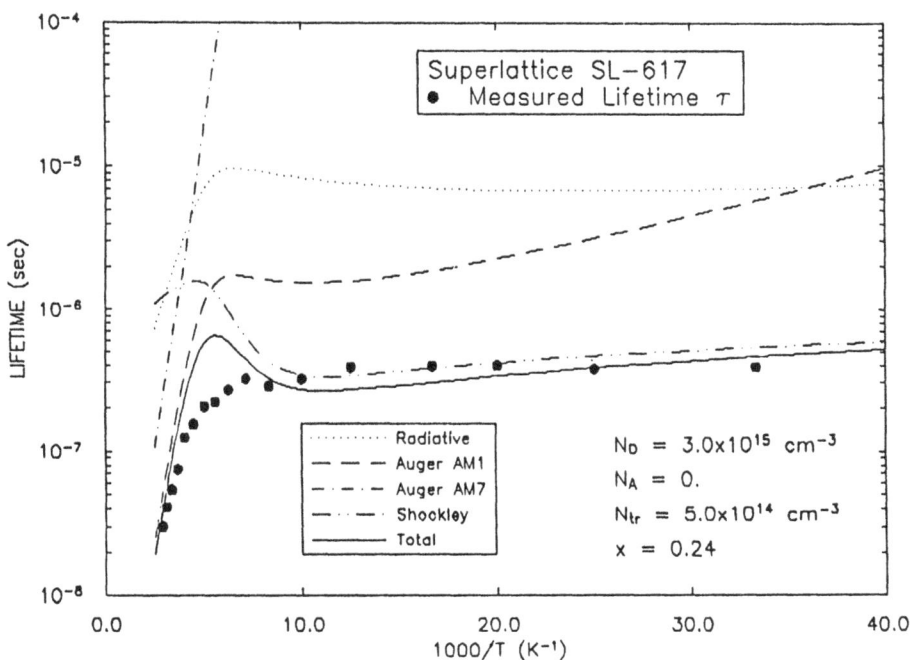

Figure 6.22 Carrier lifetime versus reciprocal temperature for an unintentionally doped ($N_D - N_A \approx 3$ $\times$ 10^{15} cm^{-3}) HgTe/CdTe superlattice with $\lambda_c \approx 8$ μm at 77 K. The lifetime due to various recombination processes for the equivalent HgCdTe alloy are shown.

Schockley-Read-Hall [130]. The observed lifetimes are about one order of magnitude shorter than those calculated based on the Auger AM1 process alone. The data are consistent with Schockley-Read-Hall recombination with a midgap trap density not in excess of 5×10^{14} cm^{-3}, which is often observed for bulk HgCdTe. This trap density may also indicate a small level of Te antisite formation occurring on the (211)B orientation. The fact that the experimental data approaches the results calculated for bulk HgCdTe testifies to the high quality of the superlattice.

For comparison, a midwave infrared superlattice with λ_c of 4.5 μm exhibited long lifetimes that exceeded the apparent limit set by the Auger AM1 process [129]. This superlattice, with an n-type carrier concentration of 8×10^{14} cm^{-3}, had a lifetime that peaked at 20 μs at 140 K and slowly decreased to 17 μs at low temperatures. This carrier lifetime is comparable to that observed in high-quality midwave infrared HgCdTe, but only for significantly lower doping levels. Recent calculations by Jiang, Teich, and Wang [131] indicate that, for HgCdTe quantum wells, Auger recombination can be significantly suppressed. They predict almost an order of magnitude increase in Auger lifetime in the quantum well case over the bulk in the midwave infrared. Their calculations did not include the nonparabolicity of the valence band, and so the quantitative predictions must be viewed with caution. Even so, these calculations may qualitatively explain the longer than expected lifetimes for this sample. Potential lifetime enhancement due to quantum confinement bodes well for the use of HgTe/CdTe or, more likely, HgCdTe/CdTe superlattice structures for near-IR injection lasers.

However, there is not yet evidence that Auger recombination suppression is present for longer λ_c. Reisinger et al. [129] also reported on a long-wavelength infrared superlattice that was intentionally doped p-type by using arsenic. Measured and calculated values of lifetime for this superlattice are shown in Figure 6.23. The superlattice had a measured 80 K lifetime of about 40 ns, which is consistent with calculations of Auger AM7 recombination in the equivalent alloy for the 2×10^{16} cm^{-3} arsenic concentration and λ_c of ≈ 8.5 μm. Certainly, more investigation of lifetimes in long-wavelength infrared superlattices is warranted. However, it must be pointed out that these values of carrier concentration and lifetime in the p-type superlattice are equivalent to that currently used in HgCdTe infrared detector fabrication.

A minority carrier lifetime of 40 ns for electrons in p-type HgTe/CdTe superlattices is also adequate for photogenerated charge collection. As discussed before, electron mobility along the z-axis should be comparable to that observed in the alloy. The minority carrier diffusion length, Λ will be about 16 μm for $\mu_{nz} \approx 10^4$ cm^2/V $\cdot$ s and $\tau_R \approx 40$ ns. Λ would still be comparable to the expected superlattice absorbing layer thickness of 5 μm if μ_{nz} was as low as 10^3 cm^2/V $\cdot$ S, resulting in a high quantum efficiency device. More importantly because the previous discussion indicates that μ_{pz} is expected to be fairly small, it is far more critical to have long minority carrier lifetimes in low-doped n-type HgTe/CdTe superlattices. The 400 ns lifetime discussed previously corresponds to a factor of seven increase in Λ over that

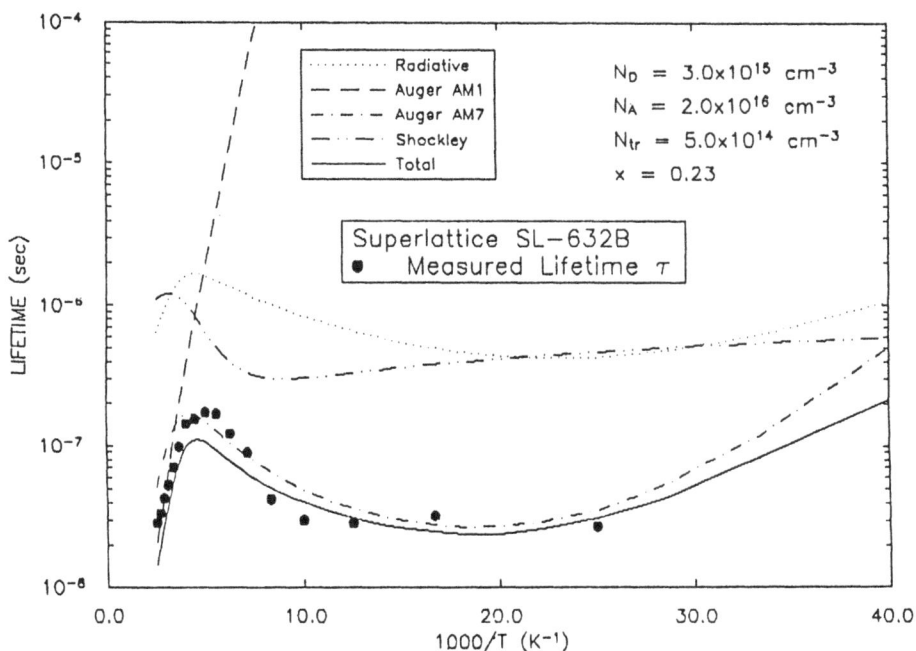

Figure 6.23 Carrier lifetime versus reciprocal temperature for an intentionally doped ($N_A - N_D \approx 2 \times 10^{16}$ cm^{-3}) HgTe/CdTe superlattice with $\lambda_c \approx 8$ μm at 77 K. The lifetimes due to various recombination processes in the equivalent HgCdTe alloy are shown.

for a lifetime of 9 ns. This increase will be critical for photogenerated charge collection from the *p*- side of an infrared detector. Certainly, the recent report of long lifetimes by Reisinger et al. [129] removes one of the last fundamental obstacles to the actual realization of HgTe/CdTe superlattice infrared detectors.

6.5 HgTe/CdTe DETECTOR FABRICATION

The recent advances in superlattice material quality indicate that many fundamental difficulties encountered with Hg-based MBE growth and doping previously limiting the studies of actual detector structures have now been solved. This is reflected by the measurement of long excess carrier lifetimes and controlled *p*-type doping. Indeed, these improvements in material properties have led to the fabrication of the first midwave photovoltaic devices based on a HgTe/CdTe superlattice with good detector characteristics. Undoubtedly, long-wavelength infrared devices using HgTe/CdTe superlattices will be fabricated in the next year that demonstrate the advantages of this materials system. The midwave infrared superlattice detector is discussed in this section, along with two device studies that proved premature for truly evaluating HgTe/CdTe superlattice detectors.

The first reported use of an HgTe/CdTe superlattice in a detector configuration was given by Goodwin, Kinch, and Koestner [77] in 1988 for a metal-insulator semiconductor device structure. The simplest metal-insulator semiconductor device structure consists of an insulator sandwiched between a metal contact and the semiconductor. Because no photovoltaic junction is formed, metal-insulator semiconductor devices do not require an extrinsic doping technology and are typically fabricated on undoped material. In principle, a preliminary investigation of superlattice devices could thus be performed without the need of an extrinsic doping technology. Three structures were grown. Two were unintentionally doped n-type superlattices on a (100) orientation and a (211)A orientation. The third structure was a p-type superlattice grown on a (111)B liquid phase epitaxy (LPE) HgCdTe layer.

The dark current measurements made on these devices indicated poor performance. Analysis of the dark currents in the n-type superlattice indicated a 77 K minority carrier lifetime of about 9 ns, a factor of 100 lower than that found in good quality bulk HgCdTe of the same λ_c with the same carrier concentration of about 2 $\times$ 10^{15} cm^{-3}. This short lifetime, which is inadequate for metal-insulator semiconductor detector fabrication on n-type material, was indicative of the problems associated with MBE growth of Hg-based material at that time and was most likely due to point defects as discussed before. Although premature for evaluating the potential of HgTe/CdTe superlattices, two important results did emerge from this study.

First, the p-type superlattice grown on LPE HgCdTe was fabricated with a semitransparent gate, allowing photoresponse and quantum efficiency measurements. Photoresponse measurements gave a λ_c of 4.5 μm, in agreement with theoretical predictions for the layer thicknesses. Analysis of the photoresponse gave a 30% quantum efficiency indicating an absorption coefficient of about 9 $\times$ 10^3 cm^{-1} and a minority carrier electron mobility large enough for appreciable photogenerated charge collection, even with the short lifetime.

The second, possibly more important, result from the metal-insulator semiconductor study was that the breakdown electric field, which is the field at which tunneling currents begin to dominate the dark currents, was very large ($E_{bd} \approx 3 \times 10^4$ V/cm) and equal to or greater than high-quality bulk HgCdTe, even with the short lifetime. This result was the first experimental demonstration of reduced tunneling currents due to the large effective masses predicted for HgTe/CdTe superlattices.

Wroge et al. [109] reported the first photovoltaic device structure based on extrinsic doping using In and Ag for n-type and p-type doping, respectively. The structure consisted of a 2 μm In-doped cap layer, with $n \approx 2 \times 10^{16}$ cm^{-3}, grown on a 5 μm Ag-doped base layer, with $p \approx 2 \times 10^{16}$ cm^{-3}. SIMS results indicated a low level of Ag diffusion, which should have allowed a rectifying junction to be formed. Evidence was presented for a weakly rectifying junction in mesa devices fabricated on this structure. However, actual photoresponse and quantum efficiency measurements were not presented, precluding a critical assessment of the detector performance. However, it should be pointed out that the superlattice in Wroge's device structure was grown based on E_g predictions using a ΔE_v of 0. Using their measured layer thicknesses of $d_B = 36$ Å and $d_W = 64$ Å with $\Delta E_v \approx 350$ meV, we would expect a bandgap of about 30 meV; that is, a wavelength of 41 μm. This alone would preclude device operation and underscores the difficulty in designing superlattice device structures when all the appropriate parameters necessary for prediction are uncertain.

As was detailed in the previous sections, combining the low-temperature technique of photon-assisted MBE with the use of the $(211)B$ orientation has allowed for the growth of multilayers exhibiting a high degree of crystalline perfection and *in-situ* *n*-type and *p*-type extrinsic doping. Using this combination, Harris et al. [66] reported the successful fabrication of midwave infrared superlattice detectors exhibiting high quantum efficiencies and uniform response.

The basic structure of the infrared homojunction superlattice devices grown by Harris consisted of a 3 μm *n*-type base layer followed by a 1 μm thick *p*-type cap layer grown on a $(211)B$ CdTe substrate. To ensure *p*-type and *n*-type behavior in the layers, As at the 10^{17} cm^{-3} level and In at the 10^{16} cm^{-3} level were used, respectively. However, typical doping levels in an optimized device structure would be about a factor of ten less than this. Dopant incorporation was verified by SIMS analysis on a small piece of the grown layer. The resultant SIMS depth profile of this homojunction superlattice, as shown previously in Figure 6.18, confirmed that the dopant profile and doping levels matched that expected from growth conditions. Mesa structures were etched from the superlattice by using standard fabrication processes. The mesas were then passivated using MBE CdTe deposited at low temperature, followed by an insulating dielectric covering. The final step was metallization to form the base and cap layer contacts.

The measured spectral response for one of these devices with a cutoff wavelength of about 4.5 μm at 140 K is shown in Figure 6.24. This value for λ_c is in excellent agreement with prediction for $\Delta E_v \approx 350$ meV. Direct measurements yielded peak quantum efficiencies as high as 66% (at 140 K) at the peak wavelength and an average over the 3–5 μm waveband of 55%. At 77 K the measured peak quantum efficiency was found to decrease to 45–50% with a cutoff wavelength of 4.9 μm. These devices represent the first photovoltaic infrared detectors based on HgTe/CdTe superlattices to exhibit significant quantum efficiencies.

Figure 6.3, shown previously, is a representative transmission electron microscopy micrograph taken from a small piece of this midwave infrared superlattice. All observed areas of the superlattice exhibited a highly regular layer spacing with a large degree of interfacial sharpness and layer uniformity. Line dislocation counts

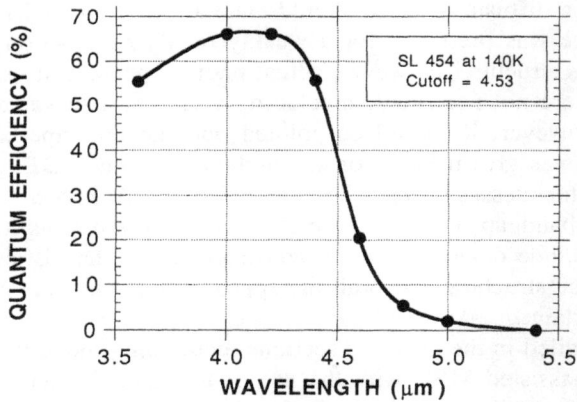

Figure 6.24 Representative spectral response of an HgTe/CdTe superlattice detector with $\lambda_c \approx 4.53$ μm at 140 K. The peak response corresponds to 66% quantum efficiency.

were consistent with a value of less than 10^6 cm^{-2} and represent the resolution of the transmission electron microscopy technique. From the transmission electron microscopy image, the layer thicknesses for this superlattice are 30 Å (HgTe) and 50 Å (CdTe). No evidence of interdiffusion was observed in the layers.

The optically sensitive area was determined by infrared spot scan, as illustrated in Figure 6.25. The response of the front-side illuminated device is uniform over the entire detector area except in the region masked by the front-side contact. The optically active area is in good agreement with that of the physical area of the mesa device.

Figure 6.26 shows a representative I–V measurement obtained for the midwave infrared superlattice detectors at 77 K. The p-on-n nature of the homojunction is clearly evident as the I–V curve is reversed from that obtained from the more conventional n-on-p diode. A standard figure of merit for a photovoltaic infrared detector is the resistance area product at zero bias, R_0A. The magnitude of R_0A is an indicator of the ultimate detectivity of the device, and the temperature variation of R_0A can be used to analyze the dark current generation mechanisms [132]. The variation of R_0A versus inverse temperature obtained for one of the superlattice devices is shown in Figure 6.27. The exponential behavior of R_0A on temperature indicates that diffusion currents dominate down to a temperature of 100 K. Below 100 K, tunneling processes begin to dominate the diffusion limited behavior, as is indicated by the very small temperature dependence of R_0A. The low-temperature behavior of R_0A may be due to both a bulk tunneling phenomenon as well as surface leakage currents [132]. As one advantage of HgTe/CdTe superlattices is to suppress bulk tunneling currents, it was important to try to determine the limiting mechanism. Additional

(a)

Contour Plot of Spot Scan

178 μm

Contact
Pad

Y — Position

25%, 50%, 75% Contours

178 μm

X — Position

(b)

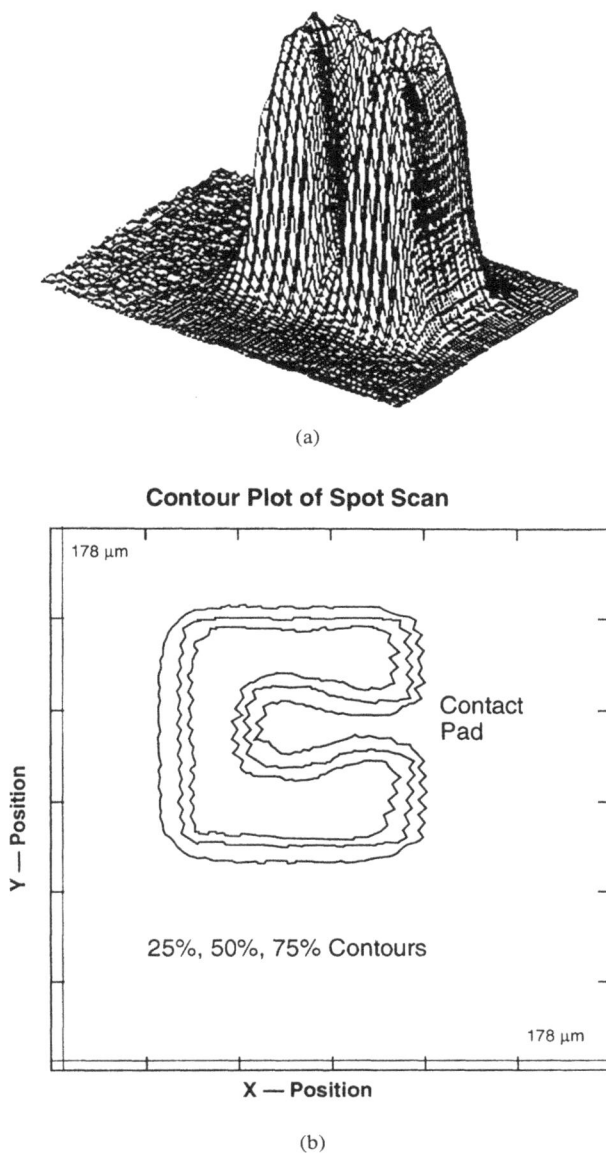

Figure 6.25 Representative infrared device response of the superlattice detectors are shown as (a) a detailed three-dimensional map of photoresponse and (b) a contour plot. The infrared photoresponse is uniform across the detector except over the region masked by the top-side metal contact.

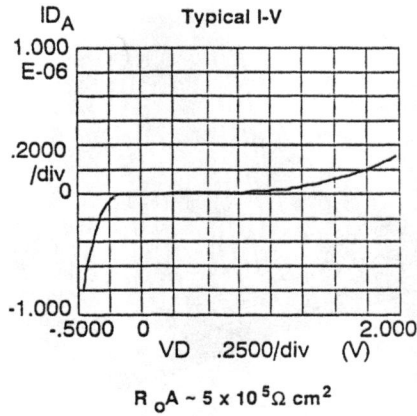

Figure 6.26 Representative I–V measurement of the midwave infrared superlattice detectors at 77 K. The *p*-on-*n* nature is evident as the I–V appears to be reversed from that obtained for the more conventional *n*-on-*p* diode. Typical R_0A values at 77 K were around 5×10^5 Ωcm^2.

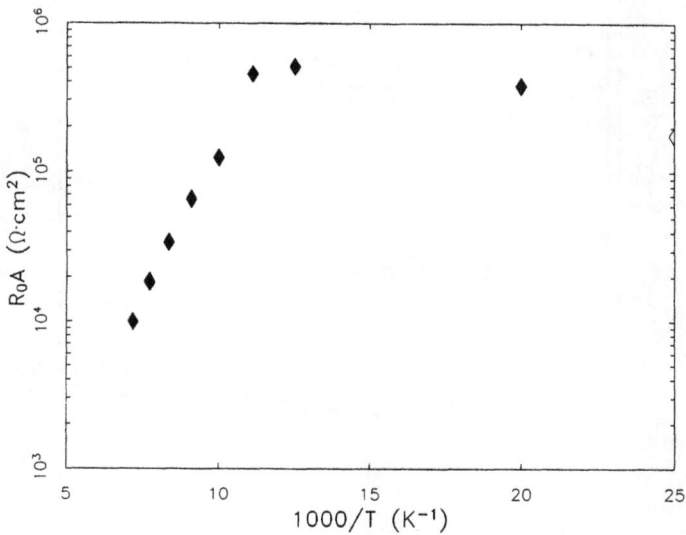

Figure 6.27 Measured R_0A as a function of temperature for a representative superlattice device. The low-temperature characteristics are indicative of tunneling processes limiting R_0A.

information was obtained by using a gated diode structure, where a metal gate was placed over the insulating passivation at the junction. The I–V curve of such a gated superlattice device is shown in Figure 6.28. The I–V curves measured as a function of gate bias indicate that the diode tunneling current is reduced by reverse biasing the guard gate. The R_0A values increase as the tunneling currents are suppressed, although the actual values are not discussed by Harris et al. [66]. In general, when surface processes are not limiting R_0A the minimum tunneling currents are observed for zero gate bias. Therefore, this trend is indicative of a surface-limiting process, with improper passivation of the surface the most likely culprit. It is conceivable that the highest performance superlattice detectors cannot be fabricated by using standard methods developed for HgCdTe, and that new processing and passivating procedures need to be developed for HgTe/CdTe superlattice detectors. Even with passivation problems, the R_0A values for superlattice detectors fabricated without gates were typically 5×10^5 Ωcm^2, comparable to that which is often achieved in the corresponding alloy.

The fabrication of high quantum efficiency midwave infrared p-on-n superlattice homojunction detectors represents a significant milestone toward the ultimate development of this material for a long-wavelength infrared or very long-wavelength infrared device technology. The results presented here clearly demonstrate the feasibility of the use of this quantum material in infrared detector structures.

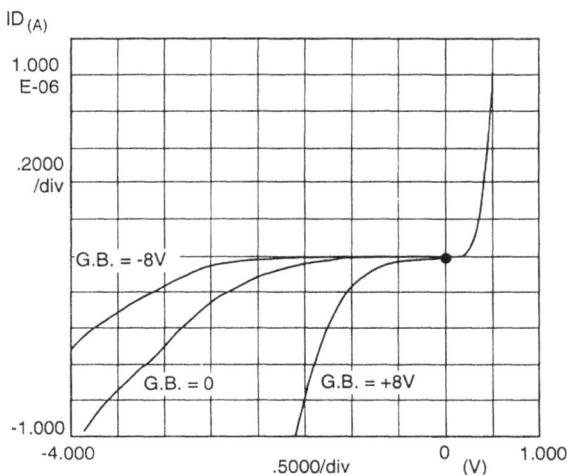

Figure 6.28 I–V curves obtained from a midwave infrared superlattice detector covered by a metal gate. The dependence of the I–V curves on bias indicate that tunneling currents are occurring near the surface and may originate from improper surface passivation. Note that this I–V curve is reversed from Figure 6.26.

6.6 CONCLUSION

HgTe/CdTe superlattices have the same basic characteristics that have made Hg-CdTe the preferred material for long-wavelength infrared and very long-wavelength infrared detection. First, its direct energy gap and large absorption coefficient allow the fabrication of high quantum efficiency detectors. The direct gap also provides a detection process that does not require complicated schemes to couple with the infrared radiation, allowing the construction of simple focal plane arrays. As with the alloy, the bandgap can be tailored over the entire wavelength range. Unlike some other quantum materials, HgTe/CdTe superlattices provide the broadband absorption of infrared radiation required by many applications involving low levels of radiation.

As detailed in this chapter, HgTe/CdTe superlattices also present distinct advantages over the HgCdTe alloy as a long-wavelength and very long-wavelength infrared detector material. Not only can the cutoff wavelength be tuned as in the alloy, HgTe/CdTe superlattices provide a higher degree of control over both cutoff wavelength and fluctuations in cutoff wavelength during growth. The small electron effective mass in the alloy causes severe problems with tunneling currents in the very long-wavelength infrared. By varying d_B, the electron effective mass in the growth direction of the superlattice can be tuned to a value large enough to effectively limit tunneling currents. In addition, the superlattice has a significantly larger absorption coefficient than in the alloy of equivalent cutoff wavelength. This allows thinner layers to be used in detector structures, limiting the volume of material contributing to diffusion dark currents.

To date, there have been only a limited number of studies of actual detector structures containing HgTe/CdTe superlattices. This has been due to problems associated with materials fabrication, as well as a lack of understanding of the basic parameters controlling superlattice properties. With respect to the latter, many so-called long-wavelength infrared and very long-wavelength infrared superlattices were fabricated with layer thicknesses based on calculations using a valence band offset of zero, leading to confusing results. Even assuming that a valid band structure approach is employed, calculation accuracy is limited by the reliability of the input parameters. In particular, it is now known that theoretical predictions for many of the fundamental optical and electronic properties of HgTe/CdTe superlattices depend quite sensitively on whether the valence band offset, ΔE_v, is taken to be small ($<$ 200 meV) or large ($>$ 200 meV). Many of the early "long-wavelength infrared" superlattices were in fact semimetallic in nature. With the current understanding that the valence band offset is large (between 350 and 400 meV), reasonably accurate predictions of superlattice properties are now obtained.

The quality of the initial HgTe/CdTe superlattices grown was limited by the difficulties encountered in Hg-based MBE growth in general. The recent advances reported in superlattice quality indicated that these problems have now been solved. A major advance in the growth of HgTe/CdTe superlattices occurred with the use of the (211)*B* orientation as a solution to the problem of defect formation due to twinning. Excess carrier lifetimes reported for (211)*B* superlattices are on the order of several hundred nanoseconds, well within the range required for infrared detector fabrication. By employing photon-assisted MBE growth, controlled *n*- and *p*-type doping are now obtainable, and high-quality superlattices exhibiting the parameters desired for detector fabrication can be produced. Indeed, these improvements have lead to the fabrication of midwave photovoltaic devices based on an HgTe/CdTe superlattice with good detector characteristics.

The ultimate decision concerning which technology is most appropriate for long-wavelength infrared detection will rest on actual device performance. Undoubtedly, long-wavelength infrared devices using HgTe/CdTe superlattices will be fabricated in the next year to demonstrate the advantages of this materials system. Based on all the calculated and measured properties presented in this chapter, long-wavelength infrared and very long-wavelength infrared HgTe/CdTe superlattice detectors with the desired performance characteristics seem quite feasible and attractive.

The authors gratefully acknowledge General Electric's Electronics Laboratory for providing many of the figures included in this chapter. Work performed at the Naval Research Laboratory was supported by the Office of Naval Research.

REFERENCES

[1]   Schulman, J.N., and T.C. McGill, *Appl. Phys. Lett.*, Vol. 34, 1979, p. 663.

[2]   Esaki, L., in *Proc. 17th Int. Conf. Physics of Semiconductors*, ed. J.D. Chadi and W.A. Harrison, Springer-Verlag, New York, 1985.

[3]   Schulman, J.N., and Y.C. Chang, *Appl. Phys. Lett.*, Vol. 46, 1985, p. 571.

[4]   McGill T.C., G.Y. Wu, and S.R. Hetzler, *J. Vac. Sci. Technol. A*, Vol. 4, 1986, p. 2091.

[5]   Zoryk, A., and M. Jaros, *Appl. Phys. Lett.*, Vol. 50, 1987, p. 1191.

[6]   Meyer, J.R., F.J. Bartoli, C.A. Hoffman, and J.N. Schulman, *Phys. Rev. B*, Vol. 38, 1988, p. 12457.

[7]   Faurie, J.P., A. Million, and J. Piaquet, *Appl. Phys. Lett.*, Vol. 41, 1982, p. 713.

[8]   Schulman, J.N., and Y.C. Chang, *Phys. Rev. B*, Vol. 33, 1986, p. 2594.

[9]   Berroir, J.M., Y. Guldner, J.P. Vieren, M. Voos, and J.P. Faurie, *Phys. Rev. B*, Vol. 34, 1986, p. 891.

[10] Ram-Mohan, L.R., K.H. Yoo, and R.L. Aggarwal, *Phys. Rev. B*, Vol. 38, 1988, p. 6151.

[11] Hoffman, C.A., J.R. Meyer, F.J. Bartoli, J.W. Han, J.W. Cook, Jr., J.F. Schetzina, and J.N. Schulman, *Phys. Rev. B*, Vol. 39, 1989, p. 5208.

[12] Johnson, N.F., H. Ehrenreich, P.M. Hui, and P.M. Young, *Phys. Rev. B*, Vol. 41, 1990, p. 3655.

[13] Wu, G.Y., and T.C. McGill, *Appl. Phys. Lett.*, Vol. 47, 1985, p. 634.

[14] Guldner, Y., G. Bastard, J.P. Vieren, M. Voos, J.P. Faurie, and A. Million, *Phys. Rev. Lett.*, Vol. 51, 1983, p. 907.

[15] Reno, J., I.K. Sou, J.P. Faurie, J.M. Berroir, Y. Guldner, and J.P. Vieren, *Appl. Phys. Lett.*, Vol. 49, 1986, p. 106.

[16] Baukus, J.P., A.T. Hunter, J.N. Schulman, and J.P. Faurie, *J. Appl. Phys.*, Vol. 64, 1988, p. 283.

[17] Yang, Z., and J.K. Furdyna, *Appl. Phys. Lett.*, Vol. 52, 1988, p. 498.

[18]   Choi, J.B., L. Ghenim, R. Mani, H.D. Drew, K.H. Yoo, and J.T. Cheung, *Phys. Rev. B*, Vol. 41, 1990, p. 10872.

[19]   Chow, D.H., T.C. McGill, I.K. Sou, J.P. Faurie, and C.W. Nieh, *Appl. Phys. Lett.*, Vol. 52, 1988, p. 54.

[20]   Reed, M.A., R.J. Koestner, M.W. Goodwin, and H.F. Schaake, *J. Vac. Sci. Technol. A*, Vol. 6, 1988, p. 2619.

[21]   Duc, T.M., C. Hsu, and J.P. Faurie, *Phys. Rev. Lett.*, Vol. 58, 1987, p. 1127.

[22]   Sporken, R., S. Sivananthan, J.P. Faurie, D.H. Ehlers, J. Fraxedas, L. Ley, J.J. Pireaux, and R. Caudano, *J. Vac. Sci. Technol. A*, Vol. 7, 1989, p. 427.

[23]   Tersoff, J., *Phys. Rev. Lett.*, Vol. 56, 1986, p. 2755.

[24]   Qteish, A., and R.J. Needs, *J. Phys. Condens. Matter*, Vol. 3, 1991, p. 617.

[25]   Meyer, J.R., C.A. Hoffman, F.J. Bartoli, J.W. Han, J.W. Cook, Jr., J.F. Schetzina, X. Chu, J.P. Faurie, and J.N. Schulman, *Phys. Rev. B*, Vol. 38, 1988, p. 2204.

[26]   Schulman, J.N., O.K. Wu, E.A. Patten, J.W. Han, Y. Lansari, L.S. Kim, J.W. Cook, Jr., and J.F. Schetzina, *Appl. Phys. Lett.*, Vol. 53, 1988, p. 2420.

[27]   Cesar, C.L., M.N. Islam, R.D. Feldman, R. Spitzer, R.F. Austin, A.E. DiGiovanni, J. Shah, and J. Orenstein, *Appl. Phys. Lett.*, Vol. 54, 1989, p. 745.

[28]   Perez, J.M., R.J. Wagner, J.R. Meyer, J.W. Han, J.W. Cook, Jr., and J.F. Schetzina, *Phys. Rev. Lett.*, Vol. 61, 1988, p. 2261.

[29]   Berroir, J.M., Y. Guldner, J.P. Vieren, M. Voos, X. Chu, and J.P. Faurie, *Phys. Rev. Lett.*, Vol. 62, 1989, p. 2024.

[30]   Meyer, J.R., C.A. Hoffman, F.J. Bartoli, T. Wojtowicz, M. Dobrowolska, J.K. Furdyna, X. Chu, J.P. Faurie, and L.R. Ram-Mohan, *Phys. Rev. B*, Vol. 44, 1991, p. 3455.

[31]   Johnson, N.F., P.M. Hui, and H. Ehrenreich, *Phys. Rev. Lett.*, Vol. 61, 1988, p. 1993.

[32]   Meyer, J.R., R.J. Wagner, F.J. Bartoli, C.A. Hoffman, M. Dobrowolska, T. Wojtowicz, J.K. Furdyna, and L.R. Ram-Mohan, *Phys. Rev. B*, Vol. 42, 1990, p. 9050.

[33]   Meyer, J.R., C.A. Hoffman, R.J. Wagner, and F.J. Bartoli, *Phys. Rev. B*, Vol. 43, 1991, p. 14715.

[34]   Meyer, J.R., D.J. Arnold, C.A. Hoffman, F.J. Bartoli, and L.R. Ram-Mohan, *J. Vac. Sci. Technol. B*, Vol. 9, 1991, p. 1818.

[35]   Reno, J., R. Sporken, Y.J. Kim, C. Hsu, and J.P. Faurie, *Appl. Phys. Lett.*, Vol. 51, 1987, p. 1545.

[36]   Ahlgren, W.L., J.B. James, R.P. Ruth, E.A. Patten, and J.L. Staudenmann, *Mater. Res. Soc. Symp. Proc.*, Vol. 90, 1987, p. 405.

[37]   Ahlgren, W.L., E.J. Smith, J.B. James, T.W. James, R.P. Ruth, E.A. Patten, R.D. Knox, and J.L. Staudenmann, *J. Crys. Growth*, Vol. 86, 1988, p. 198.

[38]   Pain, G.N., N. Bharatula, T.J. Elms, P. Gwynn, M. Kibel, M.S., Kwietniak, P. Leech, N. Petkovic, C. Sandford, J. Thompson, T. Warminski, D. Gao, S.R. Glanvill, C.J. Rossouw, A.W. Stevenson, S.W. Wilkins, and L. Wielinski, *J. Vac. Sci. Technol. A*, Vol. 8, 1990, p. 1067.

[39]   Ong, N.P., G. Kote, and J.T. Cheung, *Phys. Rev. B*, Vol. 28, 1983, p. 2289.

[40]   Chow, P.P. and D. Johnson, *J. Vac. Sci. Technol. A*, Vol. 3, 1985, p. 67.

[41]   Reno, J., and J.P. Faurie, *Appl. Phys. Lett.*, Vol. 49, 1986, p. 409.

[42]   Farrow, R.F.C., *J. Vac. Sci. Technol. A*, Vol. 3, 1985, p. 60.

[43]   Farrow, R.F.C., G.R. Jones, G.M. Williams, P.W. Sullivan, W.J.O. Boyle, and J.T.M. Wotherspoon, *J. Phys. D*, Vol. 12, 1979, p. L117.

[44]   Faurie, J.P., A. Million, R. Boch, and J.L. Tissot, *J. Vacuum, Sci. Technol. A*, Vol. 1, 1983, p. 1593.

[45]   See, for example, Harris, K.A., S. Hwang, D.K. Blanks, J.W. Cook, Jr., J.F. Schetzina, and N. Otsuka, *J. Vac. Sci. Technol. A*, Vol. 4, 1986, p. 2061.

[46]   See, for example, Wagner, B.K., R.G. Benz, II, and C.J. Summers, *J. Vac. Sci. Technol. A*, Vol. 7, 1989, p. 295.

[47]   Faurie, J.P., M. Boukerche, J. Reno, S. Sivanthanan, and C. Hsu, *J. Vac. Sci. Technol. A*, Vol. 3, 1985, p. 55.

[48]   Harris, K.A., T.H. Myers, R.W. Yanka, L.M. Mohnkern, R.W. Green, and N. Otsuka, *J. Vac. Sci. Technol. A*, Vol. 8, 1990, p. 1013.

[49]   Faurie, J.P., *J. Cryst. Growth*, Vol. 81, 1987, p. 483.

[50]   Koestner, R.J., and Schaake, H.F., *J. Vac. Sci. Technol. A*, Vol. 6, 1988, p. 2834.

[51]   Arias, J.M., S.H. Shin, and E.R. Gertner, *J. Cryst. Growth*, Vol. 86, 1988, p. 362.

[52]   Singh, J., and J. Arias, *J. Vac. Sci. Technol. A*, Vol. 7, 1989, p. 2562.

[53]   Goodwin, M.W., M.A. Kinch, R.J. Koestner, M.C. Chen, D.G. Seiler, and R.J. Justice, *J. Vac. Sci. Technol. A*, Vol. 5, 1987, p. 3110.

[54]   Myers, T.H., R.W. Yanka, K.A. Harris, A.R. Reisinger, J. Han, S. Hwang, Z. Yang, N.C. Giles, J.W. Cook, Jr., J.F. Schetzina, R.W. Green, and S. McDevitt, *J. Vac. Sci. Technol. A*, Vol. 7, 1987, p. 300.

[55]   Bicknell, R.N., N.C. Giles, and J.F. Schetzina, *Appl. Phys. Lett.*, Vol. 49, 1986, p. 1095.

[56]   Bicknell, R.N., N.C. Giles, and J.F. Schetzina, *Appl. Phys. Lett.*, Vol. 49, 1986, p. 1735.

[57]   Giles, N.C., R.L., J.W. Han, and J.F. Schetzina, *Mat. Res. Symp. Proc.*, Vol. 161, 1990, p. 227.

[58]   Schetzina, J.F., R.L. Harper, J. Han, S. Hwang, N.C. Giles, *et al.*, *Mater. Res. Symp. Proc.*, Vol. 102, 1987, p. 97.

[59]   Koestner, R.J., H.Y. Liu, H.F. Schaake, and T.R. Hanlon, *J. Vac. Sci. Technol. A*, Vol. 7, 1989, p. 517.

[60]   Arias, J.M., S.H. Shin, D.E. Cooper, M. Zandian, J.G. Pasko, E.R. Gertner, R.E. DeWames, and J. Singh, *J. Vac. Sci. Technol. A*, Vol. 8, 1990, p. 1025.

[61]   Lansari, Y., Z. Yang, S. Hwang, F.E. Reed, A.T. Sowers, J.W. Cook, Jr., and J.F. Schetzina, *J. Cryst. Growth*, in press.

[62]   See, for example, Faurie, J.P., J. Reno, S. Sivananthan, I.K. Sou, X. Chu, and M. Boukerche, *J. Vac. Sci. Technol. A*, Vol. 4, 1986, p. 2067.

[63]   See, for example, Wood, S., J. Greggi, Jr., R.F.C. Farrow, W.J. Takei, F.A. Shirland, and A.J. Noreika, *J. Appl. Phys.*, Vol. 55, 1984, p. 4225.

[64]   Million, A., L. DiCioccio, J.P. Gaillard, and J. Piaguet, *J. Vac. Sci. Technol. A*, Vol. 6, 1988, p. 2813.

[65]   DiCioccio, L., E.A. Hewat, A. Million, J.P. Gailliard, and M. Dupuy, *Inst. Phys. Conf. Ser.*, Vol. 87, Sect.3, 1987, p. 243.

[66]   Harris, K.A., T.H. Myers, R.W. Yanka, L.M. Mohnkern, and N. Otsuka, *J. Vac. Sci. Technol. B*, Vol. 9, 1991, p. 1752.

[67]   Vere, A.W., S. Cole, and D.J. Williams, *J. Electron. Mater.*, Vol. 12, 1983, p. 551.

[68]   Feldman, R.D., S. Nakahara, R.F. Austin, T. Boone, R.L. Opila, and A.S. Wynn, *Appl. Phys. Lett.*, Vol. 51, 1987, p. 1239.

[69]   Arias, J.M., S.H. Shin, J.G. Pasko, and E.R. Gertner, *Appl. Phys. Lett.*, Vol. 52, 1988, p. 39.

[70]   Goodwin, M.W., M.A. Kinch, and R.J. Koestner, *J. Vac. Sci. Technol. A*, Vol. 7, 1989, p. 523.

[71]   Arias, J.M., M. Zandian, J.G. Pasko, S.H. Shin, L.O. Bubulac, R.E. DeWames, and W.E. Tennant, *J. Appl. Phys.*, Vol. 69, 1991, p. 2143.

[72]   Koestner, R.J., M.W. Goodwin, and H.F. Schaake, *J. Vac. Sci. Technol. B*, Vol. 9, 1991, p. 1731.

[73]   Harris, K.A., R.W. Yanka, L.M. Mohnkern, T.H. Myers, Z. Zhang, Z. Yu, S. Hwang, and J.F. Schetzina, *J. Vac. Sci. Technol. B*, May–June 1992.

[74]   Faurie, J.P., HgCdTe MBE Miniworkshop, DARPA-CNVEO, Washington, DC, 1987 unpublished.

[75]   Hoffman, C.A., J.R. Meyer, E.R. Youngdale, J.R. Lindle, F.J. Bartoli, J.W. Han, K.A. Harris, J.W. Cook, Jr., and J.F. Schetzina, *J. Vac. Sci. Technol. A*, Vol. 6, 1988, p. 2785.

[76]   Hoffman, C.A., J.R. Meyer, E.R. Youngdale, J.R. Lindle, F.J. Bartoli, K.A. Harris, J.W. Cook, Jr., and J.F. Schetzina, *Phys. Rev. B*, Vol. 37, 1988, p. 6933.

[77]   Goodwin, M.W., M.A. Kinch, and R.J. Koestner, *J. Vac. Sci. Technol. A*, Vol. 6, 1988 p. 2685.

[78]   Faurie, J.P., M. Boukerche, S. Sivananthan, J. Reno, and C. Hsu, *Superlatt. and Microstruct.*, Vol. 1, 1985, p. 237.

[79]   See, for example, Zanio, K., and T. Massopust, *J. Electron. Mater.*, Vol. 15, 1986, p. 103.

[80]   Arch, D.K., J.L. Staudenmann, and J.P. Faurie, *Appl. Phys. Lett.*, Vol. 48, 1986, p. 1588.

[81]   Staudenmann, J.L., R.D. Knox, and R.D. Horning, *J. Crys. Growth*, Vol. 86, 1988, p. 436.

[82]   See, for example, Tang, M.F., and D.A. Stevenson, *Appl. Phys. Lett.*, Vol. 50, 1987, p. 1272.

[83]   Otsuka, N., Y.E. Ihm, K.A. Harris, J.W. Cook, Jr., and J.F. Schetzina, *J. Vac. Sci. Technol. A*, Vol. 56, 1987, p. 3129.

[84]   Kim, Y., A. Ourmazd, and R.D. Feldman, *J. Vac. Sci. Technol. A*, Vol. 8, 1990, p. 1116.

[85]   Ourmazd, A., D.W. Tatlor, J. Cunningham, and C.W. Tu, *Phys. Rev. Lett.*, Vol. 62, 1989, p. 933.

[86]   Kinch, M.A., and M.W. Goodwin, *J. Appl. Phys.*, Vol. 53, 1985, p. 4455.

[87]   Wu, G.Y., C. Mailhiot, and T.C. McGill, *Appl. Phys. Lett.*, Vol. 46, 1985, p. 72.

[88]   Jones, C.E., T.N. Casselman, J.P. Faurie, S. Perkowitz, and J.N. Schulman, *Appl. Phys. Lett.*, Vol. 47, 1985, p. 140.

[89]   Yang, Z., Y. Lansari, J.W. Han, Z. Yu, and J.F. Schetzina, *Mat. Res. Soc. Symp. Proc.*, Vol. 161, 1990, p. 257.

[90]   Schetzina, J.F., DARPA IR Focal Plane Array Materials and Processing Review, Washington, DC, 1988, unpublished.

[91]   Leopold, D.J., M.L. Wroge, and J.G. Broerman, *Appl. Phys. Lett.*, Vol. 50, 1987, p. 924.

[92]   Patten, E.A., K. Kosai, T.N. Casselman, J.N. Schulman, Y.C. Chang, J.L. Staudenman, *J. Vac. Sci. Technol. A*, Vol. 5, 1987, p. 3102.

[93]   Wu, O.K., J.N. Schulman, and G.S. Kamath, *J. Cryst. Growth*, Vol. 101, 1990, p. 96.

[94]   Kowalczyk, S.P., J.T. Cheung, E.A. Kraut, and R.W. Grant, *Phys. Rev. Lett.*, Vol. 56, 1986, p. 1605.

[95]   Malloy, K.J. and J.A. Van Vechten, *Appl. Phys. Lett.*, Vol. 54, 1989, p. 937.

[96]   Dobrowolska, M., T. Wojtowicz, H. Luo, J.K. Furdyna, O.K. Wu, J.N. Schulman, J.R. Meyer, C.A. Hoffman, and F.J. Bartoli, *Phys. Rev. B*, Vol. 41, 1990, p. 5084.

[97]   Yang, Z., Z. Yu, Y. Lansari, J.W. Cook, Jr., and J.F. Schetzina, *J. Vac. Sci. Technol. B*, Vol. 19, 1991, p. 1805.

[98]   Manasses, J., Y. Guldner, J.P. Vieren, M. Voos, and J.P. Faurie, *Phys. Rev. B*, Vol. 44, 1991, p. 13541.

[99]   Hoffman, C.A., J.R. Meyer, R.J. Wagner, F.J. Bartoli, X. Chu, J.P. Faurie, L.R. Ram-Mohan, and H. Xie, *J. Vac. Sci. Technol. A*, Vol. 8, 1990, p. 1200.

[100]   Hoffman, C.A., J.R. Meyer, F.J. Bartoli, J.W. Han, J.W. Cook, Jr., and J.F. Schetzina, *Phys. Rev. B*, Vol. 40, 1989, p. 3867.

[101]   Hwang, S., Y. Lansari, Z. Yang, J.W. Cook, Jr., and J.F. Schetzina, *J. Vac. Sci. Technol. B*, Vol. 9, 1991, p. 1799.

[102]   Hoffman, C.A., J.R. Meyer, F.J. Bartoli, Y. Lansari, J.W. Cook, Jr., and J.F. Schetzina, *Phys. Rev. B*, Vol. 44, 1991, p. 8376.

[103]   Meyer, J.R., D.J. Arnold, C.A. Hoffman, F.J. Bartoli, and L.R. Ram-Mohan, *Phys. Rev. B*, Vol. 46, 1992, p. 4139.

[104] Meyer, J.R., D.J. Arnold, C.A. Hoffman, and F.J. Bartoli, *Appl. Phys. Lett.*, Vol. 58, 1991, p. 2523.
[105] Dubowski, J.J., T. Dietl, W. Szymanska, and R.R. Galazka, *J. Phys. Chem. Solids*, Vol. 42, 1981, p. 351.
[106] See, for example, Capper, P., *J. Cryst. Growth*, Vol. 57, 1982, p. 280, and the references therein.
[107] Boukerche, M., P.S. Wijewarnasuriya, S. Sivanthan, I.K. Sou, Y.J. Kim, K.K. Mahavadi, and J.P. Faurie, *J. Vac. Sci. Technol. A*, Vol. 6, 1988, p. 2830.
[108] Boukerche, M., J. Reno, I.K. Sou, C. Hsu, and J.P. Faurie, *Appl. Phys. Lett.*, Vol. 48, 1986, p. 1733.
[109] Wroge, M.L., D.J. Peterman, B.J. Feldman, B.J. Morris, D.J. Leopold, and J.G. Broerman, *J. Vac. Sci. Technol. A*, Vol. 7, 1989, p. 435.
[110] Temofonte, T.A., A.J. Noreika, M.J. Bevan, P.R. Emtage, C.F. Seiler, and P. Mitra, *J. Vac. Sci. Technol. A*, Vol. 7, 1989, p. 440.
[111] Lansari, Y., Z. Yang, S. Hwang, J.W. Cook, Jr., and J.F. Schetzina, *Mat. Res. Symp. Proc.*, Vol. 161, 1990, p. 362.
[112] Hoffman, C.A., J.R. Meyer, D.J. Arnold, F.J. Bartoli, Y. Lansari, J.W. Cook, Jr., and J.F. Schetzina, *J. Vac. Sci. Technol. B*, Vol. 9, 1991, p. 1813.
[113] Hoffman, C.A., J.R. Meyer, F.J. Bartoli, Y. Lansari, J.W. Cook, Jr., and J.F. Schetzina, *Appl. Phys. Lett.*, Vol. 60, 1992, p. 2282.
[114] Brown, M., and A.F.W. Willoughby, *J. Vac. Sci. Technol. A*, Vol. 1, 1983, p. 1641.
[115] Peterman, D.J., M.L. Wroge, B.J. Morris, D.J. Leopold, and J.G. Broerman, *J. Appl. Phys.*, Vol. 63, 1988, p. 1951.
[116] Harper, R.L., Jr., S. Hwang, N.C. Giles, J.F. Schetzina, D.L. Dreifus, and T.H. Myers, *Appl. Phys. Lett.*, Vol. 54, 1989, p. 170.
[117] Han, J.W., S. Hwang, Y. Lansari, R.L. Harper, Z. Yang, N.C. Giles, J.W. Cook, Jr., J.F. Schetzina, and S. Sen, *J. Vac. Sci. Technol. A*, Vol. 7, 1989, p. 305.
[118] Lansari, Y., Z. Yang, S. Hwang, J.W. Cook, Jr., and J.F. Schetzina, *Mat. Res. Soc. Symp. Proc.*, Vol. 161, 1990, p. 363.
[119] Smith, D.L., T.C. McGill, and J.N. Schulman, *Appl. Phys. Lett.*, Vol. 43, 1983, p. 180.
[120] Guldner, Y., G. Bastard, and M. Voos, *J. Appl. Phys.*, Vol. 57, 1985, p. 1403.
[121] Arias, J., and J. Singh, *Appl. Phys. Lett.*, Vol. 55, 1989, p. 1561.
[122] Ulmer, I., N. Magnea, H. Maiette, and P. Gentile, *J. Cryst. Growth*, Vol. 111, 1991, p. 711.
[123] Smith, D.L., and V.Y. Pickhardt, *J. Appl. Phys.*, Vol. 46, 1975, p. 2366.
[124] Hansen, G.L., J.L. Schmit, and T.N. Casselman, *J. Appl. Phys.*, Vol. 53, 1982, p. 7099.
[125] Chow, D.H., and T.C. McGill, *Appl. Phys. Lett.*, Vol. 48, 1986, p. 1485.
[126] Schulman, J.N., and C.L. Anderson, *Appl. Phys. Lett.*, Vol. 48, 1986, p. 1684.
[127] Reed, M.A., R.J. Koestner, and M.W. Goodwin, *Appl. Phys. Lett.*, Vol. 49, 1986, p. 1293.
[128] Caleki, D., J.F. Palmier, and A. Chomette, *J. Phys. C*, Vol. 17, 1984, p. 5017.
[129] Reisinger, A.R., K.A. Harris, T.H. Myers, R.W. Yanka, L.M. Mohnkern, and C.A. Hoffman, *Appl. Phys. Lett.*, Vol. 61, 1992, p. 699.
[130] Bajaj, J., S.H. Shin, J.G. Pasko, and M. Khoshnevisan, *J. Vac. Sci. Technol. A*, Vol. 1, 1983, p. 1749.
[131] Jiang, Y., M.C. Teich, and W.I. Wang, *J. Appl. Phys.*, Vol. 69, 1991, p. 6869.
[132] See, for example, Nemirovsky, Y., D. Rosenfeld, R. Adar, and A. Kornfeld, *J. Vac. Sci. Technol. A*, Vol. 7, 1990, p. 528.
[133] Hoffman, C.A., D.J. Arnold, J.R. Meyer, F.J. Bartoli, Y. Lansari, J.W. Cook, Jr., J.F. Schetzina, and J.N. Schulman, *Mat. Res. Soc. Symp. Proc.*, Vol. 161, 1990, p. 413.

Index

www.ingramcontent.com/pod-product-compliance
Lightning Source LLC
Chambersburg PA
CBHW021429180326
41458CB00001B/196